# Lecture Notes on Mathematical Modelling in the Life Sciences

The rapid pace and development of new methods and techniques in mathematics and in biology and medicine creates a natural demand for up-to-date, readable, possibly short lecture notes covering the breadth and depth of mathematical modelling, mathematical analysis and numerical computations in the life sciences, at a high scientific level.

The volumes in this series are written in a style accessible to graduate students. Besides monographs, we envision the series to also provide an outlet for material less formally presented and more anticipatory of future needs due to novel and exciting biomedical applications and mathematical methodologies.

The topics in LMML range from the molecular level through the organismal to the population level, e.g. gene sequencing, protein dynamics, cell biology, developmental biology, genetic and neural networks, organogenesis, tissue mechanics, bioengineering and hemodynamics, infectious diseases, mathematical epidemiology and population dynamics.

Mathematical methods include dynamical systems, partial differential equations, optimal control, statistical mechanics and stochastics, numerical analysis, scientific computing and machine learning, combinatorics, algebra, topology and geometry, etc., which are indispensable for a deeper understanding of biological and medical problems.

Wherever feasible, numerical codes must be made accessible.

Founding Editors:

Michael C. Mackey, McGill University, Montreal, QC, Canada

Angela Stevens, University of Münster, Münster, Germany

More information about this series at http://www.springer.com/series/10049

Jinzhi Lei

# Systems Biology

## Modeling, Analysis, and Simulation

 Springer

Jinzhi Lei
Tiangong University
Tianjin, China

ISSN 2193-4789 ISSN 2193-4797 (electronic)
Lecture Notes on Mathematical Modelling in the Life Sciences
ISBN 978-3-030-73032-1 ISBN 978-3-030-73033-8 (eBook)
https://doi.org/10.1007/978-3-030-73033-8

Mathematics Subject Classification: 92B05, 92D25, 34K60

This Springer imprint is published by the registered company Springer Nature Switzerland AG
The registered company address is: Gewerbestrasse 11, 6330 Cham, Switzerland

# Preface

*The study of applied mathematics focused on physics in the twentieth century and should shift its focus to biology in the twenty-first century.*

—C. C. Lin

Biology is entering a new era in which we understand the world of organisms in a more quantitative and mechanistic way. In the past decade, there has been a rapid development of quantitative experimental techniques, especially at the single-molecule and single-cell levels. Currently, we know a substantial amount of detailed information about the microscopic world, such as the structure of proteins, the structure of genomes, the diverse processes of transcription, the transcriptomes of single cells, the imaging of chromosomes, and the molecular interactions that occur in cells. Data-informed, mechanism-based mathematical models of biological systems are fundamental approaches to discover the relationship between different components of those systems and to understand the mechanisms of cellular function and their dysfunction in diseases.

Systems biology is a biology-based interdisciplinary field of study that focuses on complex interactions within biological systems. It applies a holistic approach to decipher the complexity of biological systems from the base that the whole is greater than the sum of its parts. The field is collaborative, integrating many scientific disciplines—biology, computer science, mathematics, physics, engineering, and others—to predict how these systems change over time and under different conditions and to understand the human body and environmental issues. Different ways of thinking are involved in systems biology: biological thinking often includes the details of and the differences between subjects; physical thinking consists of the simplification and unification of mechanisms underlining different phenomena; mathematical thinking is more logical and deductive; and technological thinking usually focuses more on the practicability of controlling systems.

Mathematics, the language of science, is used to formulate and provide logical interpretations to different scientific disciplines. Once a scientific discipline becomes articulated in a mathematical structure, it goes beyond the classical view and gains predictive capability through mathematical models. This has been well verified in the domain of physical sciences and economic sciences, and today, it has been extended to the biological sciences.

The general structure of the mathematical formulation of a real-word problem includes three components: variables, relations, and equations. In biological sciences, the information fluxes from DNA sequences, mRNA, and protein to gene expression, cellular behavior, and population dynamics provide the required variables for mathematical models. The relations include different levels of interconnections, such as gene regulation networks, protein–protein interactions, cell-to-cell communications, microenvironments, and metabolism. These relations give rise to equations that are developed based on working hypotheses for the underlying mechanisms and conceptual models.

In the field of systems biology, different types of mathematical structures can be established for problems on various scales. For example, discrete mathematics for performing sequencing analysis, stochastic processes and simulations for analyzing molecular behaviors with low numbers of interacting molecules, differential equations for assessing the average behavior of a large number of interacting molecules, partial differential equations for determining spatial information and tissue growth, differential–integral equations for evaluating global interactions, and multiscale models for integrating different scale interactions. Due to the complexity of biological systems and the absence of the basic principles of biology, mathematical models are often empirical, based on available experimental facts. All models have limitations, and some models are useful for particular issues.

This lecture provides some basic techniques of mathematical modeling in systems biology, including mathematical modeling, analysis, and simulation. These techniques cover a wide range of mechanisms from molecular behavior to population dynamics. This lecture serves as a reference for readers whose backgrounds are either mathematics or biology and who wish to enter the field of systems biology.

Chapter 1 lays out an overview of the biological background that involves a basic introduction to genetics and epigenetics, cellular energy exchangers, and organism development. This chapter is intended to provide a basic introduction to readers with a nonbiological background.

Chapter 2 introduces mathematical formulations for continuous dynamics. It provides a general review of basic concepts, results, and analytic methods for different types of differential equations, including ordinary differential equations, delay differential equations, stochastic differential equations, and reaction–diffusion equations. These equations are often used to model biological processes that evolve continuously.

Chapter 3 presents stochastic models and simulation methods for biochemical reactions. These stochastic models form the basic methods for formulating the dynamics of the molecular behaviors in a cell, such as gene expression, genetic regulation networks, and protein–protein interactions.

Chapter 4 applies stochastic models to gene expression and epigenetic modifications. Gene expression is a primary cell process, and epigenetic regulation (DNA methylation and histone modification) is key to connecting gene expression regulation and cell-to-cell variance. We introduce stochastic modeling techniques for gene expression in prokaryotes and eukaryotes and propose a unified model of gene expression with epigenetic modification that extends the gene expression dynamics from one cell cycle to multiple cycles.

Chapter 5 discusses mathematical models for different types of gene regulatory networks, including positive feedback, negative feedback, and circadian regulation networks. Mathematical analyses reveal the conditions to display bistability, toggle switching, or oscillation dynamics through different structures of gene networks.

Chapter 6 demonstrates mathematical models for stem cell regeneration, which is associated with tissue growth, cancer development, and dynamic hematological diseases. We introduce a general mathematical framework for stem cell regeneration and a multiscale model that connects microscopic epigenetic variations at the single-cell level with macroscopic population dynamics. This framework studies the dynamics of tumorigenesis, in which the heterogeneity and plasticity of tumor cells play essential roles in the development of drug resistance in cancer.

Chapter 7 addresses the diffusion of morphogens, which play important roles in developmental biology. We introduce different types of reaction–diffusion equations based on different hypotheses of the process of morphogen gradient formation. The robustness of steady-state gradients is also discussed through boundary value problems.

There are many other textbooks in the field of systems biology [1–4], most of which introduce mathematical and/or computational modeling techniques for different types of examples. Many techniques are also included in this lecture. Nevertheless, these textbooks may have diverse preferences due to the various backgrounds of authors. Despite the complexity of biological systems, this lecture tries to provide logical and consistent techniques for different scale problems; hence, mathematical thinking and theoretical deductions are highlighted throughout the lecture. In contrast to this lecture, the popular textbook on systems biology written by Uri Alon [1] focuses more on the design principles of biological networks and aims for graduate and senior undergraduate students with backgrounds in biological sciences and/or engineering.

I first thought of doing this lecture in 2005, while visiting Michael Mackey at McGill University. Michael's course on mathematical biology inspired me to introduce a similar course in China. After arriving back in China, I taught a course on systems biology. In 2011, I published a textbook (in Chinese) based on the lecture notes for the course. The Chinese textbook forms the framework of this lecture. Systems biology is a burgeoning field that integrates many scientific disciplines, and this lecture can only cover a small part of the field. Nevertheless, I hope that this lecture can provoke readers to think about biological processes in a mathematical way.

I would like to give special thanks to Michael Mackey, a great mentor, colleague, and friend, who introduced me to the field of systems biology, encouraged me to write the English version of this lecture, and provided me with very important guidance and support during the process. Without his help, this lecture would never have been possible.

Finally, I express my sincere gratitude to all the people who have contributed to the lecture and assisted in checking the manuscript, especially Ming Yi, Xiulan Lai, Hao Jiang, Hongli Yang, Xiaopei Jiao, Honglei Ren, Haifeng Zhang, Meixia Zhu, Xiyin Liang, Wenxian Sun, Youyi Yang, Di Zhang, Jing Feng, Cong Liu, Yasong Chen, Yunfei Lv, and Michelle Li. After countless suggestions and revisions from my friends and colleagues, I assume full responsibility for any omissions and errors in the final draft of this lecture.

Beijing, China                                                                                            Jinzhi Lei
July 2020

# References

1. Alon, U.: An Introduction to Sytems Biology: Design Principles of Biological Circuits, 3rd edn. Chapman and Hall/CRC (2011)
2. Fall, C., Marland, E., Wagner, J. and Tyson, J., (eds.): Computational Cell Biology. Interdisciplinary Applied Mathematics, Springer, New York (2002)
3. Voit, E.: A First Course in Systems Biology. Garland Science (2017)
4. Ingalls, B. P.: Mathematical Modeling in Systems Biology. The MIT Press (2013)

# Contents

# Chapter 1
# Biological Background—Information, Energy, and Matter

*Nothing in biology makes sense except in the light of evolution.*

—Theodosius Dobzhansky

## 1.1 Introduction

The question *"what is life"* has been challenging scientists and philosophers for years due to the complexity and diversity of all known lifeforms, and any unknown lifeforms that may be different from the known organisms on Earth [1]. Currently, most definitions of *life* are descriptive. Life is considered a characteristic of something that exhibits all or most of the following traits: homeostasis, organization, metabolism, growth, adaptation, response to stimuli and reproduction [2, 3].

While it is difficult to summarize the common characteristics of all lifeforms, we focus on the essential components in all lives, which form a general system of information, energy, and matter [4]. The flux of genetic information continuously runs over the evolution of life on earth, from simple vacuoles to complex multicellular organisms. In organisms, genetic information and epigenetic information work together to guide different scales of biological processes and the transmission of information over generations. Energy is required in all processes to import negative entropy and maintain the steady state of organisms (matter) and may even reach a state of increased order and organization.

Essentially, life behaviors can be considered dynamics of information flux, and most biological processes are devoted to the interpretation, maintenance, and inheritance of information. These processes include the inheritance of genetic information, the read-out of genes, the production of protein and their conformational changes, the responses of a cell to external stimuli, cell cycling, development and evolution. In the long-term evolution of the life world, despite the random births and deaths of individual organisms, genetic information plays essential roles in maintaining the continuity of the development and evolution of lives. The storage, transition and

J. Lei, *Systems Biology*, Lecture Notes on Mathematical Modelling in the Life Sciences, https://doi.org/10.1007/978-3-030-73033-8_1

interpretation of the information of life provide the clues of how a mathematician can understand the complex biological world in a more logical way. Here, we introduce the biological background in terms of the basic information flux in all lives. Systems biology is a disciplinary science that tries to understand how the dynamics of these information fluxes are described and controlled.

## 1.2  Building Blocks of Life—Cells

### 1.2.1  Cells

The cell is a basic membrane-bound unit that contains the fundamental molecules of life and of which all living things are composed. Cells are regarded as the basic building blocks of life and are used in the elusive definition of what is means to be "alive". A single cell is often a complete organism in itself, such as a bacterium or yeast. A multicellular organism is an organism composed of many cells. The development of multicellular organisms is accompanied by cellular specialization and the division of labour; cells become efficient in one process and are dependent upon other cells for the necessities of life.

All cells are enclosed in a thin membrane, also called the plasma membrane, that delimits the cell from the environment around it. Cells keep chemical processes tidy and compartmentalized within the surrounding membrane so that individual cell processes do not interfere with others. The cell membrane serves as a barrier between the outside world and the cell's internal chemistry and selectively regulates the crossing of chemicals in and out of the cell to maintain the chemical balance necessary for the cell to live.

All living organisms can be classified into one of two groups based on the fundamental structure of their cells: prokaryotes and eukaryotes. Prokaryotes are organisms made up of cells that lack a cell nucleus or any membrane-bound organelles. Eukaryotes are organisms made up of cells that possess a membrane-bound nucleus that holds genetic material as well as membrane-bound organelles.

### 1.2.2  Prokaryotic Cells

Prokaryotes are organisms made up of cells that lack a cell nucleus or any membrane-bound organelles. There are two groups of prokaryotes, bacteria and archaea. Some bacteria, including E. coli, Salmonella, and Listeria, are found in foods and can cause disease; others are helpful to human digestion and other functions. Archaea were discovered to be unique lifeforms that are capable of living indefinitely in extreme environments such as hydrothermal vents or Arctic ice.

Prokaryotes store their DNA only in the cytoplasm. Prokaryotic cells typically have only one complete copy of their chromosome(s) and are packaged into a structure called the nucleoid. Most studied prokaryotes (such as *Escherichia coli* (*E. coli*) and *Bacillus subtilis*) have single circular chromosomes, and some prokaryotic cells have multiple chromosomes, linear chromosomes, or even both.

In prokaryotes, structural genes of related functions are often organized together on the genome and transcribed together under the control of a single promoter. The operon's regulatory region includes both the promoter and the operator. If a repressor binds to the operator, then the structural genes will not be transcribed (Sect. 4.2).

## 1.2.3 Eukaryotic Cells

Eukaryotes are organisms made up of cells that possess a membrane-bound nucleus (that holds DNA in the form of chromosomes) as well as membrane-bound organelles with a variety of functions. Multiple organelles are contained in eukaryotic cells, including mitochondria (cellular energy exchangers), the Golgi apparatus (secretory device), the endoplasmic reticulum (a canal-like system of membranes within the cell), and lysosomes (digestive apparatus within many cell types). However, there are also several exceptions to this; for example, the absence of mitochondria and a nucleus in red blood cells and the lack of mitochondria in the oxymonad *Monocercomonoides* species. Eukaryotic organisms may be multicellular or single-celled organisms. All animals are eukaryotes. Other eukaryotes include plants, fungi, and protists.

Eukaryotic organisms store most of their DNA inside the cell nucleus and some in organelles, such as mitochondria or chloroplasts. In contrast to the single loop structure of DNA in prokaryotes, in eukaryotes, DNA is organized into chromosomes. A chromosome is a string of DNA wrapped around associated proteins that give the connected nucleic acid bases a hierarchical structure. The compact structure of chromosomes plays an important role in helping to organize genetic material during cell division and enabling it to fit inside the nucleus of a cell.

All eukaryotic cells have multiple linear chromosomes. The number of chromosomes typically varies from 2 to <50, and in rare instances, it can reach thousands (e.g., in the macronucleus of the protozoa). The majority of eukaryotic cells are diploid; that is, they contain two copies of each chromosome. The two copies of a given chromosome are called homologs; one is derived from each parent. However, not all cells in a eukaryotic organism are diploid; a subset of eukaryotic cells are either haploid or polyploid. In particular, haploid cells contain a single copy of each chromosome and are involved in sexual reproduction (e.g., sperm and eggs are haploid cells).

## 1.3   Storage of Information—DNA Sequences

### 1.3.1   DNA and the Genome

In the digital world, information is encoded as digital sequences of 0 and 1 for simplicity and reliability. The life world uses a similar strategy, and the information encodes digital sequences of four nucleotides. Each nucleotide is composed of a nitrogen-containing nucleobase–either cytosine (C), guanine (G), adenine (A), or thymine (T)—as well as a sugar called deoxyribose and a phosphate group. The nucleotides are joined to one another in a chain by covalent bonds between the sugar of one nucleotide and the phosphate of the next nucleotide, resulting in an alternating sugar-phosphate backbone.

Usually, single-stranded nucleotides are unstable. To form a stable structure to store information, double-stranded deoxyribonucleic acid (DNA) is generated according to the base pairing rules (A with T and C with G) through hydrogen bonds between the separate complementary polynucleotide strands (Fig. 1.1). The DNA backbone is resistant to cleavage, and both strands of the double-stranded structure store the same biological information.

DNA strands have directionality in the information read-out. One end of a DNA polymer, the 3′ end of the molecule, contains an exposed hydroxyl group on the deoxyribose. The other end, the 5′ end, contains an exposed phosphate group. The two strands of a double helix run in opposite directions. Nucleic acid synthesis, including DNA replication and transcription, usually occurs in the 5′ → 3′ direction. This is because the exposed 3 hydroxyl group as a nucleophile is needed when new nucleotides are added via a dehydration reaction.

In a single cell or multicellular organism, each cell contains a complete DNA set that forms its genome. Each genome contains all the information needed to build and maintain the organism and hence forms the first-order information of life—the genetic information. In humans, every single cell virtually contains a complete copy of the human genome, which has approximately 3 billion DNA base pairs or letters.

Within a cell, DNA is associated with proteins, and each DNA sequence together with its proteins is called a chromosome. The packing of DNA into chromosomes has several important functions. First, chromosomes are a compact form of DNA that readily fits inside a cell. Second, packing the DNA into chromosomes protects it from damage. Third, packing DNA into chromosomes guarantees its efficient transmission to daughter cells during cell division. Finally, the chromosome confers an overall organization of each DNA sequence during the regulation of gene expression and recombination between parental chromosomes.

**Fig. 1.1** Double-stranded deoxyribonucleic acid (DNA)

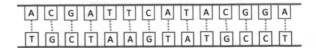

## 1.3.2 Genes

Not all regions of a DNA sequence have equal biological functions in organism behavior. There are many loci (or regions) that encode functional RNAs or protein products. Each locus forms the molecular unit of heredity and is named a gene. Each gene has two versions (or alleles), one from the father (paternal allele) and the other from the mother (maternal allele). The term homozygous refers to a gene pair in which both the maternal and paternal genes are identical. In contrast, those gene pairs in which paternal and maternal genes are different are called heterozygous. We refer to the genetic composition of a gene pair as its genotype, and the observable characteristics of an individual as its phenotype. The phenotype associated with a gene pair can be dominated by one allele or shows intermediates between the two homozygous phenotypes. A chromosome consists of a long strand of DNA that contains many genes. A human chromosome can have up to 500 million base pairs of DNA with thousands of genes. The heart of information flux in molecular biology is the process of information transfer from genes to proteins through the genetic code.

In comparison with genes (or coding sequences), the components of an organism's DNA that do not encode protein sequences are noncoding DNA sequences. Some noncoding DNA sequences are transcribed into functional noncoding RNA molecules (e.g., transfer RNAs, ribosomal RNAs, or regulatory RNAs). Other functions of noncoding DNA include the transcriptional and translational regulation of protein-coding sequences, scaffold attachment regions, origins of DNA replication, centromeres and telomeres. The amount of noncoding DNA varies greatly among species. For example, it was originally suggested that over 98% of the human genome does not encode protein sequences, while 20% of a typical prokaryote genome is noncoding [5]. Scientists once thought that noncoding DNA sequences were "junk" with unknown functions. However, it is becoming clear that at least some of these sequences are integrated into the functions of cells, particularly the control of gene activity. For example, some noncoding DNA sequences can act as regulatory elements to determine when and where genes are turned on and off [6].

## 1.4 Epigenetic Information

DNA sequences form the hereditary genetic information of life. Other information, usually associated with modifications in DNA sequences, forms the second-order information of life—epigenetic information. Epigenetics is the study of heritable changes in gene activity that are not caused by changes in DNA sequences. Such changes include, but are not limited to, DNA methylation patterns, histone modifications, regulation by noncoding RNAs, etc. Dynamic changes in epigenetic information during the development of an organism play essential roles in cell differentiation, embryo development, and the maintenance of adult homeostasis. In particular, given the same genome among all cells in an organism, the cell-to-cell variation of epigenetic information is important for different cell types.

### 1.4.1 Nucleosome and Histone Modification

Half of the molecular mass of a eukaryotic chromosome is made up of proteins. In cells, a given region of DNA with its associated proteins is called chromatin, and the majority of the associated proteins are small, basic proteins called histones. Most of the chromatin compaction in eukaryotic cells results from the regular association of DNA with histones to form structures called nucleosomes. The first step in the compaction process is the formation of nucleosomes that allow the DNA to fold into a more compact structure to reduce its linear length.

Each nucleosome contains $\sim$200 bp of DNA associated with a histone octamer, which consists of two copies of each histone (H2A, H2B, H3, and H4). They are known as the core histones. The core histones tend to form two types of subcomplexes. H3 and H4 form a stable tetramer in solution (H3$_2$-H4$_2$), and H2A and H2B most typically form a dimer (H2A-H2B).

All of the core histones are subject to numerous covalent modifications, most of which occur in the histone tails. All histones can be modified at numerous sites by methylation, acetylation, or phosphorylation (Chap. 4). Lysines in the histone tails are the most common targets of modification. Acetylation, methylation, ubiquitylation, and sumoylation all occur on the free epsilon ($\varepsilon$) amnio group of lysines. These modifications are transient and dynamic. The modification can interface the chromatin structure and gene transcription, therefore providing an additional level of information on the biological processes [7].

During DNA replication in the cell cycle, posttranslational histone modifications in the mother cells distribute onto their daughter strands in three possible ways, random distribution, semiconservative distribution, or asymmetric distribution, following which the nucleosome modifications are rebuilt with the incorporation of newly synthesized histones. The parental marks are recognized by the corresponding chromatin-binding proteins, including reader proteins that recruit chromatin modifiers and writer proteins that restore the marks at daughter strands [8]. This read-and-write process is less reliable than the maintenance of DNA sequences; hence, there are fluctuations in the nucleosome modification patterns between daughter cells and the mother cell at each cell division. The dynamic changes in nucleosome modification patterns are associated with embryo development and stem cell differentiation [9, 10].

It is less clear how prokaryotic DNA is compacted. Bacteria have no histones or nucleosomes, for example, but they do have other small basic proteins that may serve similar functions.

### 1.4.2 DNA Methylation

DNA methylation is another epigenetic mechanism that occurs by the addition of a methyl ($CH_3$) group to DNA, thereby often modifying the chromatin structure and affecting the transcription process. The most widely characterized DNA methylation process is the covalent addition of a methyl group to the 5-carbon of the cytosine ring, resulting in 5-methylcytosine (5-mC), which is primarily restricted to palindromic CpG (CG/GC) dinucleotides. These methyl groups project into the major groove of DNA and inhibit transcription.

The addition and removal of methyl groups are controlled at several different levels in cells and carried out by the families of enzymes called DNA methyltransferases (DNMTs) and the ten-eleven translocation (TET) family of 5-mC hydroxylases. These enzymes work together to regulate the patterning of DNA methylation. The biological importance of 5-mC as a major epigenetic modification in phenotype and gene expression has been widely recognized. For example, DNA hypomethylation, the decrease in global DNA methylation, is likely caused by methyl deficiency due to a variety of environmental influences and has been proposed as a molecular marker in different biological processes, such as cancer [11, 12].

Significant DNA methylation reprogramming occurs during early embryonic development in mammals [13, 14]. Upon fertilization, DNA methylation marks represent an epigenetic barrier in mammalian development and hence need to be restored and subsequently rebuilt with commitment to a particular cell fate during development. Two waves of genome-wide DNA demethylation take place in early embryonic development. The first occurs following fertilization, and the paternal pronucleus undergoes rapid demethylation in the zygote, followed by a passive loss of DNA methylation marks in the maternal genome over subsequent cell divisions. The first low point of global methylation occurs in blastocysts, followed by the reestablishment of DNA methylation patterns in the inner cell mass (ICM). Later, cells either continue to develop toward a somatic fate or are specific to primordial germ cells (PGCs). Somatically fated cells acquire distinct methylomes according to their lineage but maintain high global levels of DNA methylation. PGCs initiate a second wave of comprehensive DNA demethylation and then reestablish a unique gamete-specific methylome during gametogenesis. During fetal reprogramming, DNA methylation at imprinting control regions (ICRs) in gametes is stably maintained during embryonic development. Additionally, intracisternal A particles (IAPs) make up the sequence class that is the most highly protected against demethylation in zygotes and PGCs. Hence, the process of DNA methylation during early embryonic development defines the interesting random dynamics of the DNA methylation pattern over transgenerations.

Both histone modifications and DNA methylation are inheritable during cell division; however, they often undergo random changes due to stochastic fluctuations in biochemical reactions (Chap. 3). Hence, unlike stable genetic information, the epigenetic information of histone modification and DNA methylation is dynamically changed during cell regeneration.

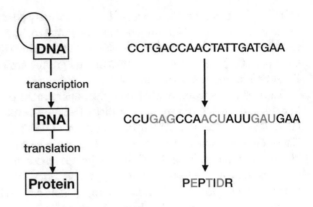

**Fig. 1.2 Central Dogma.**
DNA can be copied to DNA
(DNA replication), DNA
information can be copied
into mRNA (transcription),
and proteins can be
synthesized using the
information in mRNA as a
template (translation)

## 1.5 Expression of Genetic Information—The Central Dogma

The central dogma of molecular biology is an explanation of the flow of genetic information within a biological system [15]. The dogma is a framework for understanding the transfer of sequence information between information-carrying biopolymers, in the most common or general case, in living organisms. There are three major classes of such biopolymers: DNA, RNA, and protein. The dogma describes the normal flow of biological information among these three biopolymers: DNA can be copied to DNA (DNA replication), DNA information can be copied into mRNA (transcription), and proteins can be synthesized using the information in mRNA as a template (translation) (Fig. 1.2). Stochastic modeling of the gene expression process is discussed in Chap. 3. Reverse transcriptase (RT) is a special type of DNA polymerase that can use an RNA template to synthesize DNA and is mainly associated with retroviruses. However, non-retroviruses also use RT (for example, hepatitis B virus). Nevertheless, such sequence information cannot be transferred from protein to either protein or nucleic acid.

Ideally, the information flux described by the central dogma is a faithful, deterministic transfer, wherein one biopolymer's sequence is used as a template for the construction of another biopolymer, which is entirely dependent on the original biopolymer. The final products—proteins—are biological function executers within a cell, and the concentrations of proteins are crucial for the phenotype of a cell.

### 1.5.1 DNA Replication and Genome Inheritance

DNA replication is a biological process that produces two identical replicas of DNA from one original DNA molecule. This process occurs in all living organisms and is the basis of biological inheritance. The initiation of DNA replication is directed by specific DNA sequences called replicators. The replicator is specifically bound by a protein called the initiator, which stimulates the unwinding of the origin DNA and the recruitment of other proteins required for the initiation of replication (such as DNA helicase). The subsequent events in the initiation of DNA replication are largely driven by either protein-protein interactions or nonspecific protein-DNA interactions.

In the cell, both strands of a DNA template are duplicated simultaneously at a structure called the replication fork. Because the two strands of the DNA are antiparallel, only one of the template DNA strands can be replicated in a continuous fashion (called the leading strand). The other DNA strand (called the lagging strand) must be synthesized first as a series of short DNA fragments called Okazaki fragments. Each DNA strand is initiated with an RNA primer that is synthesized by an enzyme called primase. These primers must be removed to complete the replication process. Moreover, DNA replication cannot completely replicate the chromosome ends (called telomeres) of the lagging strand. Therefore, telomeres can be shortened after DNA replication, but there are other methods to maintain the integrity of this part of the chromosome.

The interactions between the proteins at the replication fork have an important role in DNA synthesis. The replisome, which is a complex formed by a set of proteins, is important for DNA replication. The replication reactions are most efficient when the entire array of replication proteins are present at the replication fork. These replication proteins work together to control replication efficiency, proofread the genome, fix replication errors, and regulate replication-coupled nucleosome assembly [16]. Hence, DNA replication is the intersection of genetic information inheritance and epigenetic information transition across cell cycles, which is important for the determination and maintenance of cell fate in multicellular organisms.

There are two types of cell division process in eukaryotes. Mitosis results in two daughter cells each having the same number and kind of chromosomes as the parent nucleus, which is typical for ordinary tissue growth. Alternatively, meiosis is a special type of cell division that reduces the chromosome number by half and creates four haploid cells that are genetically distinct from the parent that gave rise to them. This process occurs in all sexually reproducing single-celled and multicellular eukaryotes, including animals, plants, and fungi, and results in the production of gametes and plant spores.

## 1.5.2 Transcription and mRNA

Transcription is the process by which the information contained in a section of DNA is replicated in the form of a newly assembled piece of messenger RNA (mRNA). Enzymes facilitating the process include RNA polymerase and transcription factors. A round of transcription proceeds through three phases called initiation, elongation, and termination. Several proteins—initiation factors—are required for accurate and efficient initiation. These factors ensure that the enzyme initiates transcription only from appropriate sites on the DNA, the promoters. In bacteria, there is only one initiation factor, the sigma factor, whereas in eukaryotes, there are several, collectively called general transcription factors. In eukaryotes, DNA is wrapped within nucleosomes in vivo efficient initiation requires additional proteins to unwrap the nucleosomes. During initiation, RNA polymerase (together with the initiation factors) binds to the promoter in a closed complex. This closed complex undergoes isomerization to become an open complex. The DNA around the transcription start site (TSS) is unwound, disrupting the base pairs and forming a bubble of single-stranded DNA. This transition allows the addition of bases to the template strand, which determines the order of bases in the new RNA strand. This phase of initiation is followed by promoter escape once the enzyme has synthesized a series of short RNAs. At this point, the enzyme leaves the promoter and enters the elongation phase. During this phase, the polymerase moves along the gene in a way similar to the process of DNA replication. The polymerase stops and releases the RNA product when the full length of the gene is transcribed. The stopping process includes the addition of a 5' cap and a poly-A tail to the newly synthesized RNA chain. Transcription is terminated by signals within the RNA sequence. In bacteria, terminators come in two types: Rho-dependent and Rho-independent. In eukaryotes, termination is closely linked to an RNA processing event called 5' polyadenylation.

In eukaryotes, primary transcription from the promoter generates a pre-mRNA, which can be very long and consists of blocks of coding sequences separated from each other by blocks of noncoding sequences. The coding sequences are called exons, and the intervening sequences are called introns. Introns are removed from the pre-mRNA by a process called RNA splicing. The RNA splicing process removes introns from the pre-mRNA and converts the pre-mRNA into mature mRNA. This process must occur accurately to avoid the loss or addition of even a single nucleotide at the site at which the exons are joined. Some pre-mRNAs can be spliced in more than one way to produce alternative mRNAs. Hence, mRNAs containing different selections of exons can be generated from the same sequence of pre-mRNA. This is called alternative splicing, and the resulting products are isoforms. Therefore, a gene can give rise to more than one polypeptide product. It is estimated that up to 75% of human genes are spliced in alternative ways to generate more than one isoform.

## 1.5.3 Translation and the Genetic Code

Translation is the process by which an mRNA strand acts as a template for the assembly of amino acids that become proteins. In prokaryotes, transcription and translation simultaneously occur in the cytoplasm, and the new RNA strand is immediately ready for translation, which can start immediately and sometimes starts during transcription.

In eukaryotes, after splicing, mature mRNAs are translocated to ribosomes that direct the synthesis of proteins. Ribosomes are macromolecular machines that provide an environment for all activities involved in the translation process, including the formation of the initiation complex, the elongation of translation, and the dissoci-

**Table 1.1**  The genetic code

<table>
<tr><td></td><td></td><td colspan="4" align="center">Second letter</td><td></td><td></td></tr>
<tr><td></td><td></td><td align="center">U</td><td align="center">C</td><td align="center">A</td><td align="center">G</td><td></td><td></td></tr>
<tr>
<td rowspan="16">First letter</td>
<td rowspan="4">U</td>
<td>UUU } Phe<br>UUC }<br>UUA } Leu<br>UUG }</td>
<td>UCU }<br>UCC }<br>UCA } Ser<br>UCG }</td>
<td>UAU } Tyr<br>UAC }<br>UAA STOP<br>UAG STOP</td>
<td>UGU } Cys<br>UGC }<br>UGA STOP<br>UGG } Trp</td>
<td>U<br>C<br>A<br>G</td>
<td rowspan="16">Third letter</td>
</tr>
<tr>
<td rowspan="4">C</td>
<td>CUU }<br>CUC }<br>CUA } Leu<br>CUG }</td>
<td>CCU }<br>CCC }<br>CCA } Pro<br>CCG }</td>
<td>CAU } His<br>CAC }<br>CAA } Gln<br>CAG }</td>
<td>CGU }<br>CGC }<br>CGA } Arg<br>CGG }</td>
<td>U<br>C<br>A<br>G</td>
</tr>
<tr>
<td rowspan="4">A</td>
<td>AUU } Ile<br>AUC }<br>AUA }<br>AUG  Met</td>
<td>ACU }<br>ACC }<br>ACA } Thr<br>ACG }</td>
<td>AAU } Asn<br>AAC }<br>AAA } Lys<br>AAG }</td>
<td>AGU } Ser<br>AGC }<br>AGA } Arg<br>AGG }</td>
<td>U<br>C<br>A<br>G</td>
</tr>
<tr>
<td rowspan="4">G</td>
<td>GUU }<br>GUC }<br>GUA } Val<br>GUG }</td>
<td>GCU }<br>GCC }<br>GCA } Ala<br>GCG }</td>
<td>GAU } Asp<br>GAC }<br>GAA } Glu<br>GAG }</td>
<td>GGU }<br>GGC }<br>GGA } Gly<br>GGG }</td>
<td>U<br>C<br>A<br>G</td>
</tr>
</table>

ation of the ribosome from the mRNA. During translation, components of the translational machines (mRNAs, tRNAs, aminoacyl-tRNA synthetases, and ribosomes) work together to accomplish the extraordinary task of translating a code written in a four-base alphabet into a second code written in the language of 20 amino acids. mRNAs provide the information that is interpreted by the translation machinery.

The genetic code is a set of rules used by living cells to translate the information encoded within genetic material (DNA or mRNA sequences) into proteins. The strategy of the genetic code involves the triplet codons listed in Table 1.1. There are 64 permutations, with the left-hand column indicating the base at the 5' end of the triplet, the row across the top specifying the middle base, and the right-hand column identifying the base in the 3' position. Here, we note that 61 of the 64 possible triplets specify an amino acid, with the remaining three triplets being chain-terminating signals. Many amino acids are specified by more than one codon, which is a phenomenon called degeneracy. Codons specifying the same amino acid are synonyms. The genetic code is rather universal and is used by every organism from bacteria to humans, with small differences in mitochondria (AUA → Met, {TGA, AGA, AGC}→ STOP) [17]. tRNAs provide the physical interface between the codons in the mRNA and the amino acids being added to the growing polypeptide chain. There are many types of tRNA molecules, each of which is attached to a specific amino acid and recognizes a particular codon, or codons, in the mRNA [18].

The newly synthesized proteins undergo further conformational changes and posttranslational modifications and then perform their specific functions. The functional properties of proteins depend upon their three-dimensional structures. Protein folding is a physical process by which a linear amino acid polypeptide acquires its compact three-dimensional structure that is usually biologically functional, in an expeditious and reproducible manner. Hence, protein folding further converts linear chain information to three-dimensional information. The problem of predicting the three-dimensional structure from the amino acid sequence and understanding the mechanisms of how proteins fold, collectively known as "the protein folding problem", has been a grand challenge in molecular biology for over half a century [19, 20]. Breakthrough of the challenge was made using methods of artificial intelligence (AI) [21].

The central dogma involves the transcription of a DNA segment into an RNA form, which is then used as a template for translation into protein. However, not all genes are expressed in all cells at all times. Indeed, the success of life depends on the ability of cells to express their genes in different combinations at different times and in different places. The development of multicellular organisms offers an example of this differential gene expression. Essentially, all cells in a human contain identical genes; however they are expressed heterogeneously. The transcription of genes is well regulated through multiple regulators, and the gene expression products further regulate the expression of other genes to form gene regulatory networks (Chap. 5).

## 1.6   Variation of the Information

In addition to the well-controlled information flux, there are many variations that occur throughout the life of organisms. The variations could be harmful to lifeforms but are also necessary for their adaptation of environmental changes and evolutions [22].

Mutations are permanent alterations in the nucleotide sequence of the genome of an organism, virus, or extrachromosomal DNA. Mutations result from errors during DNA replication or other types of damage to DNA, which may then undergo error-prone repair, cause an error during other forms of repair, or cause an error during replication. Mutations also include the insertion or deletion of segments of DNA due to mobile genetic elements. Mutations may or may not produce discernible changes in phenotypes.

Another type of variation involves recombinant DNA sequences. Genetic exchange often works constantly to blend and rearrange chromosomes, most obviously during meiosis when homologous chromosome pairing occurs prior to the first nuclear division. During this pairing, genetic exchange between the chromosomes occurs. This exchange, classically termed crossover, is caused by homologous recombination, which involves the physical exchange of DNA sequences between chromosomes. Homologous recombination is an essential cellular process catalyzed by specific enzymes. In addition to providing genetic variation, recombination allows cells to retrieve sequences that were lost through DNA damage by replacing the damaged section with an undamaged DNA strand from a homologous chromosome. Recombination also provides a mechanism to restart stalled or damaged replication forks. Furthermore, special instances of recombination can regulate the expression of some genes.

Gene order can also be changed occasionally. For example, transposons a movable DNA segments that occasionally jump over chromosomes and promote DNA rearrangements and other chromosomal organization. The transposition of genes often results in the duplication of the same genetic material. Transposons make up a large fraction of the genome and are responsible for much of the mass of DNA in eukaryotic cells.

Somatic recombination is important for adaptive immunity in response to specific antigens. The best example is V(D)J recombination, which occurs in the cells of the vertebrate immune system. V(D)J recombination is a specialized mechanism of genetic recombination that occurs only in developing lymphocytes during the early stages of T and B cell maturation. It involves somatic recombination and results in the highly diverse repertoire of antibodies/immunoglobulins and T cell receptors in B cells and T cells, respectively. The process is a defining feature of the adaptive immune system.

## 1.7   Cellular Energy Exchangers—Mitochondria

Mitochondria are membrane-bound organelles found in the cytoplasm of almost all eukaryotic cells, and their primary function is to generate large quantities of energy in the form of adenosine triphosphate (ATP). Mitochondria are typically round to oval in shape and range in size from 0.5 to 1.0 μm. In addition to producing energy, mitochondria store calcium for cell signaling activities, generate heat, and mediate cell growth and death. The number of mitochondria per cell varies widely; for example, in humans, erythrocytes (red blood cells) do not contain any mitochondria, whereas liver cells and muscle cells may contain hundreds or even thousands of mitochondria. The only eukaryotic organism known to lack mitochondria is the oxymonad *Monocercomonides* species. Mitochondria have two distinct membranes and a unique genome and are reproduced by binary fission. These features indicate that mitochondria share an evolutionary past with prokaryotes. Currently, it is well accepted that mitochondria originate from an alphaproteobacteria-like ancestor [23].

Mitochondrial DNA (mtDNA), is contained in a small circular chromosome found inside mitochondria and comprises another set of genomic information in eukaryotic cells. Mitochondrial DNA is different from DNA in the nucleus [24, 25]. It has only 16,500 or so base pairs and is small and circular. It encodes proteins that are specific for the mitochondria for producing energy, using a set of genetic codes slightly different from that in the nucleus. Mitochondrial DNA is critically important for many of the pathways that produce energy within the mitochondria. Defects in some of those mitochondrial DNA bases, such as mutations, may result in mitochondrial disease, which involves the inability to produce sufficient energy in important parts such as the muscle, brain, and kidney [26]. Mitochondrial DNA, unlike nuclear DNA, is inherited from the mother, while nuclear DNA is inherited from both parents. The mutation rates of mtDNA genes are exceptionally high, perhaps 100- to 1000-fold higher than those of nuclear DNA genes [27–29]. Hence, mtDNA genes have a very high sequence evolution rate, on the order of 10-20 times that of comparable nuclear DNA genes [24].

## 1.8   Organism Development—From Genotype to Phenotype

Organism development involves the progressive changes in size, shape, and function during the life of an organism by which its genetic potential (genotype) is translated into functioning mature systems (phenotype). Phenotype is the set of observable characteristics of an individual resulting from the interaction of its genotype with the environment. During organism development, the process in which the shape and structure of an organism develop is termed as morphogenesis. The process in which significant changes occur in the internal and external organs, tissues, and cells of the body of an organism is known as differentiation.

There is a marked difference in the development of plants and animals. In a plant, certain groups of cells are retained throughout the whole life of the plant, and their embryonic capability gives rise to many types of cells. These regions, known as meristems, occur at the growing tips of branches and roots and as cylindrical sheaths around the stem. They consist of rapidly dividing cells capable of assembling into groups that form buds from which new stems, leaves, flowers, or roots may arise.

Animal development refers to the processes that eventually lead to the formation of a new animal starting from cells derived from one or more parent individuals. Most animals have no special regions that retain an embryonic character. In most forms, the whole egg and the whole collection of cells immediately derived from it take part in the development processes and form parts of the developing embryo.

The life cycle of an organism during development from a zygote to an adult is characterized by phases of reprogramming and differentiation. Cells of the early mammalian embryo are epigenetically dynamic and heterogeneous. During early development, this heterogeneity in epigenetic states is associated with the stochastic expression of lineage-determining factors that establish intimate crosstalk with epigenetic modifiers [30]. Generally, somatic cells progressively acquire an increasing array of epigenetic marks that are important for cell specification. By contrast, germ cells and early embryos can reset (or reprogram) epigenetic marks [31–33]. This epigenetic reprogramming of germ cells is an essential characteristic for their immortality and occurs in preparation for the eventual acquisition of totipotency; for example, epigenetic marks in primordial germ cells (PGCs) are reset so that they are in place at the time of fertilization.

The epigenetic landscape is a concept representing embryonic development. It was proposed by Conrad Hal Waddington to illustrate the various developmental pathways a cell might take toward differentiation [34]. The idea has been depicted as a ball rolling down from the top of Waddington's mountain to the bottom of a valley. Waddington's epigenetic landscape has been used for decades to illustrate the progressive restriction of cell differentiation during development. In recent years, a series of landmark experiments showed that cell fate is flexible and reversible and largely extended our understanding of the Waddington landscape. It is now known that cells can transition from a differentiated state to a pluripotent state during reprogramming by the introduction of specific genes, such as the transcription factors Oct3/4, Sox2, Klf4, and c-Myc [35]. Further studies suggested that the ectopic expression of tissue-specific transcription factors can convert a differentiated cell to a cell of another lineage, a process known as transdifferentiation [36]. Conceptually, this process is depicted as moving from one valley to another valley across the ridge of Waddington's landscape.

Pattern formation is one of the classic problems in developmental biology regarding how complex patterns in spatial organization are formed from an apparently uniform field of cells. In 1969, Lewis Wolpert proposed the concept of positional information in which cells acquire positional identities as in a coordinate system and then interpreted their positions to give rise to spatial patterns [37]. Positional information implies that cells have a graded set of values that varies continuously along an axis, and under the influence of the local positional information, a cell obtains a positional value that they interpret in their position by developing in particular ways. One of the main mechanisms for setting up positional information is based on a gradient of morphogens, which are diffusible molecules produced by the cells at a localized site, whose concentration specifies the position [38, 39]. Some of the best evidence for positional information comes from regeneration experiments and the pattering of the leg and antenna in Drosophila and the vertebrate limb [40–42]. The central problems are how positional information is set up, how it is recorded, and then how it is interpreted by the cells. For detailed discussions on position information and pattern formation, we refer the readers to [43–45], and mathematical models of morphogen gradients are introduced in Chap. 7.

## 1.9   Summary

Life is a complex system of information, energy, and matter. Information flux continuously runs over the evolution of life and throughout the whole life of an organism. Different types of information, genetic codes, epigenetic modifications, molecules, and position information work together to determine cell behaviors and cell fate under certain microenvironments. The information is transferred to daughter cells or offspring through different types of cell division, with possible variations or random fluctuations during inheritance. Mathematically, life can be understood as a system of information flux.

From a mathematical point of view, the concentrations of functional molecules, RNA or proteins, can be represented by a vector $\mathbf{x}$. The dynamics of the concentration of functional molecules, $\mathbf{x}(t)$, can often be described by a stochastic process (Chap. 3) or a deterministic chemical rate equation

$$\frac{d\mathbf{x}}{dt} = \mathbf{F}(\mathbf{x}). \qquad (1.1)$$

The function $\mathbf{F}$ describes the regulatory relationships among functional molecules, such as the genetic network, protein-protein interactions, or modifications of molecules, and is usually determined by the information in the genome sequence. The regulations may affect the biochemical interactions involved in the production and degradation/dilution of the molecules. The reaction rates are often represented by parameters $\mathbf{p}$ that are involved in the equation. Hence, the Eq. (1.1) can be rewritten as

$$\frac{d\mathbf{x}}{dt} = \mathbf{F}(\mathbf{x}; \mathbf{p}), \qquad (1.2)$$

with parameters $\mathbf{p}$ explicitly included. In Eq. (1.2), the reaction rates, such as the transcription rate or rate constants of protein-protein interactions, often depend on the epigenetic state or microenvironment of a cell, which are dynamically changing. Hence, the parameters $\mathbf{p}$ are often time dependent and represented as $\mathbf{p}(t)$, which yield the non-autonomous equation (Chap. 4)

$$\frac{d\mathbf{x}}{dt} = \mathbf{F}(\mathbf{x}; \mathbf{p}(t)). \qquad (1.3)$$

During cell division, epigenetic information (histone modification, DNA methylation, etc.) redistributes to the two daughter cells, followed by the re-establishment of epigenetic marks. Meanwhile, the proteins and mRNA are also redistributed to the two daughter cells. Mathematically, cell division brings discontinuous boundary conditions in variables $\mathbf{x}$ and parameters $\mathbf{p}$. Hence, the above equation is extended below over cell divisions.

$$\begin{cases} \dfrac{d\mathbf{x}}{dt} = \mathbf{F}(\mathbf{x}; \mathbf{p}(t)) & \text{Between cell divisions} \\ (\mathbf{x}, \mathbf{p}) \mapsto (\mathbf{x}', \mathbf{p}') \sim (\mathscr{P}(\psi(\mathbf{x}, \mathbf{p})), \mathscr{P}(\phi(\mathbf{x}, \mathbf{p}))) & \text{Cell divison} \end{cases} \qquad (1.4)$$

Here, $\mathscr{P}(f(\mathbf{x}, \mathbf{p}))$ $(f = \psi, \phi)$ represents the random number with a probability density function given by $f(\mathbf{x}, \mathbf{p})$ (Sect. 4.7). The equations of form (1.4) provide a general framework for describing the dynamics of single-cell cross-cell divisions and will be the topics in Chaps. 3–5. In Chap. 7, we show the mathematical models used when the spatial information of molecules is involved, which result in reaction-diffusion equations.

When we try to model the information flux among cell populations across cell divisions, cell behaviors such as cell division, apoptosis, and differentiation/aging must be taken into account. Moreover, we should also consider the heterogeneity of cells due to the variance in epigenetic states, as well as the transition between epigenetic states during cell division. These considerations result in a differential–integral equation

$$
\begin{cases}
\dfrac{\partial Q(t, \mathbf{x})}{\partial t} = -Q(t, \mathbf{x})(\beta(c, \mathbf{x}) + \kappa(\mathbf{x})) + 2\int_\Omega \beta(c, \mathbf{y})Q(t, \mathbf{y})e^{-\mu(\mathbf{y})}p(\mathbf{x}, \mathbf{y})d\mathbf{y}, \\
\qquad c = \int_\Omega Q(t, \mathbf{x})\zeta(\mathbf{x})d\mathbf{x},
\end{cases}
$$

$$(1.5)$$

as described in Chap. 6. Here, $Q(t, \mathbf{x})$ represents the number of cells with epigenetic state $\mathbf{x}$, and $\beta$, $\kappa$, and $\mu$ represent the rates of proliferation, differentiation/senescence, and apoptosis, respectively, during the proliferating phase; $c$ stands for the concentration of growth factors secreted by all cells; $p(\mathbf{x}, \mathbf{y})$ quantifies the transition probability of epigenetic states during cell division. The variable $\mathbf{x}$ can also include the position information of cells. This type of equation provides a general framework for understanding the behavior of heterogeneous stem cell regeneration, including tissue development, degeneration, and abnormal growth [46].

# References

1. Schrodinger, E.: What is Life. Cambridge University Press (1944)
2. McKay, C.P.: What is life-and how do we search for it in other worlds? PLoS Biol. **2**, E302 (2004)
3. Koshland, D.E.: Special essay. The seven pillars of life. Science **295**, 2215–2216 (2002)
4. von Bertalanffy, L.: General System Theory: Foundations, Development, Applications. George Braziller, New York (1968)
5. Elgar, G., Vavouri, T.: Tuning in to the signals: noncoding sequence conservation in vertebrate genomes. Trends Genet. **24**, 344–352 (2007)
6. Maston, G.A., Evans, S.K., Green, M.R.: Transcriptional regulatory elements in the human genome. Ann. Rev. Genom. Hum. Genet. **7**, 29–59 (2006)
7. Probst, A.V., Dunleavy, E., Almouzni, G.: Epigenetic inheritance during the cell cycle. Nat. Rev. Mol. Cell. Biol. **10**, 192–206 (2009)
8. Ruthenburg, A.J., Allis, C.D., Wysocka, J.: Methylation of Lysine 4 on Histone H3: intricacy of writing and reading a single epigenetic mark. Mol. Cell **25**, 15–30 (2006)
9. Cui, K., Zang, C., Roh, T.-Y., Schones, D.E., Childs, R.W., Peng, W., Zhao, K.: Chromatin signatures in multipotent human hematopoietic stem cells indicate the fate of bivalent genes during differentiation. Stem Cell **4**, 80–93 (2009)
10. Zheng, H., Huang, B., Zhang, B., Xiang, Y., Du, Z., Xu, Q., Li, Y., Wang, Q., Ma, J., Peng, X., Xu, F., Xie, W.: Resetting epigenetic memory by reprogramming of histone modifications in mammals. Mol. Cell **63**, 1066–1079 (2016)
11. ...Hao, X., Luo, H., Krawczyk, M., Wei, W., Wang, W., Wang, J., Flagg, K., Hou, J., Zhang, H., Yi, S., Jafari, M., Lin, D., Chung, C., Caughey, B.A., Li, G., Dhar, D., Shi, W., Zheng, L., Hou, R., Zhu, J., Zhao, L., Fu, X., Zhang, E., Zhang, C., Zhu, J.-K., Karin, M., Xu, R.-H., Zhang, K.: DNA methylation markers for diagnosis and prognosis of common cancers. Proc. Natl. Acad. Sci. USA **114**, 7414–7419 (2017)
12. Yang, X., Gao, L., Zhang, S.: Comparative pan-cancer DNA methylation analysis reveals cancer common and specific patterns. Brief Bioinform. **18**, 761–773 (2016)
13. Seisenberger, S., Peat, J.R., Reik, W.: Conceptual links between DNA methylation reprogramming in the early embryo and primordial germ cells. Curr. Opinion Cell Biol. **25**, 281–288 (2013)
14. Wu, H., Zhang, Y.: Reversing DNA methylation: mechanisms, genomics, and biological functions. Cell **156**, 45–68 (2014)
15. Crick, F.: Central dogma of molecular biology. Nature **227**, 561–563 (1970a)

16. Serra-Cardona, A., Zhang, Z.: Replication-coupled nucleosome assembly in the passage of epigenetic information and cell identity. Trends Biochem. Sci. **43**, 136–148 (2018)
17. King, R.C., Stansfield, W.D., Mulligan P.K.: A Dictionary of Genetics, 7th edn. Oxford University Press, Oxford (2007)
18. Fluitt, A., Pienaar, E., Viljoen, H.: Ribosome kinetics and aa-tRNA competition determine rate and fidelity of peptide synthesis. Comput. Biol. Chem. **31**, 335–346 (2007)
19. Dill, K.A., MacCallum, J.L.: The protein-folding problem, 50 years on. Science **338**, 1042–1046 (2012)
20. Bian, L., Michaela, F., Sten, H., Jens, M.: Finding the needle in the haystack: towards solving the protein-folding problem computationally. Crit. Revi. Biochem. Molecul. Biol. 1–28 (2020)
21. Senior, A.W., Evans, R., Jumper, J., Kirkpatrick, J., Sifre, L., Green, T., Qin, C., Žídek, A., Nelson, A.W.R., Bridgland, A., Penedones, H., Petersen, S., Simonyan, K., Crossan, S., Kohli, P., Jones, D.T., Silver, D., Kavukcuoglu, K., Hassabis, D.: Improved protein structure prediction using potentials from deep learning. Nature **577**, 706–710 (2020)
22. Gralka, M., Hallatschek, O.: Environmental heterogeneity can tip the population genetics of range expansions. eLife **8**, 4087 (2019)
23. Fan, L., Wu, D., Goremykin, V., Xiao, J., Xu, Y., Garg, S., Zhang, C., Martin, W.F., Zhu, R.: Phylogenetic analyses with systematic taxon sampling show that mitochondria branch within Alphaproteobacteria. Nat. Ecol. Evol. **4**, 1213–1219 (2020)
24. Wallace, D.C., Chalkia, D.: Mitochondrial DNA genetics and the heteroplasmy conundrum in evolution and disease. Cold Spring Harbor Perspect. Biol. **5**, (2013)
25. Lawless, C., Greaves, L., Reeve, A.K., Turnbull, D.M., Vincent, A.E.: The rise and rise of mitochondrial DNA mutations. Open Biol. **10** (2020)
26. Stewart, J.B., Chinnery, P.F.: The dynamics of mitochondrial DNA heteroplasmy: implications for human health and disease. Nat. Rev. Genet. **16**, 530–542 (2015)
27. Wallace, D.C., Ye, J.H., Neckelmann, S.N., Singh, G., Webster, K.A., Greenberg, B.D.: Sequence analysis of cDNAs for the human and bovine ATP synthase beta subunit: mitochondrial DNA genes sustain seventeen times more mutations. Curr. Genet. **12**, 81–90 (1987)
28. Neckelmann, N., Li, K., Wade, R.P., Shuster, R., Wallace, D.C.: cDNA sequence of a human skeletal muscle ADP/ATP translocator: lack of a leader peptide, divergence from a fibroblast translocator cDNA, and coevolution with mitochondrial DNA genes. Proc. Natl. Acad. Sci. USA **84**, 7580–7584 (1987)
29. Brown, W.M., Prager, E.M., Wang, A., Wilson, A.C.: Mitochondrial DNA sequences of primates: tempo and mode of evolution. J. Molec. Evol. **18**, 225–239 (1982)
30. Hemberger, M., Dean, W., Reik, W.: Epigenetic dynamics of stem cells and cell lineage commitment: digging Waddington's canal. Nat. Rev. Mol. Cell Biol. **10**, 526–537 (2009)
31. Hajkova, P., Erhardt, S., Lane, N., Haaf, T., El-Maarri, O., Reik, W., Walter, J., Surani, M.A.: Epigenetic reprogramming in mouse primordial germ cells. Mech. Dev. **117**, 15–23 (2002)
32. Reik, W., Dean, W., Walter, J.: Epigenetic reprogramming in mammalian development. Science **293**, 1089–1093 (2001)
33. Seisenberger, S., Peat, J.R., Hore, T.A., Santos, F., Dean, W., Reik, W.: Reprogramming DNA methylation in the mammalian life cycle: building and breaking epigenetic barriers. Philos. Trans. R Soc. Lond. B Biol. Sci. **368**, 20110330–20110330 (2013)
34. Waddington, C.H.: The Strategy of the Genes. Geo Allen & Unwin, London (1957)
35. Takahashi, K., Yamanaka, S.: Induction of pluripotent stem cells from mouse embryonic and adult fibroblast cultures by defined factors. Cell **126**, 663–676 (2006)
36. Cho, Y.-D., Ryoo, H.-M.: Trans-differentiation via epigenetics: a new paradigm in the bone regeneration. J. Bone Metab. **25**, 9–13 (2018)
37. Wolpert, L.: Positional information and the spatial pattern of cellular differentiation. J. Theor. Biol. **25**, 1–47 (1969)
38. Crick, F.: Diffusion in embryogenesis. Nature **225**, 420–422 (1970b)
39. Lander, A.D., Nie, Q., Wan, F.Y.M.: Do morphogen gradients arise by diffusion? Dev. Cell **2**, 785–796 (2002)

40. Grimm, O., Coppey, M., Wieschaus, E.: Modelling the Bicoid gradient. Development **137**, 2253–2264 (2010)
41. Gurdon, J.B., Bourillot, P.Y.: Morphogen gradient interpretation. Nature **413**, 797–803 (2001)
42. Lecuit, T., Cohen, S.M.: Dpp receptor levels contribute to shaping the Dpp morphogen gradient in the Drosophila wing imaginal disc. Development **125**, 4901–4907 (1998)
43. Wolpert, L.: Positional information and patterning revisited. J. Theor. Biol. **269**, 359–365 (2011)
44. Lander, A.D.: Morpheus unbound: reimagining the morphogen gradient. Cell **128**, 245–256 (2007)
45. Lander, A.D.: Pattern, growth, and control. Cell **144**, 955–969 (2011)
46. Lei, J.: A general mathematical framework for understanding the behavior of heterogeneous stem cell regeneration. J. Theor. Biol. **492** (2020)

# Chapter 2
# Mathematical Preliminary–Continuous Dynamics

*God used beautiful mathematics in creating the world.*

—Paul Dirac

## 2.1 Introduction

There are two types of mathematical models in biology: data-driven discrete mapping models and mechanism-driven dynamic models. These two types of models reflect different logics in understanding the world.

The mapping models try to establish a mapping relationship between different objects. The mapping relationship provides a dictionary to understand the language of nature, e.g., the genetic code in Table 1.1. For a long time, many biologists (or naturalists) have tried to identify different species and find the relationships among them. More recently, the logic of mapping models has been extensively applied to different aspects of biological and medical studies, including mapping models between DNA sequences and genes, genotypes and phenotypes, transcriptomes and cell types, gene mutants and complex diseases, *etc.* They have been successful in many studies, especially when the objects are static, and hence there, intrinsic mapping relationships exist between them. However, *nothing in biology makes sense except in light of dynamics*. When studying dynamic biological processes, there are no simple maps, and hence, dynamic models are required.

Dynamic models are often used to describe processes that change with time, such as oscillations, growth, production and degradation, and state transitions. Dynamic models are often established based on the mechanisms of how different components in the system interact with each other in governing system behaviors. Our understanding of these mechanisms comes from empirical knowledge from either basic principles or working hypotheses derived from experimental observations. The basic principles often include fundamental roles in science and well-established facts, such as Newton's law of motion, the law of mass action, the conservation law of flux,

J. Lei, *Systems Biology*, Lecture Notes on Mathematical Modelling in the Life Sciences,
https://doi.org/10.1007/978-3-030-73033-8_2

the conservation of energy, and the central dogma of genetics. For example, Isaac Newton (1642–1727) deduced the law of universal gravitation from Newton's second law and Kepler's laws of planetary motion. In physiology, the Hodgkin-Huxley equation is a good example of a dynamic model. The equation was established based on a hypothesis of the gating mechanism of ion channels and a series of dynamic information of neuronal electrical signals under various conditions [1]. The Hodgkin-Huxley equation nicely describes the mechanism of action potential and predicts a series of experimental results and has become a basic equation in computational neuronscience.

Dynamic models can either be continuous or discrete. In this lecture, we mainly introduce continuous models. Mathematically, continuous dynamic models are often formulated by differential equations, which describe the dynamics of observable variables through the rules of how these variables change at an infinitesimal time step. These types of models enable us to predict the long-term dynamics from an initial state. In this chapter, we introduce a mathematical preliminary of different types of differential equations. Here, we mainly focus on the basic concepts and results of related topics. For further details, please refer to mathematics textbooks, such as [2] and [3] for ordinary differential equations, [4] for stochastic differential equations, [5] and [6] for delay differential equations.

## 2.2  Ordinary Differential Equations

Differential equations are an important mathematical tool to describe our natural dynamic processes. An ordinary differential equation (ODE) is an equation that relates a function $x(t)$ of one variable to its ordinary derivatives (as opposed to partial derivatives). In applications, the functions usually represent physical quantities, the derivatives represent their states of change, and the differential equations define the relationships between the two.

The derivative, written as $x'$ or $\frac{dx}{dt}$ (in this lecture, we usually take time $t$ as the argument), of a function $x(t)$ expresses its rate of change at each point—that is, how fast the value $x$ increases or decreases as the value of the argument $t$ increases or decreases (Fig. 2.1). For the linear function $x(t) = at + b$, the rate of change is simply its slope, expressed as $x'(t) = a$. For a general function, the rate of change varies along the curve of the function, and the precise way to define and calculate the derivative is based on the subject of differential calculus.

In general, the derivative of a function is again a function of the variable $t$, and therefore, the derivative of the derivative can also be calculated, namely, $x''$ or $\frac{d^2x}{dt^2} = \frac{d}{dt}\left(\frac{dx}{dt}\right)$, and is called the second-order derivative of the original function. Higher-order derivatives can be similarly defined. We usually use the notation $x^{(n)}$ for higher-order derivatives, i.e., $x^{(n)} = \frac{d^n x}{dt^n} = \frac{d}{dt}\left(\frac{d^{n-1}x}{dt^{n-1}}\right)$.

**Fig. 2.1**  Illustration of the derivative

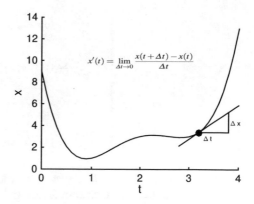

A differential equation gives the relationship that connects a function to its derivatives. The order of a differential equation is defined as that of the highest-order derivative it contains. The general form of an $n$th order ODE is given as

$$F(t, x, x', \ldots, x^{(n)}) = 0, \tag{2.1}$$

or

$$x^{(n)} = f(t, x, x', \ldots, x^{(n-1)}). \tag{2.2}$$

When a differential equation does not depend on the variable $t$, we call the equation an autonomous differential equation.

For an $n$th order ODE of form (2.2), we can introduce new functions $x_0(t) = x(t), x_1(t) = x'(t), x_2(t) = x''(t), \ldots, x_{n-1}(t) = x^{(n-1)}(t)$, so that Eq. (2.2) can be rewritten as a system of $n$ differential equations

$$\begin{cases} \dfrac{dx_0}{dt} = x_1 \\[2mm] \dfrac{dx_1}{dt} = x_2 \\[2mm] \cdots \quad \cdots \\[2mm] \dfrac{dx_{n-1}}{dt} = f(t, x_0, x_1, \ldots, x_{n-1}) \end{cases} \tag{2.3}$$

More generally, a system of $n$ differential equations is written as

$$\begin{cases} \dfrac{dx_0}{dt} = f_0(t, x_0, x_1, \ldots, x_{n-1}) \\[2ex] \dfrac{dx_1}{dt} = f_1(t, x_0, x_1, \ldots, x_{n-1}) \\[2ex] \cdots \quad\quad \cdots \\[2ex] \dfrac{dx_{n-1}}{dt} = f_{n-1}(t, x_0, x_1, \ldots, x_{n-1}) \end{cases} \quad\quad (2.4)$$

The functions $f_0, f_1, \ldots, f_{n-1}$ give the relations between the derivatives $x_0'(t), x_1'(t),$ $\ldots, x_{n-1}'(t)$ and the unknown functions $x_0(t), x_1(t), \ldots, x_{n-1}(t)$. A system of form (2.4) is also an $n$th order differential equation.

### 2.2.1  First-Order Differential Equations

Considering a quantity $x$ that changes with time $t$ (represented by a function $x(t)$), the rate of change is given by the derivative $x'(t)$. When $t$ increases to $t + \Delta t$, the quantity $x$ changes from $x(t)$ to $x(t + \Delta t)$. Hence, the change in $x$ in $\Delta t$ gives the average changing rate

$$\text{average changing rate} = \frac{x(t + \Delta t) - x(t)}{\Delta t}.$$

Letting $\Delta t$ approach to 0, the limitation of the above average changing rate (mathematically, we can prove that the limit exists when $x(t)$ is a smooth function) gives the instant changing rate of $x(t)$ at time $t$, which is denoted as $x'(t)$ or $\frac{dx}{dt}$, i.e.,

$$x'(t) = \lim_{\Delta t \to 0} \frac{x(t + \Delta t) - x(t)}{\Delta t}. \quad\quad (2.5)$$

When the time $t$ varies, $x'(t)$ defines a function from $x(t)$. We call $x'(t)$ the derivative of $x(t)$. Hence, the derivative of a function measures the rate of change of the original function.

Now, we introduce several simple forms of differential equations through the modeling of cell growth. Let $N(t)$ represent the number of cells in a cell population at time $t$. The change rate of $N(t)$ includes an increase due to self-renewal and a decrease due to cell death, hence

cell number change = increase due to self-renewal − decrease due to cell death

We assume that each cell undergoes self-renewal at a rate of $\beta$ (cells/day), and death at a rate of $\gamma$ (day$^{-1}$). In a time interval $\Delta t$, the cell number increases by $\beta N(t) \Delta t$

due to self-renewal and decreases by $\gamma N(t)\Delta t$ due to cell death. Thus, the net change of the cell number is

$$N(t + \Delta t) - N(t) = \beta N(t)\Delta t - \gamma N(t)\Delta t.$$

Dividing both sides by $\Delta t$ and taking the limit with $\Delta t$ approaching 0, we obtain an exponential growth model

$$\frac{dN}{dt} = \alpha N(t), \tag{2.6}$$

where $\alpha = \beta - \gamma$ is the net growth rate. Equation (2.6) gives a simple differential equation model for the exponential growth (or decay) of the cell number.

When the net growth rate $\alpha$ is a constant and given the initial cell number $N(0) = N_0$, we can solve (2.6) to obtain the cell number

$$N(t) = N_0 e^{\alpha t} \tag{2.7}$$

at any time $t$. In particular, the cell number exponentially increases when $\alpha > 0$, and exponentially decreases to 0 when $\alpha < 0$. Taking the logarithm to (2.7) gives

$$\log N(t) = \log N_0 + \alpha t. \tag{2.8}$$

This formula gives the method to estimate the net growth rate through the cell population number, which is given by the slope of $\log N(t)$ with respect to time $t$.

The above model of exponential growth is a linear differential equation. Based on exponential growth, when $t \to \infty$, the cell number either tends to infinity or tends to 0. This is not realistic in most situations. In many situations, we may need to consider saturation due to resource limitations. Hence, we can introduce a maximum capacity $N_{\max}$, so the net growth rate is limited by the ratio between the cell number $N$ and the maximum capacity $N_{\max}$, and hence

$$\alpha = \alpha_0 \left( 1 - \frac{N}{N_{\max}} \right).$$

Thus, we obtain the following logistic growth model [7]:

$$\frac{dN}{dt} = \alpha_0 \left( 1 - \frac{N}{N_{\max}} \right) N. \tag{2.9}$$

Given an initial cell number $N(0) = N_0$, Eq. (2.9) can be solved analytically, which gives

$$N(t) = N_{\max} \frac{(N_0/N_{\max})e^{\alpha_0 t}}{1 + (N_0/N_{\max})(e^{\alpha_0 t} - 1)}. \tag{2.10}$$

Thus, when $\alpha_0 > 0$, $N(t)$ approaches the maximum capacity $N_{\max}$ when $t$ goes to infinity.

The logistic model is widely used in modeling cell population dynamics when cell divisions are limited by external resources. Nevertheless, the limitation can also come from cell-to-cell interactions, such as the inhibition of cell proliferation through cytokines (Chap. 6). In this case, the cell proliferation rate $\beta$ may depend on the total cell number $N$ through a nonlinear function $\beta = \beta(N)$. Thus, we have a nonlinear cell growth model

$$\frac{dN}{dt} = \beta(N)N - \gamma N. \tag{2.11}$$

Given an initial condition $N(0) = N_0$, the population dynamics $N(t)$ can be obtained by solving Eq. (2.11).

Both of the Eqs. (2.9) and (2.11) are nonlinear differential equations. In general, nonlinear differential equations are analytically unsolvable (expect for specific exceptions); hence, numerical schemes are required to solve the equation. The simplest numerical method was originally devised by Euler and is called Euler's method. From the Eq. (2.11), we write

$$\frac{dN}{dt} = \frac{N(t + \Delta t) - N(t)}{\Delta t}$$

when $\Delta t$ is small enough, then

$$N(t + \Delta t) = N(t) + (\beta(N)N - \gamma N)\Delta t. \tag{2.12}$$

Equation (2.12) gives the iterative scheme of solving (2.11); this scheme gives an application of Euler's method. For readers who are not familiar with numerical methods, XPP-AUT is a great tool to solve differential equations and perform bifurcation analysis [8].

In applications, we are interested in long-term behaviors when the time $t$ approaches infinity, e.g., the stable steady states. Given a differential equation

$$\frac{dx}{dt} = f(x). \tag{2.13}$$

A steady state of the equation, denoted by $x(t) \equiv x^*$, is given by a constant $x^*$ that vanishes the right-hand side $f(x)$, i.e.

$$f(x^*) = 0.$$

An equation may have multiple steady states. For example, when the proliferation rate is given by a Hill function (Sect. 2.8)

$$\beta(N) = \beta_0 \frac{\theta^n}{\theta^n + N^n}, \tag{2.14}$$

Equation (2.11) always has a zero steady state $N_1^* = 0$, and when $\beta_0 > \gamma$, it has a positive steady state $N_2^* = \theta(\frac{\beta_0}{\gamma} - 1)^{1/n}$.

In applications, we are often interested in stable steady states. The stability of (2.13) at the steady state $x = x^*$ represents how the steady state resists small perturbations. Mathematically, the steady state $x = x^*$ is stable if for any positive constant $\varepsilon$, there exists $\delta > 0$, so that for any $z^*$ satisfies $|z^*| < \delta$, the solution of (2.13) with initial condition $x(0) = x^* + z^*$ satisfies $|x(t) - x^*| < \varepsilon$ for all $t > 0$. More specifically, if $\lim_{t \to \infty} |x(t) - x^*| = 0$, the steady state $x(t) \equiv x^*$ is asymptotically stable.

To obtain the condition for the stability of the state $x = x^*$, let $z(t) = x(t) - x^*$, then $z(t)$ satisfies a differential equation

$$\frac{dz}{dt} = f(x^* + z). \tag{2.15}$$

The displacement $z(t)$ represents a perturbation of the original equation at $x = x^*$. When the perturbation is small, we can take the Taylor expansion of $f(x^* + z)$ around $z = 0$ and note that $f(x^*) = 0$,

$$f(x^* + z) \approx f'(x^*)z + o(z)$$

Here, $o(z)$ means higher-order small values compared to $z$. Thus, up to the first-order approximation, we have

$$\frac{dz}{dt} = f'(x^*)z.$$

This approximation gives the exponential growth of the perturbation. Particularly, when $z(0) = z^*$, we have

$$z(t) = z^* e^{f'(x^*)t}.$$

Hence, when $f'(x^*) > 0$, the perturbation tends to infinity, and the original steady state $x = x^*$ is unstable; when $f'(x^*) < 0$, the perturbation tends to zero, and the steady state $x = x^*$ is stable. Thus, the stability of (2.13) at the steady state $x = x^*$ is determined by the sign of the derivative $f'(x^*)$.

For the nonlinear growth model (2.11), when there is a zero solution $N_1^* = 0$ and a positive steady state $N_2^* > 0$, we have

$$\frac{d}{dN}(\beta(N)N - \gamma N)\bigg|_{N=0} = \beta(0) - \gamma$$

and

$$\frac{d}{dN}(\beta(N)N - \gamma N)\bigg|_{N=N_2^*} = \beta'(N_2^*)N_2^*.$$

Thus, the zero solution is stable if and only if $\beta(0) < \gamma$, and the positive solution $N_2^*$ is stable if and only if $\beta'(N_2^*) < 0$. In particular, when the proliferation rate is a decreasing function (for example, the Hill function (2.14)), the zero solution is stable if and only if there is no positive steady state, and when a positive steady state exists, the positive steady state is always stable.

The concept of steady state and the corresponding stability can easily be extended to general $n$th order differential equations. Consider an $n$th order autonomous differential equation

$$\frac{d\mathbf{x}}{dt} = \mathbf{f}(\mathbf{x}),\tag{2.16}$$

where $\mathbf{x} = (x_1, x_2, \ldots, x_n)^T$, $\mathbf{f} = (f_1, f_2, \ldots, f_n)^T$, with $x_i \in C^1(\mathbb{R}, \mathbb{R})$ and $f_i \in C^1(\mathbb{R}^n, \mathbb{R})$. The steady state of (2.16), or the steady state solution $\mathbf{x}(t) = \mathbf{x}^*$, means a vector $\mathbf{x}^* \in \mathbb{R}^n$ that satisfies

$$\mathbf{f}(\mathbf{x}^*) = \mathbf{0}.$$

Let $\mathbf{x}(t) = \mathbf{x}^*$ be a steady-state solution of (2.16); if for any $\varepsilon > 0$, there exists $\delta > 0$, so that for any $\mathbf{z}$ such that $\|\mathbf{z} - \mathbf{x}^*\| < \delta$, the solution of (2.16) with initial condition $\mathbf{x}(0) = \mathbf{z}$ satisfies $\|\mathbf{x}(t) - \mathbf{x}^*\| < \varepsilon$ for all $t > 0$. Moreover, if $\lim_{t \to \infty} \|\mathbf{x} - \mathbf{x}^*\| = 0$, the steady-state solution $\mathbf{x}(t) = \mathbf{x}^*$ is asymptotically stable. Mathematically, the steady state $\mathbf{x}(t) = \mathbf{x}^*$ is stable if all eigenvalues of the Jacobin matrix

$$A = \left.\frac{\partial \mathbf{f}}{\partial \mathbf{x}}\right|_{\mathbf{x}=\mathbf{x}^*} = \left.\begin{bmatrix} \frac{\partial f_1}{\partial x_1} & \frac{\partial f_1}{\partial x_2} & \cdots & \frac{\partial f_1}{\partial x_n} \\ \frac{\partial f_2}{\partial x_1} & \frac{\partial f_2}{\partial x_2} & \cdots & \frac{\partial f_2}{\partial x_n} \\ \cdots & \cdots & \cdots & \cdots \\ \frac{\partial f_n}{\partial x_1} & \frac{\partial f_n}{\partial x_2} & \cdots & \frac{\partial f_n}{\partial x_n} \end{bmatrix}\right|_{\mathbf{x}=\mathbf{x}^*}$$

have negative real parts [2, 3].

### 2.2.2  Second-Order Differential Equations

Here, we briefly review the mathematical results often used in the discussion of oscillation solutions.

Consider a system of second-order differential equations

$$\frac{dx}{dt} = P(x, y), \quad \frac{dy}{dt} = Q(x, y),\tag{2.17}$$

where $x$ and $y$ are (nonnegative) concentrations of two components. In (2.17), the right-hand side $\mathscr{X} = (P(x, y), Q(x, y))$ defines a vector field in the $x$–$y$ plane, and hence, we can apply quantitative analysis to obtain the properties of the equations. Here, we always assume that both $P(x, y)$ and $Q(x, y)$ have continuous derivatives over a domain $\Omega \subseteq \mathbb{R}^+ \times \mathbb{R}^+$, and hence, $\mathscr{X}$ defines a smooth vector field. If there is a point $(x_0, y_0) \in \Omega$ such that $P(x_0, y_0) = Q(x_0, y_0) = 0$, the point $(x_0, y_0)$ is called the *singular point* (or the *fixed point*). In this case, $(x(t), y(t)) \equiv (x_0, y_0)$ is a constant solution of (2.17). We often call the fixed point $(x_0, y_0)$ the *steady state* of equation (2.17).

Let $\varphi(t) = (x(t), y(t))$ be a solution of (2.17) defined by a smooth vector field; then, $\varphi(t)$ gives a curve in the phase plane, and the tangent at each point $(x(t), y(t))$ satisfies

$$\left.\frac{d\varphi(t)}{dt}\right|_{(x(t),y(t))} = \left.\left(\frac{dx}{dt}, \frac{dy}{dt}\right)\right|_{(x(t),y(t))} = (P(x, y), Q(x, y)).$$

In particular, any periodic solution $\varphi(t)$ defines a closed orbit in the phase plane. Hence, geometrically, to determine the periodic solution of (2.17), we only need to find the closed orbit from the vector field $(P(x, y), Q(x, y))$.

Let $(P, Q)$ be a smooth vector field in $\Omega$ and $\Gamma$ be a simple closed curve (not necessarily a closed orbit) that contains no singular point. When a point $M$ moves along $\Gamma$ anticlockwise for one cycle, the vector $(P, Q)$ rotates $j$ full circles, i.e., the vector $(P, Q)$ rotates $2\pi j$ ($j$ is an integer). We call this integer $j$ the *rotation number* of the vector field $(P, Q)$ along the closed curve $\Gamma$ and denote it by $J(\Gamma)$ [3]. For a smooth vector field, the rotation number is calculated by the integral along the circle $\Gamma$

$$J(\Gamma) = \frac{1}{2\pi} \oint_\Gamma \frac{PdQ - QdP}{P^2 + Q^2}. \tag{2.18}$$

The following are the basic properties of the rotation number:

(1) If $\Gamma$ is a closed orbit of (2.17), then $J(\Gamma) = 1$.
(2) If there is no singular point of $(P, Q)$ inside the region surrounded by $\Gamma$, then $J(\Gamma) = 0$.
(3) The rotation number $J(\Gamma)$ is unchanged when $\Gamma$ changes continuously and does not encounter singular points.

Let $S_0 = (x_0, y_0)$ be a singular point of $(P, Q)$; we take any closed curve $\Gamma$ surrounding $S_0$, and there is no other singular point inside $\Gamma$. We define the *index* of $S_0$ (in the vector field $(P, Q)$) as the rotation number along $\Gamma$, i.e., $J_{S_0} = J(\Gamma)$.

The above properties yield a criterion for a closed orbit, if it exists, of the vector field $(P, Q)$: any closed orbit $\Gamma$ must contain singular points within the surrounding region. Moreover, if there are only isolated singular points inside a closed orbit, the summation of the indexes of these singular points equals 1.

The Bendixson criterion below is a straightforward way to determine the nonexistence of a closed orbit [3].

**Bendixson Criterion**: Consider Eq. (2.17); if in a simply connected domain $G$, the divergence

$$\text{div}(P, Q) = \frac{\partial P}{\partial x} + \frac{\partial Q}{\partial y}$$

has a constant sign, there is no closed trajectory in the domain $G$.

This criterion was generalized by Dulac as follows [3]: If $G$ is a simply connected domain and there is a function $B(x, y) \in C^1(G)$ so that

$$\frac{\partial(BP)}{\partial x} + \frac{\partial(BQ)}{\partial y}$$

has a constant sign in $G$, the domain $G$ does not contain any simple closed trajectory.

The Poincaré-Bendixon theorem is often used to determine the existence of a closed orbit in a smooth vector field [3].

**Poincaré-Bendixon Theorem**:  Let $R$ be a region of the plane that is closed and bounded. Consider a dynamical system in $R$ where the vector field $\mathscr{X} = (P(x, y), Q(x, y))$ is at least $C^1$. Assume that $R$ contains no fixed point of $\mathscr{X}$, and that there exists a trajectory $\gamma$ of $\mathscr{X}$ starting in $R$ that will stay in $R$ for all future times. Then, either $\gamma$ is a closed orbit or asymptotically approaches a closed orbit; in other words, there exists a limit cycle in $R$.

Practically, we have a scheme to determine the existence of a periodic solution by applying the Poincaré-Bendixson theorem: construct an annular trapping region $R$ in the plane so that the vector field points into $R$ on all boundaries of $R$. If we can construct a trapping region $R$ that contains no fixed points, every trajectory of $\mathscr{X}$ starting in $R$ will stay in $R$ for all future times, and hence asymptotically approach a closed orbit.

For a vector field defined by a biological system, the solutions are often bounded. Hence, for sufficiently large constants $M_1$ and $M_2$, the vector field points into the region bounded by a curve

$$\Gamma_1 : (0, 0) \rightarrow (M_1, 0) \rightarrow (M_1, M_2) \rightarrow (0, 0).$$

If there is only one steady state within the region bounded by $\Gamma_1$ and the steady state is unstable, we can always define a circle $\Gamma_0$ that centers on the steady state. Thus, it is easy to see that the region $R$ bounded by $\Gamma_1$ and $\Gamma_0$ is a trapping region and contains no fixed points, and there exists a limit cycle in $R$ according to the Poincaré-Bendixson theorem.

To analyze the stability of a fixed point $S_0 = (x^*, y^*)$, we linearize equation (2.17) at $S_0$ to obtain the coefficient matrix

$$A = \begin{bmatrix} a & b \\ c & d \end{bmatrix} = \left. \begin{bmatrix} \dfrac{\partial P}{\partial x} & \dfrac{\partial P}{\partial y} \\ \dfrac{\partial Q}{\partial x} & \dfrac{\partial Q}{\partial y} \end{bmatrix} \right|_{(x,y)=(x^*,y^*)}.$$

The corresponding eigenvalues are

$$\lambda_{1,2} = \frac{-p \pm \sqrt{p^2 - 4q}}{2},$$

**Fig. 2.2** Classification of
stabilities for the steady state
of planar systems

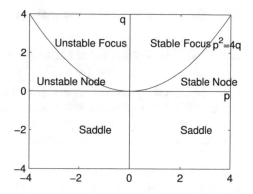

where

$$p = -\mathrm{tr}(A) = -(a+d), \quad q = \det(A) = ad - bc.$$

Here, $p$ is the negative trace of the coefficient matrix $A$, and $q$ is the determinant. Here, we only consider the simple fixed points so that $q \neq 0$; otherwise, we need to consider higher-order approximations. We can classify different types of fixed points according to the eigenvalues: saddle ($\lambda_1$ and $\lambda_2$ are real numbers with different signs), node ($\lambda_1$ and $\lambda_2$ are real numbers with the same sign), focus ($\lambda_1$ and $\lambda_2$ are complex numbers with nonzero real parts), and center ($\lambda_1$ and $\lambda_2$ are pure imaginary numbers). The fixed point is stable when all eigenvalues have negative real parts; otherwise, the fixed point is unstable (Fig. 2.2). Thus, only when $p, q > 0$, the steady state is stable.

We can further obtain the indexes of the fixed points. The index of a node or a focus is 1, and the index of a saddle is $-1$. Hence, if the system has a stable limit cycle and there is a single fixed point within the limit cycle, the fixed point must have index 1 and hence is either a node or a focus ($q > 0$). In particular, from the Poincaré-Bendixon theorem, if all solutions of the system defined by (2.17) are bounded, the system has only one fixed point, and the trace of the linearized coefficient matrix around the fixed point is larger than zero ($\mathrm{tr}(A) > 0$), the fixed point is unstable and there is a stable periodic solution. For a two-component biochemical reaction or genetic network system, $\partial P/\partial x$ and $\partial Q/\partial y$ are usually negative and then $\mathrm{tr}(A) < 0$. Hence, the above implies that a two-component system should have an autocatalytic interaction (or positive feedback) ($\partial P/\partial x > 0$ or $\partial Q/\partial y > 0$) to exhibit sustained oscillations. Nevertheless, a system with only positive feedback is a monotone dynamical system and would not have oscillation solutions. Hence, combinations of positive feedback with negative feedback are required to exhibit sustained oscillations in a two-component system. This conclusion provides a principle to design genetic oscillators [9].

## 2.3  Delay Differential Equations

### 2.3.1  Delay Differential Equations

Many biological systems involve feedbacks with time delays (lags) between the
sensor and the corresponding changes, such as the physiological control systems of
$CO_2$ elimination or cell production [10]. Delay differential equations are often used
to model the dynamics of these feedback systems with lag time. Delay differential
equations differ from ordinary differential equations in that the derivative at any time
depends on the solution at prior times. For the basic concepts and notions of delay
differential equations, please refer to [5, 6].

A simple discrete delay equation has the form

$$\frac{dx}{dt} = f(t, x(t), x(t - \tau_1), x(t - \tau_2), \ldots, x(t - \tau_k)), \tag{2.19}$$

where the time delays (lags) $\tau_j$ are positive constants. Sometimes the lags can depend
on the state, i.e. $\tau_i = \tau_i(t, x(t))$, and the equations are said to have state-dependent
delays. In many applications, the delays are not deterministic but are random numbers. When the lag time is given by a random number with probability density $g(\tau)$,
we have a distributed delay differential equation

$$\frac{dx}{dt} = \int_0^\infty f(t, x(t), x(t - s))g(s)ds. \tag{2.20}$$

An example of a delay differential equation model can be derived from feedback in
gene expression, in which the expression products regulate the transcription process,
and the delay comes from multiple steps of protein modification (Sect. 5.5.3). In this
case, the concentration $X$ of the gene expression product evolves following a delay
differential equation

$$\frac{dX}{dt} = f(Z) - bX, \quad Z(t) = X(t - \tau). \tag{2.21}$$

Here, $Z(t)$ represents the concentration of effective molecules at time $t$, which
depends on the past of $X$ due to the multiple modification process (e.g., Fig. 5.27).

The delay differential equation model can also be derived from an age-structured
model of cell regeneration of form (see Sect. 6.2.1)

$$\nabla s(t, a) = -\mu s(t, a), \quad (t > 0, 0 \le a \le \tau) \tag{2.22}$$

$$\frac{dQ}{dt} = 2s(t, \tau) - (\beta(Q) + \kappa)Q, \quad (t > 0). \tag{2.23}$$

Here, $\nabla = \partial/\partial t + \partial/\partial a$. The boundary condition at $a = 0$ is given by

$$s(t, 0) = \beta(Q(t))Q(t). \tag{2.24}$$

We can solve (2.22) using the characteristic line method and substitute the obtained $s(t, \tau)$ into (2.23) to obtain a delay differential equation (when $t > \tau$) (detailed in Sect. 6.2.2)

$$\frac{dQ}{dt} = -(\beta(Q) + \kappa)Q + 2e^{-\mu\tau}\beta(Q(t - \tau))Q(t - \tau). \tag{2.25}$$

To determine the solution of a delay differential equation of form (2.19) or (2.20), we need to provide an initial history function to specify the value of the solution before $t = 0$, i.e., the initial function

$$x(\theta) = \varphi(\theta), \theta \le 0. \tag{2.26}$$

Hence, the solution of a delay differential equation depends on an infinite dimensional initial function.

### 2.3.2  Stability Analysis of Delay Differential Equations

Consider a first-order delay differential equation

$$\frac{dx}{dt} = f(x, x_\tau) \tag{2.27}$$

where $x_\tau = x(t - \tau)$. We assume that $x(t) \equiv x^*$ is a steady-state solution of (2.27). To analyze the stability of the steady state, let $y(t) = x(t) - x^*$ and linearize the equation around the steady state $x^*$ to obtain

$$\frac{dy}{dt} = -ay(t) + by(t - \tau) + \text{higher-order terms}, \tag{2.28}$$

where

$$a = -\frac{\partial f(x^*, x^*)}{\partial x}, \quad b = \frac{\partial f(x^*, x^*)}{\partial x_\tau}.$$

Here, we note a difference in the sign of the definitions in $a$ and $b$. In many biological systems, $\partial f(x, x_\tau)/\partial x < 0$ due to the process of dissipation; hence, we put a negative sign so that the coefficient $a$ is positive and biologically more straightforward, e.g., the degradation rate of molecules or death rate of cells.

We omit the higher-order terms and find the solution of (2.28) of form $y(t) = y_0 e^{\lambda t}$; the exponent $\lambda$ satisfies the characteristic equation

$$\lambda + a = b e^{-\lambda \tau}. \tag{2.29}$$

Equation (2.29) gives the equation for the eigenvalues of the delay equation (2.28). From the stability theory of delay-differential equations, the zero solution of (2.28) is stable (and hence the steady state $x(t) = x^*$ of (2.27) is stable) only when all eigenvalues have negative real parts.

When $\tau = 0$, there is only one eigenvalue $\lambda = b - a$, and hence, the zero solution of (2.28) is stable only when $a - b > 0$. When $\tau > 0$, Eq. (2.29) cannot be solved explicitly. Here we assume that $a - b > 0$.

Substituting $\lambda = i\omega$ (here, $\omega$ is a positive real number) into (2.29) and separating the real part and the imaginal part, we have (here, we note $e^{-i\omega\tau} = \cos \omega\tau - i \sin \omega\tau$)

$$\begin{cases} a - b \cos \omega\tau = 0 \\ \omega + b \sin \omega\tau = 0. \end{cases} \tag{2.30}$$

For $\omega \in \mathbb{R}^+$, Eq. (2.30) gives a curve in the $(a, b)$ plane, which corresponds to the Hopf bifurcation of the original Eq. (2.27). For the equilibrium state $x^*$, this curve gives the boundary of the parameter region $S$ for the stable equilibrium state (Fig. 2.3).

From (2.30), it is easy to have $a = b$ when $\omega = 0$, and

$$a^2 + \omega^2 = b^2$$

when $\omega \neq 0$, which implies $|b| > |a|$. When $\omega > 0$, the equation (2.30) yields

$$\omega = -a \tan \omega\tau, \quad b = a \sec \omega\tau. \tag{2.31}$$

**Fig. 2.3** The region $S$ in the $(a, b)$ plane for the stability of the zero solution of (2.28)

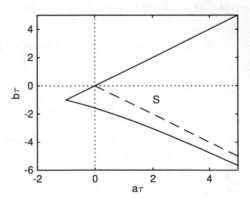

We note that when $b = 0$, the zero solution of (2.28) is stable when $a > 0$. Hence, for any $\tau \geq 0$, the parameter region in the $(a, b)$ plane that stabilizes the zero solution of (2.28) is given by

$$S = \{(a, b) \in \mathbb{R}^2 | a \sec \omega\tau < b < a, \text{ where } \omega = -a \tan \omega\tau, a > -\frac{1}{\tau}, \omega \in (0, \frac{\pi}{\tau})\}. \tag{2.32}$$

Figure 2.3 shows the region $S$.

## 2.4 Stochastic Differential Equations

### 2.4.1 Stochastic Differential Equations and Stochastic Integral

Ordinary differential equations are usually used to describe deterministic dynamics. When there are random perturbations to the deterministic dynamics, we need to apply stochastic differential equations in the modeling. Most contents here referred to [4] and [11].

First, we introduce the Wiener process. A standard Wiener process (often called Brownian motion) is a random process $W_t(\omega)$ (hereafter, we omit the random variable $\omega \in \Omega$) that depends continuously on $t \in \mathbb{R}$ and satisfies the following:

(1) $W_0 = 0$,

(2) For any $t, \tau \geq 0$,
$$W_{t+\tau} - W_t \sim \sqrt{\tau}\mathcal{N}(0, 1),$$

where $\mathcal{N}(0, 1)$ represents the standard normal distribution with zero mean and unit variance.

(3) For $0 \leq t_1 < t_2 < t_3 < t_4$, $W_{t_2} - W_{t_1}$ and $W_{t_4} - W_{t_3}$ are independent of each other.

From (2) and (3), it is easy to have

$$\langle (W_{t+\tau} - W_t) \rangle = 0, \langle (W_{t+\tau} - W_t)^2 \rangle = \tau \tag{2.33}$$

and

$$\langle (W_{t_4} - W_{t_3})(W_{t_2} - W_{t_1}) \rangle = 0. \tag{2.34}$$

Hereafter, $\langle \cdot \rangle$ denotes the expectation of a random variable over a proper statistical space.

Through the standard Wiener process $W_t$, a general form of a one-dimensional stochastic differential equation is given by

$$dx = f(x, t)dt + g(x, t)dW_t, \qquad (2.35)$$

where $dW_t = W_{t+dt} - W_t$ and $f(x, t), g(x, t)$ are continuous functions of $x$ and $t$. From the definition of the Wiener process, $dW_t$ has a normal distribution with zero mean, is independent of the time $t$, and has the variance

$$\langle (dW_t)^2 \rangle = dt.$$

Hence, we can consider $dW_t$ as an infinitesimal unit of order $(dt)^{1/2}$.

In many literatures, the stochastic differential equation of form (2.35) is also written as

$$\frac{dx}{dt} = f(x, t) + g(x, t)\xi(t), \qquad (2.36)$$

where $\xi(t)$ is a standard white noise. The expression of form (2.36) is mathematically not precise because the solution $x(t)$ is usually not differentiable with respect to $t$, and hence, the left-hand side $\frac{dx}{dt}$ is not well defined. The equation of form (2.36) should be understood as the standard form (2.35).

The solution of (2.35) is a stochastic process $x(t)$ that satisfies

$$x(t) = x(0) + \int_0^t f(x(s), s)ds + \int_0^t g(x(s), s)dW_s. \qquad (2.37)$$

Writing the integrals in (2.37) in the form of summations, we have

$$x(t) = x(0) + \lim_{\Delta t \to 0} \sum_{0 < t_i < t} f(x(t_i'), t_i')\Delta t_i + \lim_{\Delta s \to 0} \sum_{0 < s_i < t} g(x(s_i'), s_i')\Delta W_{s_i},$$

where $\Delta t_i = t_{i+1} - t_i$, $\Delta W_{s_i} = W_{s_{i+1}} - W_{s_i}$, $0 = t_0 < t_1 < \cdots < t_n = t$ and $0 \le s_0 < s_1 < \cdots < s_n = t$ are subdivisions of the interval $[0, t]$, and $t_i \le t_i' \le t_{i+1}$, $s_i \le s_i' \le s_{i+1}$. When $\Delta t_i \to 0$, the first summation converges to the Riemann integral $\int_0^t f(x(t), t)dt$ and is independent of the selection of $t_i'$. The second summation, however, may depend on how we select $s_i'$ from the interval $[s_i, s_{i+1}]$.

To show the dependence on $s_i'$, we calculate the following integral:

$$\int_0^t W_s dW_s.$$

If we take $s_i'$ as the left point $s_i' = s_i$,

$$\left\langle \int_0^t W_s dW_s \right\rangle = \left\langle \lim_{\Delta s \to 0} \sum_{i=0}^{n-1} W_{s_i}(W_{s_{i+1}} - W_{s_i}) \right\rangle$$

$$= \left\langle \lim_{\Delta s \to 0} \sum_{i=0}^{n-1} \left( \frac{1}{2}(W_{s_{i+1}} + W_{s_i}) - \frac{1}{2}(W_{s_{i+1}} - W_{s_i}) \right) (W_{s_{i+1}} - W_{s_i}) \right\rangle$$

$$= \frac{1}{2} \left\langle \lim_{\Delta s \to 0} \sum_{i=0}^{n-1}(W_{s_{i+1}}^2 - W_{s_i}^2) \right\rangle - \frac{1}{2} \left\langle \lim_{\Delta s \to 0} \sum_{i=0}^{n-1}(W_{s_{i+1}} - W_{s_i})^2 \right\rangle$$

$$= \frac{1}{2}(\langle W_t^2 \rangle - \langle W_0^2 \rangle) - \frac{1}{2}t$$

$$= 0.$$

If we take $s_i'$ so that $W_{s_i'} = (W_{s_{i+1}} + W_{s_i})/2$,

$$\left\langle \int_0^t W_s dW_s \right\rangle = \left\langle \lim_{\Delta s \to 0} \sum_{i=0}^{n-1} \frac{1}{2}(W_{s_{i+1}} + W_{s_i})(W_{s_{i+1}} - W_{s_i}) \right\rangle$$

$$= \frac{1}{2} \left\langle \sum_{\Delta s \to 0} \sum_{i=0}^{n-1}(W_{s_{i+1}}^2 - W_{s_i}^2) \right\rangle$$

$$= \frac{1}{2}(\langle W_t^2 \rangle - \langle W_0^2 \rangle)$$

$$= \frac{t}{2}.$$

Different results are obtained with different selections of $s_i'$.

From the above discussions, we may have different results under different definitions of the stochastic integral $\int_0^t g(x(s), x) dW_s$. In particular, there are two types of definitions of the stochastic integral: the Itô integral and the Stratonovich integral. The Itô integral is defined by taking $s_i'$ as the left point $s_i' = s_i$, while the Stratonovich integral is defined by taking $s_i'$ so that

$$g(x(s_i'), s_i') = \frac{1}{2}(g(x(s_i), s_i) + g(x(s_{i+1}), s_{i+1})).$$

For the stochastic differential equation (2.35), we have two interpretations when different types of integrals are applied, and the Eq. (2.35) is termed the "Itô equation" or the "Stratonovich equation" appropriately. Sometimes, to avoid confusion, the Stratonovich equation is written as

$$dx = f(x, t)dt + g(x, t) \circ dW_t. \tag{2.38}$$

Mathematically, the Stratonovich equation (2.38) is equivalent to the Itô equation of form [4]

$$dx = \left( f(x, t) + \frac{1}{2} \frac{\partial g(x, t)}{\partial x} g(x, t) \right) dt + g(x, t) dW_t. \tag{2.39}$$

Hence, we usually use the term stochastic differential equation for the Itô equation. The Itô integral satisfies the following properties [4]:

(1) It $X(t)$ and $Y(t)$ are two simple random processes,

$$\int_0^t (\alpha X(s) + \beta Y(s)) dW_s = \alpha \int_0^t X(s) dW_s + \beta \int_0^t Y(s) dW_s,$$

where $\alpha$ and $\beta$ are constants.

(2)

$$\int_0^t I_{[a,b]} dW_s = W_{\min\{b,t\}} - W_{\max\{a,0\}},$$

where $I_{[a,b]}(t)$ means the characteristic function that takes value 1 when $a \leq t \leq b$ and 0 otherwise.

(3) If $\langle X(t)^2 \rangle < \infty$ ($\forall t \geq 0$),

$$\left\langle \int_0^t X(s) dW_s \right\rangle = 0.$$

(4) If $\langle X(t)^2 \rangle < \infty$ ($\forall t \geq 0$),

$$\left\langle \left( \int_0^t X(s) dW_s \right)^2 \right\rangle = \int_0^t \langle X^2(s) \rangle ds.$$

Similar to first-order stochastic differential equations, a general equation for $n$th-order stochastic differential equation is written as

$$dX^j = a^j(\mathbf{X}, t) dt + \sum_{k=1}^m b_k^j(\mathbf{X}, t) dW_t^k, \quad (j = 1, 2, \ldots, n). \tag{2.40}$$

Here, $\mathbf{X} = (X^1, X^2, \ldots, X^n)$, and $W_t^k$ means the $k$th Wiener process; different Wiener processes are independent of each other.

We can apply the Stratonovich integral to Eq. (2.40). In this case, the Stratonovich equation can be transformed to the following form of the Itô equation:

$$dX^j = \left( a^j(\mathbf{X}, t) + \frac{1}{2} \sum_{k=1}^m \sum_{l=1}^n \frac{\partial b_k^j(\mathbf{X}, t)}{\partial X^l} b_k^l(\mathbf{X}, t) \right) dt + \sum_{k=1}^m b_k^j(\mathbf{X}, t) dW_t^k \tag{2.41}$$

$$(j = 1, 2, \ldots, n).$$

The Itô-Stratonovich dilemma is important in applications, and we need to choose a proper interpretation according to the source of the noise perturbation. Usually, we apply the Stratonovich integral to external noise and the Itô integral to intrinsic noise [11].

### 2.4.2  Itô Formula

Let $x(t)$ satisfies the Itô equation (2.35), a stochastic process $V(x, t)$ depending on $x$ satisfies the following Itô formula [4]

$$dV = \left( \frac{\partial V}{\partial t} + \frac{\partial V}{\partial x} f(x, t) + \frac{1}{2} \frac{\partial^2 V}{\partial x^2} g(x, t)^2 \right) dt + \frac{\partial V}{\partial x} g(x, t) dW_t. \qquad (2.42)$$

The Itô formula is an important tool in the study of stochastic differential equations.

Now, we examine the Itô formula. The process $x(t)$ depends on the time $t$ and the Wiener process $W_t$. Hence, we can consider $x(t)$ as a function $x(t, W_t)$. Taking the Taylor expansion of $V(x, t)$, we have

$$
\begin{aligned}
dV(x, t) &= V(x(t + dt), t + dt) - V(x(t), t) \\
&= \frac{\partial V}{\partial t} dt + \frac{\partial V}{\partial x} dx + \frac{1}{2} \frac{\partial^2 V}{\partial x^2} (dx)^2 + o(dt) \\
&= \frac{\partial V}{\partial t} dt + \frac{\partial V}{\partial x} (f(x, t) dt + g(x, t) dW_t) \\
&\quad + \frac{1}{2} \frac{\partial^2 V}{\partial x^2} (f(x, t) dt + g(x, t) dW_t)^2 + o(dt) \\
&= \frac{\partial V}{\partial t} dt + \frac{\partial V}{\partial x} (f(x, t) dt + g(x, t) dW_t) + \frac{1}{2} \frac{\partial^2 V}{\partial x^2} (g(x, t)(dW_t))^2 + o(dt) \\
&= \left( \frac{\partial V}{\partial t} + \frac{\partial V}{\partial x} f(x, t) + \frac{1}{2} \frac{\partial^2 V}{\partial x^2} g(x, t)^2 \right) dt + \frac{\partial V}{\partial x} g(x, t) dW_t + o(dt).
\end{aligned}
$$

Here, we note $\langle (dW_t)^2 \rangle = dt$ in the last step.

Similarly, we have the Itô formula for the $n$th order stochastic differential equation (2.40), which is given by

$$dV(\mathbf{X}, t) = \left[\frac{\partial V(\mathbf{X}, t)}{\partial t} + \sum_{j=1}^{n} \frac{\partial V(\mathbf{X}, t)}{\partial X^j} a^j(\mathbf{X}, t)\right.$$

$$\left. + \frac{1}{2} \sum_{k=1}^{m} \left(\sum_{i,j=1}^{n} b_k^i(\mathbf{X}, t) \frac{\partial^2 V(\mathbf{X}, t)}{\partial X^i \partial X^j} b_k^j(\mathbf{X}, t)\right)\right] dt$$

$$+ \sum_{k=1}^{m} b_k^j(\mathbf{X}, t) dW_t^k \tag{2.43}$$

To show an application of the Itô formula, we solve the stochastic differential equation

$$dX = a(t)Xdt + b(t)XdW_t. \tag{2.44}$$

Let $V = \log X$, the Itô formula yields

$$dV = \left(\frac{\partial V}{\partial X} a(t)X + \frac{1}{2} \frac{\partial^2}{\partial X^2}(b(t)X)^2\right) dt + \frac{\partial V}{\partial X} dW_t$$

$$= \left(\frac{1}{X} a(t)X - \frac{1}{2} \frac{1}{X^2}(b(t)X)^2\right) dt + \frac{1}{X} b(t)XdW_t$$

$$= (a(t) - \frac{1}{2}b^2(t))dt + b(t)dW_t.$$

Hence, $V(t) = V(X(t))$ is given by the integral

$$V = V(0) + \int_0^t (a(s) - \frac{1}{2}b^2(s))ds + \int_0^t b(s)dW_s.$$

Thus, we obtain the solution of (2.44)

$$X(t) = X(0) \exp\left(\int_0^t (a(s) - \frac{1}{2}b^2(s))ds + \int_0^t b(s)dW_s\right).$$

### 2.4.3   Fokker-Planck Equation

Consider the $n$th order stochastic differential equation (2.41). Let $P(\mathbf{x}, t)$ be the probability density of the solution, so that $P(\mathbf{x}, t)d\mathbf{x}dt$ means the probability of having $x^i < X^i < x^i + dx^i$ in the time interval $(t, t + dt)$. Here $\mathbf{x} = (x^1, x^2, \ldots, x^n)$ and $d\mathbf{x} = dx^1 \cdots dx^n$. It is easy to have the normalization condition

$$\int_{\mathbf{x} \in \mathbb{R}^n} P(\mathbf{x}, t)d\mathbf{x} = 1 \tag{2.45}$$

for any $t$.

Now, we derive the equation for $P(\mathbf{x}, t)$. From the Itô integral, the solution $\mathbf{X}(t)$ of (2.41) satisfies

$$X^i(t + \Delta t) = X^i(t) + a^i(\mathbf{x}, t)\Delta t + \sum_{k=1}^{m} b_k^i(\mathbf{x}, t)\Delta W_t^k, \quad (i = 1, \ldots, n) \quad (2.46)$$

where

$$\Delta W_t^k = W_{t+\Delta t}^k - W_t^k.$$

Thus, assuming $\mathbf{X}(t) = \mathbf{x}$, the displacement $\Delta \mathbf{x} = \mathbf{X}(t + \Delta t) - \mathbf{X}(t)$ satisfies

$$\langle \Delta x^i \rangle = a^i(\mathbf{x}, t)\Delta t \quad (2.47)$$

and

$$\langle \Delta x^i \Delta x^j \rangle = \sum_{k=1}^{m} b_k^i(\mathbf{x}, t)b_k^j(\mathbf{x}, t)\Delta t + o(\Delta t). \quad (2.48)$$

Let $W(\Delta \mathbf{x}, \Delta t; \mathbf{x}, t)$ be the transition probability from $\mathbf{X}(t) = \mathbf{x}$ to $\mathbf{X}(t + \Delta t) = \mathbf{x} + \Delta \mathbf{x}$; the probability density $P(\mathbf{x}, t)$ satisfies

$$P(\mathbf{x}, t + \Delta t) - P(\mathbf{x}, t) = \int_{\Delta \mathbf{x} \in \mathbb{R}^n} P(\mathbf{x} - \Delta \mathbf{x}, t)W(\Delta \mathbf{x}, \Delta t; \mathbf{x} - \Delta \mathbf{x}, t)d\Delta \mathbf{x}$$

$$- \int_{\Delta \mathbf{x} \in \mathbb{R}^n} P(\mathbf{x}, t)W(\Delta \mathbf{x}, \Delta t; \mathbf{x}, t)d\Delta \mathbf{x}. \quad (2.49)$$

We note that

$$\int_{\Delta \mathbf{x} \in \mathbb{R}^N} \Delta x^i W(\Delta \mathbf{x}, \Delta t; \mathbf{x}, t)d\Delta \mathbf{x} = \langle \Delta x^i \rangle, \quad (2.50)$$

and

$$\int_{\Delta \mathbf{x} \in \mathbb{R}^N} \Delta x^i \Delta x^j W(\Delta \mathbf{x}, \Delta t; \mathbf{x}, t)d\Delta x = \langle \Delta x^i \Delta x^j \rangle. \quad (2.51)$$

Applying the Taylor expansion to the function

$$h(\mathbf{x} - \Delta \mathbf{x}) \equiv P(\mathbf{x} - \Delta \mathbf{x}, t)W(\Delta \mathbf{x}, \Delta t; \mathbf{x} - \Delta \mathbf{x}, t)$$

in (2.49), we have

$$h(\mathbf{x} - \Delta \mathbf{x}) = h(\mathbf{x}) - \sum_{i=1}^{n} \Delta x^i \frac{\partial}{\partial x^i} h(\mathbf{x}, t)$$

$$+ \frac{1}{2} \sum_{1 \leq i,j \leq n} \Delta x^i \Delta x^j \frac{\partial^2}{\partial x^i \partial x^j} h(\mathbf{x}, t) + o(\|\Delta \mathbf{x}\|^2).$$

Hence, from (2.50) and (2.51), we obtain

$$P(\mathbf{x}, t + \Delta t) - P(\mathbf{x}, t) = -\sum_{i=1}^{n} \frac{\partial}{\partial x^i} \left[ P(\mathbf{x}, t) \langle \Delta x^i \rangle \right]$$

$$+ \frac{1}{2} \sum_{1 \leq i,j \leq n} \frac{\partial^2}{\partial x^i \partial x^j} \left[ P(\mathbf{x}, t) \langle \Delta x^i \Delta x^j \rangle \right].$$

From (2.47) and (2.48), dividing both sides by $\Delta t$ and letting $\Delta t \to 0$, we obtain

$$\frac{\partial P(\mathbf{x}, t)}{\partial t} = -\sum_{i=1}^{n} \frac{\partial}{\partial x^i} (P(\mathbf{x}, t) a^i(\mathbf{x}, t))$$

$$+ \frac{1}{2} \sum_{1 \leq i,j \leq n} \frac{\partial^2}{\partial x^i \partial x^j} (P(\mathbf{x}, t) b_k^i(\mathbf{x}, t) b_k^j(\mathbf{x}, t)).$$

Thus, we obtain the *Fokker-Planck equation*

$$\frac{\partial}{\partial t} P(\mathbf{x}, t) + \sum_{i=1}^{n} \frac{\partial}{\partial x^i} J_i(\mathbf{x}, t) = 0 \qquad (2.52)$$

where $J_i(\mathbf{x}, t)$ is the probability flux defined as

$$J_i(\mathbf{x}, t) = a^i(\mathbf{x}, t) P(\mathbf{x}, t) - \frac{1}{2} \sum_{j=1}^{n} \frac{\partial}{\partial x^j} (G_{ij}(\mathbf{x}, t) P(\mathbf{x}, t)),$$

with

$$G_{ij}(\mathbf{x}, t) = \sum_{k=1}^{m} b_k^i(\mathbf{x}, t) b_k^j(\mathbf{x}, t).$$

The Fokker-Planck equation is a partial differential equation, and boundary conditions are required to determine the solution $P(\mathbf{x}, t)$. We are only interested in the nonnegative solutions, i.e.,

$$P(\mathbf{x}, t) \geq 0, \quad \forall \mathbf{x} \in \mathbb{R}^n. \qquad (2.53)$$

Moreover, the normalization condition (2.45) is always required, which implies

$$\lim_{|\mathbf{x}| \to +\infty} P(\mathbf{x}, t) = 0. \qquad (2.54)$$

Hence, (2.45), (2.53), and (2.54) are natural conditions for the Fokker-Planck equation (2.52).

In practical studies, the boundary condition should be given by the physical problem under consideration. For example, when we assume an absorption boundary of a region $\Omega$, we should have

$$\frac{\partial}{\partial t}P(\mathbf{x}, t) + \sum_{i=1}^{n} \frac{\partial}{\partial x^i} J_i(\mathbf{x}, t) = 0, \quad \mathbf{x} \in \Omega \subset \mathbb{R}^n, \tag{2.55}$$

$$P(\mathbf{x}, t) = 0, \quad \mathbf{x} \in \partial\Omega, \tag{2.56}$$

$$P(\mathbf{x}, t) \geq 0, \quad \mathbf{x} \in \bar{\Omega}, \tag{2.57}$$

$$\int_{\Omega} P(\mathbf{x}, t)d\mathbf{x} = 1. \tag{2.58}$$

For a one-dimensional stochastic differential equation

$$dx = f(x)dt + g(x)dW_t, \tag{2.59}$$

the Fokker-Planck equation is

$$\frac{\partial P(x, t)}{\partial t} + \frac{\partial}{\partial x}\left(f(x)P(x, t) - \frac{1}{2}\frac{\partial}{\partial x}(g(x)^2 P(x))\right) = 0. \tag{2.60}$$

The solution of the Fokker-Planck equation gives the evolution of the distribution of the stochastic process defined by (2.59). The stationary distribution is given by $P_{ss}(x) \geq 0$, which satisfies

$$\frac{d}{dx}\left(f(x)P_{ss}(x) - \frac{1}{2}\frac{\partial}{\partial x}(g(x)^2 P_{ss}(x))\right) = 0, \quad \int_{-\infty}^{+\infty} P_{ss}(x)dx = 1. \tag{2.61}$$

Specifically, when $g(x) = \sigma$ and $f(x) = -\varphi'(x)$, we obtain a Boltzmann distribution $P_{ss}(x) \propto e^{-\frac{\varphi(x)}{\sigma^2/2}}$.

## 2.5   Reaction-Diffusion Equations

Diffusion is a basic form of spreading molecules in biological media that arises from the passive movement of molecules from a region of higher concentration to a region of lower concentration. Mathematically, the diffusion process can be described by reaction-diffusion equations. Here, we introduce the mathematical formulation and basic results of the reaction-diffusion equation.

### 2.5.1  One-Dimensional Conservation Law

In physics, the conservation law states that a particular measurable property of an isolated physical system does not change as the system evolves over time. In the diffusion process of molecules, the conservation law equation describes how molecule concentration evolves due to production, degradation, and transportation. Here, we first introduce the equation for one-dimensional transportation and then extend the equation to higher dimensions.

Consider an amount of molecules C distributed in a long tube (denoted by $x$ for the location along the tube) with a cross-sectional area $A$. Let $c(x, t)$ be the concentration of C (per unit area) at time $t$ at $x$ (Fig. 2.4). For any region $R$ ($x_a < x < x_b$), changes in the total amount of molecules C are mainly due to the net effect of transportation and local production/degradation. Hence, the net variation rate of the concentration of C in the region $R$ is given by

Net variation rate of C in $R$ = In flux rate $-$ Out flux rate

+ Production rate in $R$ $-$ Degradation rate in $R$.

The total amount of $C$ within the region $R$ ($x_a < x < x_b$) is given by the integral as

$$\text{Total amount of C in } [x_a, x_b] = \int_{x_a}^{x_b} c(x, t)A\,dx.$$

Assuming that C can freely move inside the tube, both the fluxes into and out of the $R$ region are obtained by moving across the $x = x_a$ and $x_b$ boundaries. Let $J(x, t)$ be the flux (from left to right) of C at time $t$ through the boundary $x$; then, at time $t$, the net flux of C into the region $R$ is

$$\text{Net flux at } t = AJ(x_a, t) - AJ(x_b, t).$$

Let $f(x, t, c)$ be the net production rate (production rate $-$ degradation rate) of C at location $x$ at time $t$; then the net production rate of C in $R$ is

**Fig. 2.4** One-dimensional conservation law

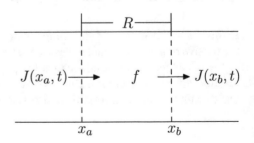

$$\text{Net production rate in } R = \int_{x_a}^{x_b} f(x, t, c(x, t)) A dx.$$

Now, we can write the above relationship for the net variation rate of C in R as

$$\frac{d}{dt} \int_{x_a}^{x_b} c(x, t) dx = J(x_a, t) - J(x_b, t) + \int_{x_a}^{x_b} f(x, t, c(x, t)) dx. \qquad (2.62)$$

Here, we assume that the cross-sectional area A is independent of the location x and hence can be eliminated on both sides. Equation (2.62) gives the integral form of the one-dimensional conservation law of the molecules.

Now, we can write the net flux in the form of an integral,

$$J(x_a, t) - J(x_b, t) = \int_{x_a}^{x_b} \frac{\partial}{\partial x} J(x, t) dx.$$

Hence, the Eq. (2.62) is rewritten as

$$\frac{\partial}{\partial t} \int_{x_a}^{x_b} c(x, t) dx = - \int_{x_a}^{x_b} \frac{\partial}{\partial x} J(x, t) dx + \int_{x_a}^{x_b} f(x, t, c(x, t)) dx. \qquad (2.63)$$

When the function $c(x, t)$ is smooth enough, the derivative and integral in (2.63) can be exchanged, and hence, (2.63) is further rewritten as

$$\int_{x_a}^{x_b} \left[ \frac{\partial}{\partial t} c(x, t) + \frac{\partial}{\partial x} J(x, t) - f(x, t, c(x, t)) \right] dx = 0. \qquad (2.64)$$

Since Eq. (2.64) is valid for any region $x_a < x < x_b$, the integrand is identical to zero, i.e.,

$$\frac{\partial c}{\partial t} + \frac{\partial J}{\partial x} = f(x, t, c). \qquad (2.65)$$

This equation gives the differential form of one-dimensional conservation law.

## 2.5.2 Various Forms of Flux

The conservation law equation (2.65) includes two unknown functions $J(x, t)$ and $c(x, t)$ and hence is not in a closed-form equation. Usually, we need to establish the constitutive relation between the flux $J(x, t)$ and the concentration $c(x, t)$ based on different physical problems. The constitutive relation is often derived from empirical observations instead of the basic principles. Here, we introduce a few forms of the constitutive relation between the flux and concentration.

A common constitutive relation is given by Fick's law of diffusion—the flux goes from a region of high concentration to a region of low concentration, with a magnitude that is proportional to the concentration gradient. Mathematically, Fick's law is represented as

$$J(x, t) = -D\frac{\partial}{\partial x}c(x, t),$$  (2.66)

where the coefficient $D$ (area per unit time) represents the diffusion coefficient. The diffusion coefficient can be either a constant or a function of the spatial variable $x$ and is usually dependent on the physical properties of the solute $C$ and the solution. Here, the negative sign means that the solute molecules move from a region of high concentration to a region of low concentration.

From Fick's law, the conservation law equation is written in the form of the reaction-diffusion equation

$$\frac{\partial c}{\partial t} - \frac{\partial}{\partial x}\left(D\frac{\partial c}{\partial x}\right) = f(x, t, c).$$  (2.67)

Fick's law can be used to describe molecular movements that follow Brownian motion, which is the basic form of molecular transportation. In some applications, the diffusion coefficient $D$ depends on the concentration $c$ so that we have a nonlinear diffusion equation.

When the medium moves along $x$ with a velocity $v$, then in a time $\Delta t$, the molecules $C$ within the region $x_a - v\Delta t < x < x_a$ will move across the boundary $x = x_a$ into the region $x > x_a$. Hence, an amount of $Avc(x_a, t)\Delta t$ (here, we assume that $\Delta t$ is small enough so that the concentration $c$ is independent of $x$ within the region $x_a - v\Delta t < x < x_a$) will flux into the region $x > x_a$ per unit time. Thus, the flux due to the movement with velocity $v$ is given by

$$J(x, t) = vc(x, t).$$

This flux is termed the advection flux. Accordingly, we have the reaction-advection equation

$$\frac{\partial c}{\partial t} + \frac{\partial(vc)}{\partial x} = f(x, t, c).$$  (2.68)

Here, we note that the reaction-advection equation can also be used for the age-structure model of population dynamics. In this case, the time evolution automatically results in the flux of age, and hence, the corresponding velocity of age is $v = 1$. This gives the age-structured operator

$$\nabla = \frac{\partial}{\partial t} + \frac{\partial}{\partial a}$$

in (2.22) and Chap. 6.

When we consider a situation with both advection and diffusion, then

$$J(x, t) = vc(x, t) - D\frac{\partial}{\partial x}c(x, t).$$

In this case, the conservation law equation has the form of the reaction-advection-diffusion equation

$$\frac{\partial c}{\partial t} + \frac{\partial}{\partial x}\left(vc - D\frac{\partial c}{\partial x}\right) = f(x, t, c). \tag{2.69}$$

## 2.5.3  Random Walk and the Diffusion Equation

In the reaction-diffusion equation (2.67), the diffusion term comes from Fick's law (2.66), which describes the macroscopic flux of how particles diffuse under random thermal motion tend to spread from a region of higher concentration to a region of lower concentration to equalize the concentration in both regions. Alternatively, Fick's law of diffusion can also be derived from the microscopic Brownian motion of particles or random walks [7].

The one-dimensional motion of a Brownian particle can be described by the Langevin equation of form

$$dv = -\gamma v + \sigma dW_t, \tag{2.70}$$

where $v$ is the velocity of the particle, $\gamma$ is the coefficient of viscosity, and $\sigma = \sqrt{2\gamma k_B T}$ according to the Einstein relation [12]. Let $P(v, t)$ be the probability density of having velocity $v$ at time $t$ and let $P(v, t)$ satisfies the Fokker-Planck equation (see (2.52))

$$\frac{\partial P(v, t)}{\partial t} + \frac{\partial}{\partial v}\left(-\gamma v P(v, t) - \frac{1}{2}\sigma^2\frac{\partial}{\partial v}P(v, t)\right) = 0, \tag{2.71}$$

or the diffusion equation

$$\frac{\partial P(v, t)}{\partial t} = \frac{1}{2}\sigma^2\frac{\partial^2}{\partial v^2}P(v, t) + \gamma v\frac{\partial}{\partial v}P(v, t) = 0. \tag{2.72}$$

Thus, the diffusion coefficient $D = \frac{1}{2}\sigma^2 = \gamma k_B T$.

Here, we note that

$$J = -\gamma v P(v, t) - \frac{1}{2}\sigma^2\frac{\partial}{\partial v}P(v, t)$$

gives the probability flux due to the random walk of Brownian particles. We can further generalize this concept of probability flux to biochemical reactions (Chap. 3) or random processes described by stochastic differential equations (Sect. 2.4), which gives the Fokker-Planck equations.

Now, let us consider the drunken walker. We assume a substantial population of drunken walkers distributed in space and time $n(x, t)$, where $x$ and $t$ take discrete values $x = 0, \pm \Delta x, \pm 2\Delta x, \ldots$, and $t = 0, \Delta t, 2\Delta t, \ldots$. Let each walker move to the right with a probability $p$ and to the left with a probability $q$ $(p + q = 1)$; then, the population distribution $n(x, t)$ satisfies

$$n(x, t + \Delta t) = pn(x - \Delta x, t) + qn(x + \Delta x, t). \tag{2.73}$$

We rewrite (2.73) as

$$n(x, t + \Delta t) - n(x, t) = -p(n(x, t) - n(x - \Delta x, t)) + q(n(x + \Delta x, t) - n(x, t)).$$

Since

$$n(x, t + \Delta t) - n(x, t) = \frac{\partial n(x, t)}{\partial t} \Delta t + o(\Delta t),$$

$$n(x, t) - n(x - \Delta x, t) = \frac{\partial n(x, t)}{\partial x} \Delta x - \frac{1}{2} \frac{\partial^2 n(x, t)}{\partial x^2} (\Delta x)^2 + o((\Delta x)^2),$$

$$n(x + \Delta x, t) - n(x, t) = \frac{\partial n(x, t)}{\partial x} \Delta x + \frac{1}{2} \frac{\partial^2 n(x, t)}{\partial x^2} (\Delta x)^2 + o((\Delta x)^2),$$

we have

$$\frac{\partial n(x, t)}{\partial t} = D \frac{\partial^2 n(x, t)}{\partial x^2} - V \frac{\partial n(x, t)}{\partial x}, \tag{2.74}$$

where $D = (\Delta x)^2/(2\Delta t)$ and $V = (p - q)\Delta x/\Delta t$. Now, we obtain Fourier-Fick's law for the transport flux

$$J(x, t) = -D \frac{\partial n(x, t)}{\partial x} + Vn(x, t). \tag{2.75}$$

When $p = q = \dfrac{1}{2}$, we say that the random walk is unbiased and obtain the diffusion equation (or heat equation)

$$\frac{\partial n(x, t)}{\partial t} = D \frac{\partial^2 n(x, t)}{\partial x^2}. \tag{2.76}$$

### 2.5.4  Initial and Boundary Conditions

Now, we have established an equation for the concentration $c(x, t)$ from the conservation law and the constitutive relation. However, the equation itself is not enough to determine the evolution of the concentration, and we need to identify the initial and boundary conditions to determine the solution. Usually, the initial condition $c(x, 0)$ can be obtained by experimental measurements. The boundary condition is usually defined according to the physical problem under consideration. Boundary conditions are important for the determination of the solution, and different conditions often give rise to different solutions. Hence, in applications, it is essential to determine the proper boundary conditions based on the particular physical problem. Here, we introduce the formulations of a few common boundary conditions.

In the above example of one-dimensional transportation, if the concentration at the boundary $x = x_a$ can be determined for a given function $g(t)$, the corresponding boundary condition is $c(x_a, t) = g(t)$. We call this type of boundary condition the Dirichlet condition. If the flux on the boundary can be controlled by external conditions so that $-D\frac{\partial c}{\partial x}(x_a, t) = g(t)$, we have the Neumann condition. If the flux on the boundary is associated with the concentration, we have the Robin condition, for example, $-D\frac{\partial c}{\partial x}(x_a, t) = g(t) - ac(x_a, t)$. These three types of conditions are often used in different problems. Here, we omit the discussions of the boundary conditions and refer the readers to textbooks of partial differential equations, e.g., [13] or [14].

### 2.5.5 High Dimensional Conservation Laws

Now, we briefly introduce the formulation for high-dimensional conservation laws. Let $c(x, y, z, t)$ be the concentration (per unit volume) of C at time $t$ at spatial location $(x, y, z)$; the total amount of C within volume $V$ is

$$\text{Total amount of C} = \int_V c(x, y, z, t) dV.$$

Let $S$ be the boundary of $V$, and $\mathbf{J}(x, y, z, t)$ be the flux on $S$ towards the outward normal direction $\mathbf{n}(x, y, z, t)$. The net flux into $V$ per unit time is expressed as

$$\text{Net flux} = -\int_S \mathbf{J}(x, y, z, t) \cdot \mathbf{n}(x, y, z, t) dA,$$

where $dA$ is the area of the differential element on $S$. Let $f(x, y, z, t, c)$ be the net production rate at $(x, y, z)$. Similar to the above discussions, the integral form of the conservation law is expressed as

$$\frac{d}{dt} \int_V c dV = -\int_S \mathbf{J} \cdot \mathbf{n} dA + \int_V f dV. \tag{2.77}$$

Applying the Gaussian formula, we replace the above surface integral $\int_S \mathbf{J} \cdot \mathbf{n} dA$ with the volume integral $\int \nabla \cdot \mathbf{J} dV$ and obtain the following integral form conservation law:

$$\frac{d}{dt} \int_V c \, dV = - \int_V \nabla \cdot \mathbf{J} \, dV + \int_V f \, dV, \tag{2.78}$$

where $\nabla = (\frac{\partial}{\partial x}, \frac{\partial}{\partial y}, \frac{\partial}{\partial z})$ represents the gradient operator. If the concentration $c(x, y, z, t)$ is smooth enough, we can exchange the derivative with the integral operator and obtain the following conservation law equation:

$$\frac{\partial c}{\partial t} + \nabla \cdot \mathbf{J} = f. \tag{2.79}$$

When Fick's law is applied to the flux,

$$\mathbf{J}(x, y, z, t) = -D \nabla c(x, y, z, t),$$

we obtain a high-dimensional reaction-diffusion equation

$$\frac{\partial c}{\partial t} - \nabla \cdot (D \nabla c) = f. \tag{2.80}$$

In particular, if the diffusion coefficient is homogeneous, i.e., $D$ is independent of the spatial location, we have

$$\nabla \cdot (D \nabla c) = D \nabla \cdot (\nabla c) = D \Delta c,$$

where $\Delta$ is the Laplacian operator defined as

$$\Delta c = \frac{\partial^2 c}{\partial x^2} + \frac{\partial^2 c}{\partial y^2} + \frac{\partial^2 c}{\partial z^2}, \tag{2.81}$$

so the Eq. (2.80) becomes

$$\frac{\partial c}{\partial t} - D \Delta c = f. \tag{2.82}$$

Here, we omit the discussions of other forms of constitutive relations and the boundary conditions.

## 2.6  Discrete Dynamical Systems

In previous sections, we introduced various forms of differential equations used in modeling continuous dynamical systems. Alternatively, the evolution of quantities can occur over discrete time steps, which can be described by discrete dynamical systems.

When we model a system as a discrete dynamical system, we imagine that we take a snapshot of the system at a sequence of times. Snapshots can occur once a year, once every millisecond, or even irregularly, such as once every time a cell is undergoing mitosis.

When we take these snapshots, the idea is that we are recording whatever variable determines the state of the system: our chosen state variables $\mathbf{x}$ that evolve through the state space $\Omega$. To complete the description of the dynamical system, we need to specify a rule $f : \Omega \to \Omega$ that determines, given an initial snapshot, what the resulting sequence of future snapshots should be. A general mathematical formulation of a discrete dynamical system is given by a map

$$\mathbf{x}_{t+1} = f(\mathbf{x}_t), \tag{2.83}$$

where $\mathbf{x}_t$ represents the state variable at the $t$'th snapshot. Given an initial state $\mathbf{x}_0$, the state at the $t$'th snapshot can be obtained by iterations

$$\mathbf{x}_t = f(\mathbf{x}_{t-1}) = f(f(\mathbf{x}_{t-2})) = \cdots = f(f(\cdots f(\mathbf{x}_0))) = f^t(\mathbf{x}_0).$$

We use $f^t(\mathbf{x})$ to represent the iteration that $f$ composes with itself $t$ times.

One important application of discrete dynamical systems in the field of life sciences is describing the size of populations. For example, suppose a species of fish lays eggs every spring. Putting aside the issue of random variation, it is reasonable to expect the population next year to be a function of the population of this year. Letting $x_t$ be the size of the population in year $t$, we have $x_{t+1} = f(x_t)$. Various forms of $f(x_t)$ have been used in different biological situations; see, for example, the list in [15]. One model is often used to describe the discrete analogue of the continuous logistic model, in which $x_{t+1}$ is given in terms of $x_t$ as

$$x_{t+1} = f(x_t) = rx_t \left(1 - \frac{x_t}{K}\right), \quad r > 0, \ K > 0, \tag{2.84}$$

where $r$ is a constant measuring the growth rate and $K$ is the maximum size of the population (Fig. 2.5). This model (2.84) is sometimes referred to as the Verhulst process.

An obvious drawback of the logistic model is that if $x_t > K$, then $x_{t+1} < 0$. A more realistic model would be such that for large $x_t$, there should be a reduction in the growth rate b $x_{t+1}$ that should remain nonnegative. One such frequently used model is

$$x_{t+1} = g(x_t) = x_t \exp\left[r(1 - \frac{x_t}{K})\right], \quad r > 0, \ K > 0, \tag{2.85}$$

which is a modification of the logistic model (2.84).

Given an initial state $\mathbf{x}_0$ and the model (2.83), the sequence $\{\mathbf{x}_t\}_{t=0}^{\infty}$ gives the system states at each snapshot. The fixed points of the iteration sequence give steady states $x^*$, which are solutions of

$$\mathbf{x}^* = f(\mathbf{x}^*).$$

The logistic growth model (2.84) has a trivial steady state $x_1^* = 0$, and a nontrivial positive steady state $x_1^* = \dfrac{K}{r}(r-1)$ when $r > 1$.

Given an equilibrium state $\mathbf{x}^* \in \Omega \subset \mathbb{R}^n$, the dynamics of a vector of small deviations $\mathbf{y} = \mathbf{x} - \mathbf{x}^*$ from this equilibrium are given by

$$\mathbf{y}_{t+1} = J \cdot \mathbf{y}_t, \quad \mathbf{y}_t \in \mathbb{R}^n \tag{2.86}$$

where $J$ is the Jacobian matrix of partial derivatives evaluated at the equilibrium

$$J = \left.\frac{\partial f(\mathbf{x}^*)}{\partial \mathbf{x}}\right|_{\mathbf{x}=\mathbf{x}^*} = \begin{bmatrix} \frac{\partial f_1(\mathbf{x}^*)}{\partial x_1} & \frac{\partial f_1(\mathbf{x}^*)}{\partial x_2} & \cdots & \frac{\partial f_1(\mathbf{x}^*)}{\partial x_n} \\ \vdots & \vdots & \ddots & \vdots \\ \frac{\partial f_n(\mathbf{x}^*)}{\partial x_1} & \frac{\partial f_n(\mathbf{x}^*)}{\partial x_2} & \cdots & \frac{\partial f_n(\mathbf{x}^*)}{\partial x_n} \end{bmatrix}.$$

This dynamical equation holds approximately as long as $\|\mathbf{y}_t\| \ll 1$ is small.

In most cases, the matrix $J$ has $n$ eigenvalues $\lambda_1, \ldots, \lambda_n$ with corresponding eigenvectors $\mathbf{w}_1, \ldots, \mathbf{w}_n$:

$$J\mathbf{w}_i = \lambda_i \mathbf{w}_i.$$

The eigenvectors are linearly independent, and any state variable $\mathbf{y}_0$ can be expressed as a linear combination of the eigenvectors:

$$\mathbf{y}_0 = b_1 \mathbf{w}_1 + \cdots + b_n \mathbf{w}_n.$$

Thus,

$$\mathbf{y}_1 = J\mathbf{y}_0 = b_1\lambda_1\mathbf{w}_1 + \cdot + b_n\lambda_n\mathbf{w}_n.$$

The general form of the dynamical system (2.86) is given by

$$\mathbf{y}_t = b_1\lambda_1^t\mathbf{w}_1 + \cdots + b_n\lambda_n^t\mathbf{w}_n.$$

In the long term, i.e., for $t \to \infty$, there are two qualitatively distinct cases.

Let $\lambda_1$ be the eigenvalue of $J$ with the largest absolute value, i.e., the dominant eigenvalue, and let $\mathbf{w}_1$ be the corresponding eigenvector. Then, if $\lambda_1$ is a real number, the long-term behavior is given by

$$\mathbf{x}_t = b_1\lambda_1^t\mathbf{w}_1 \quad \text{for } t \to \infty.$$

Thus, if the eigenvalue with the largest absolute value $\lambda_1$ is real, the dynamic variables will eventually grow (or decrease) at a rate $\lambda_1$, and the vector of variables will be a multiple of the corresponding eigenvector $\mathbf{w}_1$.

If $\lambda_1$ is not real, $\lambda_1 = r_1 + is_1 = c(\cos\varphi_1 + i\sin\varphi_1)$ with $\varphi_1 \neq 0$ (hence $s_1 \neq 0$), then $\mathbf{w}_1 = \mathbf{u}_1 + i\mathbf{v}_1$ with $\mathbf{v}_1 \neq 0$. In this case, the complex conjugate $\bar{\lambda}_1 = r_1 - is_1$ ($= \lambda_2$) is also an eigenvalue of $J$ with corresponding eigenvector $\mathbf{w}_2 = \mathbf{u}_1 - i\mathbf{v}_1$. The

real valued solution of (2.86) will eventually converge to oscillating behavior with
exponentially growing (or decreasing) amplitude:

$$\mathbf{y}_t = c^t(b_1(\cos(t\varphi_1)\mathbf{u}_1 - \sin(t\varphi_1)\mathbf{v}_1) + b_2(\sin(t\varphi_1)\mathbf{u}_1 + \cos(t\varphi_1)\mathbf{v}_1)) \quad \forall t \to \infty.$$

The equilibrium is called locally stable if and only if $\mathbf{y}_t \to \mathbf{0}$ for $t \to \infty$ for any
initial condition $\mathbf{y}_0$ with $\|\mathbf{y}_0\| \ll 1$. The above analysis shows that the equilibrium is
locally stable if and only if the dominant eigenvalue of the Jacobian matrix $J$ at this
equilibrium has absolute value $\|\lambda_1\| < 1$.

We apply the above argument to a one-dimensional discrete dynamical system
$x_{t+1} = f(x_t)$. Suppose a function $y = f(x)$ has an equilibrium point at $x = a$. Then,
the equilibrium point is stable if $|f'(a)| < 1$ and unstable if $|f'(a)| > 1$.

For the logistic growth model, we have

$$f'(x_0^*) = r, \quad f'(x_1^*) = 2 - r.$$

Thus, the zero solution is unstable when $r > 1$, i.e., there exists a positive steady
state, and the positive steady state is stable only when $1 < r < 3$. In particular,
when $r > 3$, the logistic growth model may have periodic obits or chaotic dynamics
(Fig. 2.5).

Given the initial state $\mathbf{x}_0$, the discrete dynamical system (2.83) gives a deterministic sequence $\{\mathbf{x}_0, \mathbf{x}_1, \ldots\}$. Alternatively, many biological processes involve opera-
tions of chance, which yield a stochastic process. For example, epigenetic inheritance
occur during cell division. Let the sequence $\{\mathbf{x}_t\}$ represent snapshots of the state variable of the stochastic process, with $\mathbf{x}_t$ being random variables in a probability space
$(\Omega, \mathscr{F}, P)$; then, the probability (or prediction) of $\mathbf{x}_{t+1}$ may depend on the history
$\{\mathbf{x}_0, \ldots, \mathbf{x}_t\}$. Particularly, a sequence $\mathbf{x}_0, \mathbf{x}_1, \ldots$ of random variables is called Markov
if, for any $t$,

$$P(\mathbf{x}_{t+1}|\mathbf{x}_t, \ldots, \mathbf{x}_0) = P(\mathbf{x}_{t+1}|\mathbf{x}_t),$$

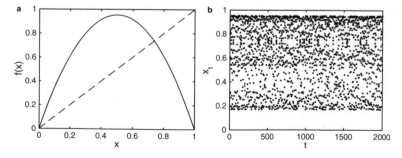

**Fig. 2.5** Logistic growth model. **a.** The logistic map $y = f(x)$ defined by (2.84). **b.** The chaotic
dynamics $x_{t+1} = f(x_t)$. Here $r = 3.81232$, $K = 1$

**Fig. 2.6 Random logistic growth model**. Dynamic sequences of deterministic (red) and random (black) logistic growth models. The deterministic sequence is generated by $x_{t+1} = g(x_t)$, and the random sequence is generated following the random dynamics (2.87) with $g(x_t)$ given by (2.85), and parameters $r = 2.6, K = 1, \sigma = 0.1$

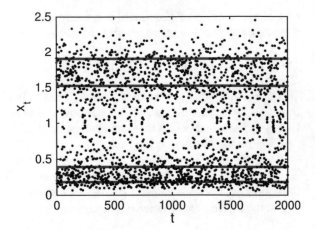

i.e., if the conditional probability of $\mathbf{x}_{t+1}$ assuming $\mathbf{x}_t, \mathbf{x}_{t-1}, \ldots, \mathbf{x}_0$ equals the conditional probability of $\mathbf{x}_{t+1}$ assuming only $\mathbf{x}_t$.

As an example, we consider a random perturbation to the modified logistic growth model (2.85) so that $\frac{x_{t+1}}{g(x_t)}$ is a lognormal distribution, i.e.,

$$ P(x_{t+1}|x_t) = \mathcal{N} \left( \log \frac{x_{t+1}}{g(x_t)}; 0, \sigma \right) = \frac{1}{\sigma \sqrt{2\pi}} \exp \left[ -\frac{(\log \frac{x_{t+1}}{g(x_t)})^2}{2\sigma^2} \right]. \qquad (2.87) $$

Figure 2.6 shows the sequences obtained from the deterministic and random modified logistic growth models. Here, the deterministic system shows a period-4 sequence; however, random perturbation to the iteration process yields a chaotic dynamic that switches between large and small population sizes.

For more discussions on discrete dynamical system and their applications to biological systems, please refer the textbook written by [7].

## 2.7 Multi-timescale Analysis

In the last two sections of this chapter, we introduce two techniques often used in the mathematical modeling of systems biology.

Here, we consider an example of the opening and closing process of ion channels. Ion channels are pore-forming membrane proteins that allow ions to pass through the channel pore. The process of opening and closing a pore depends on the membrane potential. When the membrane potential changes, an ion channel is stimulated and open so that ions pass through the channel and generate an ion current, which in turn may drive the membrane potential to reach an equilibrium state and close the channel. Here, we consider calcium channels, as shown in Fig. 2.7. The extracellular calcium

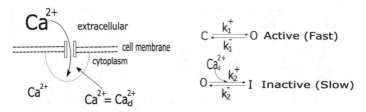

**Fig. 2.7 Illustration of the model of calcium channel activation and inactivation.** See the text for details

concentration (2 mM) is much higher than the intracellular calcium concentration (0.1 μM). Calcium ion channels are active and open quickly when stimulated by the membrane potential. Extracellular calcium ions transported into the cell when ion channels are open. However, due to the slow process of calcium ion diffusion inside a cell, there is a buffer area with a high level calcium concentration (200–500 μM) inside the cell near the region of ion channels. This activation process is fast and is completed in a few microseconds. Here, the reason for the slow diffusion process is because calcium ions may form complexes with non-diffusive substances. The calcium ions in the buffer area can directly bind to the ion channels and inactivate them. These processes are shown in Fig. 2.7.

Let $C$, $O$, and $I$ represent the three states of the ion channels: closed, open, and inactive. The process of ion channel activation can be separated into two steps: the fast step process from close to open, and the slow process from open to inactivation. Here, ion channel inactivation results from the accumulation of high-concentration calcium ions in the buffer area. Hence, the channel must be open before it becomes inactivated.

Applying the chemical rate equation in Chap. 3, the above process can be modeled by the following differential equation. There are many calcium ion channels on the cell membrane, and we assume that all channels follow the same process and are independent of each other. Let $x_C, x_O$ and $x_I$ denote the fraction of channels with different states (i.e., closed, open, and inactive, respectively), then

$$\frac{dx_C}{dt} = -V_1 \tag{2.88}$$

$$\frac{dx_O}{dt} = V_1 - V_2, \tag{2.89}$$

where $x_I = 1 - x_C - x_O$, and $V_1$ and $V_2$ represent the rates of activation and inactivation, which are given by

$$V_1 = k_1^+ x_C - k_1^- x_O$$
$$V_2 = k_2^+ [Ca_d^{2+}] x_O - k_2^- (1 - x_C - x_O).$$

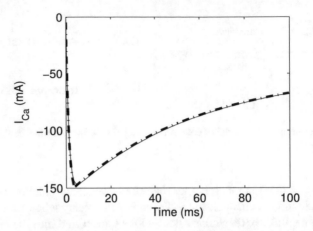

**Fig. 2.8 Simulated calcium ion current**. The solid line shows the current obtained from the original model (2.88)–(2.89), and the dashed line shows the result obtained from the simplified model in Sect. 2.7.1. The parameters used in the simulations are $k_1^+ = 0.7\,\mathrm{ms}^{-1}$, $k_1^- = 0.2\,\mathrm{ms}^{-1}$, $k_2^+ = 0.05\,\mathrm{mM}^{-1}\,\mathrm{ms}^{-1}$, $k_2^- = 0.005\,\mathrm{ms}^{-1}$, and $[\mathrm{Ca}_\mathrm{d}^{2+}] = 0.3\,\mathrm{mM}$. The calcium current is given by $I_{\mathrm{Ca}} = gx_O(V - V_{\mathrm{Ca}})$, with $g = 5\,\mathrm{nS}$, $V = 20\,\mathrm{mV}$, and $V_{\mathrm{Ca}} = 60\,\mathrm{mV}$

Here, $[\mathrm{Ca}_\mathrm{d}^{2+}]$ represents the extracellular calcium concentration and is assumed to be independent of the state of ion channels.

We can numerically solve the above equations. Figure 2.8 shows the simulation results. When the membrane potential changed, the calcium ion current increased immediately, followed by a slow decrease, in agreement with experimental observations [16]. In the following discussions, we introduce two methods, the quasi-steady-state assumption and the perturbation method, to establish simplified models for the fast open process and slow inactivation process, respectively, and obtain approximate expressions of the calcium current for the two processes.

### 2.7.1 Quasi-Equilibrium Assumption

To model how the ion channels reach a state of equilibrium after the process of fast opening and closing following stimuli with changes in membrane potential, we first omit the inactivation process and let $V_2 = 0$. Since all ion channels are closed at the initial stage, the inactivation process can only occur when the channel is open and hence can be omitted at the early stage. The channels are closed initially, hence $x_C(0) = 1$. Moreover, $x_I = 0$ and $x_C + x_O = 1$ at the early stage. Hence, the Eq. (2.89) can be approximated as

$$\frac{dx_O}{dt} = -\frac{1}{\tau_{\mathrm{act}}}\left(x_O - \frac{1}{1 + K_1}\right), \tag{2.90}$$

where $K_1 = k_1^- / k_1^+$ is the equilibrium constant of the opening and closing process of an ion channel, and $\tau_{\text{act}} = 1/(k_1^+ + k_1^-)$ is the time constant of the activation process. From (2.90), the fraction of open state ion channels at the early stage is

$$x_O(t) = \frac{1}{1+K_1}(1 - e^{-t/\tau_{\text{act}}}). \tag{2.91}$$

After an ion channel is active, the opening and closing process is a fast process relative to the inactivation process (approximately a few microseconds, a fast process relative to the other process). Hence, the transitions between the closing and opening of an ion channel can be considered a quasi-steady-state process, i.e., $V_1 \approx 0$. This quasi-steady-state assumption (QSSA) is often used in modeling biochemical reactions and allows us to generate reaction-rate expressions that can capture the details of chemical reactions through a minimum number of rate constants. Hence, we have an approximate relation

$$x_C = (k_1^- / k_1^+) x_O = K_1 x_O. \tag{2.92}$$

Here, we note that it is not correct to set $V_1 = 0$ in equations (2.88)–(2.89). Because the fast equilibrium between closing and opening ($V_1 \approx 0$) implies the relation (2.92), it does not mean that the number of ion channels in the closed state is unchanged ($dx_C/dt = 0$). In fact, when the opening and closing process is fast, the total number of ion channels at state $C$ or $O$ is of interest. The process of changes in this total number is a slow process.

Let $y = x_C + x_O$, then

$$\frac{dy}{dt} = -V_2. \tag{2.93}$$

Based on the relation (2.92), we have

$$x_O = \frac{1}{1+K_1}y, \quad x_C = \frac{K_1}{1+K_1}y. \tag{2.94}$$

Hence, we obtain the equation

$$\frac{dy}{dt} = k_2^-(1-y) - \frac{k_2^+[\text{Ca}_d{}^{2+}]}{1+K_1}y. \tag{2.95}$$

Let

$$\tau([\text{Ca}_d{}^{2+}]) = \frac{1+K_1}{k_2^+[\text{Ca}_d{}^{2+}] + k_2^-(1+K_1)}, \tag{2.96}$$

$$y_\infty([\text{Ca}_d{}^{2+}]) = k_2^- \tau([\text{Ca}_d{}^{2+}]), \tag{2.97}$$

we obtain the following equation

$$\frac{dy}{dt} = \frac{y_\infty([Ca_d^{2+}]) - y}{\tau([Ca_d^{2+}])}. \tag{2.98}$$

Here, $y_\infty([Ca_d^{2+}])$ represents the value of $y$ at the equilibrium state, and $\tau([Ca_d^{2+}])$ is the timescale of approaching the equilibrium state.

Now, we need to determine the initial condition for (2.98). From the above analysis, (2.98) is applied to the situation after the ion channel is active. Since early activation is a fast process, the equilibrium between the opened and closed states is essential for the initial condition of (2.98). When the membrane potential changes, we have $x_C(0) = 1$. Hence, after the early fast activation process, some channels are open and change from the state $C$ to the state $O$ (but not the state $I$ yet); therefore, $y = x_C + x_O = 1$. Thus, we can set the initial condition

$$y(0) = 1 \tag{2.99}$$

for the Eq. (2.98). Now, we solve Eq. (2.98) with the initial condition (2.99) to obtain

$$y(t) = y_\infty([Ca_d^{2+}]) + (1 - y_\infty([Ca_d^{2+}]))e^{-t/\tau([Ca_d^{2+}])},$$

and the fraction of opening state channels is

$$x_O(t) = \frac{1}{1 + K_1}\left(y_\infty([Ca_d^{2+}]) + (1 - y_\infty([Ca_d^{2+}]))e^{-t/\tau([Ca_d^{2+}])}\right). \tag{2.100}$$

The calcium ion currents at different stages can be approximately obtained from equations (2.91) and (2.100) (the current is given by $I_{Ca}(t) = gx_O(t)(V - V_{Ca})$). Figure 2.8 shows the currents obtained from both the original model and the simplified equations, which are consistent. Hence, the above simplifications based on the quasi-steady-state assumption provide a reasonable approximation of the original dynamical system.

Here, we note that the current obtained from the approximation solution is slightly higher than that obtained from the original model because in the initial condition (2.99), we assume that there are no inactive channels, which is slightly different from the original model. In the region of exponential decay, (2.100) provides a good approximation of the exact solution because the time scale of activation (approximately 1 ms) is much faster than the time scale of inactivation (approximately 45 ms), and hence, the quasi-steady-state assumption is valid.

### 2.7.2  Perturbation Method

We have obtained an approximation of the calcium current through the method of quasi-steady-state assumption. Now, we introduce another method: the perturbation method. Perturbation methods include a group of methods for studying various problems employed in many branches of mathematics, mechanics, physics, and technology. Perturbation methods are often used to find an approximation solution to a problem, usually by starting from an exact solution of a related, simpler problem. A critical feature of the technique is a step to break the problem into solvable and perturbed parts. For multi-timescale systems, we need to identify different time scales and simplify the system at different timescales into a proper simplified problem that can be approximately solved.

In the above example of calcium ion channel activation, there are two time scales corresponding to the timing of ion channel activation $\tau_{act} = 1/(k_1^+ + k_1^-)$ and inactivation $\tau([Ca_d^{2+}])$. Here, we perform nondimensional analysis in accordance with the two time scales. The variables $x_C, x_O$ and $x_I$ are nondimensional, and hence, we only need to perform nondimensional analysis for the time.

First, we consider the slow process after ion channel activation. We apply the slow time scale to perform the nondimensionalization and consider $\tau([Ca_d^{2+}])$ as the unit of time. Since channel inactivation is mainly determined by the kinetic rates $k_2^+[Ca_d^{2+}]$ and $k_2^-$, $\tau([Ca_d^{2+}]) \sim k_2^-$. For simplicity, we take $k_2^-$ as the time scale for nondimensionalization and $\hat{t} = k_2^- t$ as the nondimensional time. Applying the nondimensional time $\hat{t}$, Eqs. (2.88)–(2.89) are rewritten as

$$\varepsilon \frac{dx_C}{d\hat{t}} = -\frac{k_1^+}{k_1^-}x_C + x_O, \tag{2.101}$$

$$\varepsilon \frac{dx_O}{d\hat{t}} = \frac{k_1^+}{k_1^-}x_C - x_O - \varepsilon \frac{k_2^+}{k_2^-}[Ca_d^{2+}]x_O + \varepsilon(1 - x_C - x_O), \tag{2.102}$$

where $\varepsilon = k_2^-/k_1^-$ is a small parameter.

Here, we note that $\varepsilon$ is a small parameter, and $dx_C/d\hat{t} \sim O(1)$ since we are considering a slow process. Hence, when $\varepsilon \to 0$, we should have

$$-\frac{k_1^+}{k_1^-}x_C + x_O = 0,$$

which gives the relation (2.94). Now, we add (2.101)–(2.102) and eliminate $\varepsilon$ to obtain an equation that is independent of $\varepsilon$

$$\frac{d(x_C + x_O)}{d\hat{t}} = -\frac{k_2^+}{k_2^-}[Ca_d^{2+}]x_O + (1 - x_C - x_O). \tag{2.103}$$

This gives the previous Eq. (2.95). Thus, we obtain the simplified Eqs. (2.94) and (2.95) under the region of slow time scale.

Similarly, in the region of fast time scale, we define the nondimensional time $\tilde{t} = k_1^- t$ and obtain the equations

$$\frac{dx_C}{d\tilde{t}} = -\frac{k_1^+}{k_1^-}x_C + x_O, \tag{2.104}$$

$$\frac{d(x_C + x_O)}{d\tilde{t}} = \varepsilon \left( -\frac{k_2^+}{k_2^-}[Ca_d^{2+}]x_O + (1 - x_C - x_O) \right). \tag{2.105}$$

Under the limit $\varepsilon \to 0$, (2.105) gives

$$\frac{d(x_C + x_O)}{d\tilde{t}} = 0.$$

Therefore, we have $x_C + x_O = 1$ (the initial condition (2.99)) during the fast time scale of ion channel activation, and $x_C(t)$ satisfies

$$\frac{dx_C}{d\tilde{t}} = -\frac{k_1^+}{k_1^-}x_C + 1 - x_C,$$

which is equivalent to Eq. (2.90). Thus, we reobtain the previous simplified model with the perturbation method.

## 2.8  Michaelis-Menten Function and Hill Function

Here, we introduce two important functions often used in modeling gene regulatory networks: Michaelis-Menten function and Hill function.

### 2.8.1  Michaelis-Menten Function

The Michaelis-Menten kinetic model is one of the best-known models of enzyme kinetics. It provides a general explanation of the velocity and gross mechanism of enzyme-catalyzed reactions [17].

Consider the reaction system that involves an enzyme, E, binding to a substrate, S, to form a complex, ES, which in turn releases a product, P, regenerating the original enzyme. This process may be represented schematically as

$$S + E \underset{k_1^-}{\overset{k_1^+}{\rightleftharpoons}} ES \overset{k_2}{\longrightarrow} EP \overset{k_3}{\longrightarrow} P + E.$$

The dynamics of the molecule concentrations can be described by the following chemical rate equation:

$$
\begin{cases}
\dfrac{d[S]}{dt} = -k_1^+[S][E] + k_1^-[ES] \\[2mm]
\dfrac{d[ES]}{dt} = k_1^+[S][E] - k_1^-[ES] - k_2[ES] \\[2mm]
\dfrac{d[EP]}{dt} = k_2[ES] - k_3[EP] \\[2mm]
\dfrac{d[P]}{dt} = k_3[EP].
\end{cases}
\tag{2.106}
$$

Hereafter, the square brackets [·] represent the concentration of molecules. Since the total amount of enzyme is unchanged, the total enzyme concentration

$$
[E] + [ES] + [EP] = E_{total}
\tag{2.107}
$$

is a constant. We assume that the enzymatic process is fast, i.e., $k_2, k_3 \gg k_1^+, k_1^-$. Applying the quasi-steady-state assumption, the concentrations of the enzyme-substrate complexes [ES] and [EP] can be considered constants, and hence,

$$
0 = \frac{d[ES]}{dt} = k_1^+[S][E] - k_1^-[ES] - k_2[ES]
\tag{2.108}
$$

and

$$
0 = \frac{d[EP]}{dt} = k_2[ES] - k_3[EP].
\tag{2.109}
$$

From (2.108) and (2.109),

$$
[ES] = \frac{k_3}{k_2}[EP], \quad [E] = \frac{k_1^- + k_2}{k_1^+[S]}[ES].
$$

Substituting the above equations into (2.107), we obtain

$$
[EP] = \frac{(C/k_3)[S]}{K + [S]} E_{total},
$$

where

$$
C = \frac{k_2 k_3}{k_2 + k_3}, \quad K = \frac{k_1^- + k_2}{k_1^+(1 + k_2/k_3)}.
$$

Thus, from (2.106), the production rate of the outcome P from the substrate S is approximately given by

$$
v = k_3[EP] = \frac{C[S]}{K + [S]} E_{total}.
\tag{2.110}
$$

The Eq. (2.110) gives the relation between the production rate of the outcomes and the substrate [S] in enzymatic reactions.

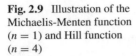

**Fig. 2.9** Illustration of the Michaelis-Menten function ($n = 1$) and Hill function ($n = 4$)

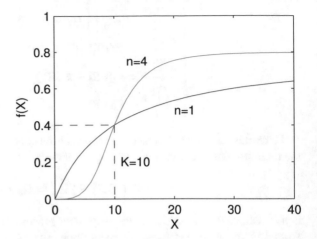

From (2.110), the production rate reaches the maximum value of $CE_{total}$ when the concentration of the substrate is high enough. The maximum rate is proportional to the enzyme concentration. When the substrate concentration [S] equals $K$, the production rate reaches half of the maximum rate. Hence, the value $K$ corresponds to the substrate concentration that reaches 50% activity (Fig. 2.9) and is named the half-maximal effective concentration (EC50).

The function of form (2.110) is called the Michaelis-Menten function, named by German biochemist Lenor Michaelis and Canadian physician Maud Menten, who investigated the kinetics of the enzymatic reaction mechanism of invertase, which catalyzes the hydrolysis of sucrose into glucose and fructose [18]. The Michaelis-Menten function describes the relation between the production rate and the substrate concentration.

## 2.8.2 Hill Function

In the above example of an enzymatic reaction, each enzyme can bind to only one substrate molecule. If there are multiple binding sites, one enzyme can bind to multiple substrate molecules, and the above chemical reaction becomes

$$nS + E \underset{k_1^-}{\overset{k_1^+}{\rightleftharpoons}} ES_n \overset{k_2}{\longrightarrow} EP_n \overset{k_3}{\longrightarrow} nP + E.$$

Here, the intermediate steps are omitted, and we assume that all $n$ molecules bind to an enzyme simultaneously. The corresponding chemical rate equation becomes

$$\begin{cases} \dfrac{d[S]}{dt} = -nk_1^+[S]^n[E] + nk_1^-[ES_n] \\[2mm] \dfrac{d[ES_n]}{dt} = k_1^+[S]^n[E] - k_1^-[ES_n] - k_2[ES_n] \\[2mm] \dfrac{d[EP_n]}{dt} = k_2[ES_n] - k_3[EP_n] \\[2mm] \dfrac{d[P]}{dt} = nk_3[EP_n]. \end{cases} \tag{2.111}$$

Here, we assume that the number of substrate molecules is large enough; hence, we can use the term $[S]^n$ in the propensity function (refer to Chap. 3).

Similar to the above argument, we apply the quasi-steady-state assumption to $[ES_n]$ and $[EP_n]$ and have

$$\frac{d[ES_n]}{dt} = \frac{d[EP_n]}{dt} = 0.$$

These assumptions together with (2.107) give the concentration of $[EP_n]$ at equilibrium state

$$[EP_n] = \frac{1}{nk_3} \frac{C[S]^n}{K^n + [S]^n} E_{\text{total}},$$

where

$$C = \frac{nk_2 k_3}{k_2 + k_3}, \quad \text{and } K^n = \frac{k_1^- + k_2}{k_1^+(1 + k_2/k_3)}.$$

Hence, the production rate of the outcome P is approximately

$$v = nk_3[EP_n] = \frac{C[S]^n}{K^n + [S]^n} E_{\text{total}}. \tag{2.112}$$

Here, $C$ and $K$ have the same meanings as in (2.110), and $n$ is the Hill coefficient.

The function (2.112) gives the production rate when an enzyme can bind to multiple substrates. The function of form (2.112) is called the Hill function (sometimes referred to as the Hill-Langmuir equation), which was originally formulated by Archibald Hill in 1910 to describe the sigmoidal $O_2$ binding curve of hemoglobin [19].

From (2.112), let the production rate be

$$v = \frac{v_{\max}[S]^n}{K^n + [S]^n},$$

then

$$\log\left(\frac{v/v_{\max}}{1 - v/v_{\max}}\right) = n\log[S] - n\log K.$$

This gives a linear relation of $\log\left(\dfrac{v/v_{\max}}{1 - v/v_{\max}}\right)$ versus $\log[S]$, which is called a Hill plot. The slope of the Hill plot equals the Hill coefficient.

In (2.112), the Hill coefficient $n$ describes cooperativity (or other biochemical properties depending on the context). Usually, the value $n$ means the following process of cooperativity in the processes of ligand binding:

- $n > 1$–Positive cooperative binding: Once one ligand molecule is bound to the enzyme, its affinity for other ligand molecules increases.
- $n < 1$–Negative cooperative binding: Once one ligand molecule is bound to the enzyme, its affinity for other ligand molecules decreases.
- $n = 1$–Noncooperative binding: The affinity of the enzyme for a ligand molecule is not dependent on whether other ligand molecules are already bound.

In particular, when $n = 1$, we obtain a model of Michaelis-Menten kinetics.

Figure 2.9 gives plots of the Michaelis-Menten function and Hill function. Here, we note an inflection point at the EC50 concentration $X = K$. Similar to the above increasing Hill function, the decreasing Hill function is given by

$$f(X) = \frac{CK^n}{K^n + X^n}. \tag{2.113}$$

Here, $n = 1$ gives the decreasing form of the Michaelis-Menten function.

### 2.8.3 Hill-Type Function Transcription Rate

In the above examples, we derived the Hill function for the dependence of the production rate on the substrate concentration in the enzymatic reaction. Now, we consider the situation of transcription with regulation through transcription factors (activators or repressors). In this case, the transcription factor binds to DNA (enhancers or operators) to regulate the transcription process (Chap. 5).

During transcription, regulatory proteins (such as transcription factors) bind to DNA sequences to regulate the activity of gene expression (please refer to Chap. 4). This transcription regulation process can be described in a simple form by the following chemical reactions:

$$D + nX \underset{k_1^-}{\overset{k_1^+}{\rightleftharpoons}} D^*, \quad D \xrightarrow{k_2} R, \quad D^* \xrightarrow{k_3} R, \quad R \xrightarrow{k_4} \emptyset.$$

Here, $X$ represents the regulators, $D$ represents the free DNA sequence (such as the enhancer, promoter, or operator), $D^*$ indicates the situation in which the DNA sequence is binding to the regulator ($n$ proteins $X$), and $R$ represents the transcriptional product mRNA. The coefficients $k_2$ and $k_3$ are the transcription rates when the gene sequence is either free or bound to the regulator, respectively. If $k_2 < k_3$, the protein $X$ activates transcription; in contrast, if $k_2 > k_3$, the protein $X$ inhibits transcription.

The above process can be formulated as chemical rate equations as

$$\begin{cases} \dfrac{d[D]}{dt} = -k_1^+[D]X^n + k_1^-[D^*] \\ \dfrac{d[D^*]}{dt} = k_1^+[D]X^n - k_1^-[D^*] \\ \dfrac{d[R]}{dt} = k_2[D] + k_3[D^*] - k_4[R]. \end{cases} \qquad (2.114)$$

Here, we assume that the protein number $X$ is large enough.

We assume that the protein-DNA interaction is a fast process, and hence, the quasi-equilibrium assumption yields

$$\frac{d[D]}{dt} = \frac{d[D^*]}{dt} = 0.$$

Moreover, the gene sequence is either free or bound to the transcription factor, and hence

$$[D] + [D^*] = 1.$$

Thus, we have

$$[D] = \frac{K^n}{K^n + X^n}, \quad [D^*] = \frac{X^n}{K^n + X^n},$$

where $K^n = k_1^-/k_1^+$. When $k_2 < k_3$, letting $\rho = k_2/k_3 < 1$, the production rate of mRNA is

$$k_2[D] + k_3[D^*] = k_3 \left( \rho + (1 - \rho)\frac{X^n}{K^n + X^n} \right), \qquad (2.115)$$

and when $k_3 < k_2$, letting $\rho = k_3/k_2 < 1$, the production rate of mRNA is

$$k_2[D] + k_3[D^*] = k_2 \left( \rho + (1 - \rho)\frac{K^n}{K^n + X^n} \right). \qquad (2.116)$$

The production rates (2.115) and (2.116) are Hill-type functions that determine how the effective transcription rates depend on the concentration of transcription factor $X$. Here, the maximum transcription rate is $k_{\max} = \max\{k_2, k_3\}$, and the minimum rate is $\rho k_{\max}$. When $X = 0$, we have the basal transcription level, and when $X = K$, the transcription rate reaches the half-maximum level $(1 + \rho)k_{\max}/2$.

Now, we have obtained Hill-type functions from enzymatic and protein-DNA interactions. These types of functions are widely used in simplifying mathematical formulations for protein interactions. Nevertheless, in real applications, one should not be limited to the form of the expressions given here and should choose suitable expressions for the problems under consideration.

## 2.9   Summary

In the history of mathematics, the discovery of calculus by Isaac Newton and Gottfried Wilhelm Leibniz is no doubt a remarkable step toward the understanding of our natural world. Differential equations are direct consequences following the discovery of calculus. Through differential equations, we are able to understand the long-term dynamics of a complex system based on the rules behind them. In biological systems, differential equation models are often established based on our understanding of the mechanisms or assumptions behind the observations. These types of mathematical models can be powerful in discovering the amazing world of biological dynamics.

## References

1. Hodgkin, A.L., Huxley, A.F.: A quantitative description of membrane current and its application to conduction and excitation in nerve. J. Physiol. **117**, 500–544 (1952)
2. Adkins, W.A., Davidson, M.G.: Ordinary Differential Equations. Springer, New York (2012)
3. Hale, J.K.: Ordinary Differential Equations. Dover Publications (2009)
4. Øksendal, B.: Stochastic Differential Equations, 6th edn. Springer, Berlin (2005)
5. Smith, H.: An Introduction to Delay Differential Equations. Springer, New York (2010)
6. Kuang, Y.: Delay Differential Equations. Academic Press, Boston (1993)
7. Murray, J.D.: Mathematical Biology. Springer, New York (1993)
8. Ermentrout, B., Mahajan, A.: Simulating, analyzing, and animating dynamical systems: a guide to XPPAUT for researchers and students. Appl. Mech. Rev. **56**, 53 (2003)
9. Tsai, T.Y.-C., Choi, Y.S., Ma, W., Pomerening, J.R., Tang, C., Ferrell, J.E.: Robust, tunable biological oscillations from interlinked positive and negative feedback loops. Science **321**, 126–129 (2008)
10. Mackey, M.C., Glass, L.: Oscillation and chaos in physiological control systems. Science **197**, 287–289 (1977)
11. van Kampen, N.: Stochastic Processes in Physics and Chemistry. North-Holland, Amsterdam (2007)
12. Einstein, A.: Eine neue Bestimmung der Moleküldimensionen. Ann. Phys. **19**, 289 (1906)
13. Olver, P.J.: Introduction to Partial Differential Equations. Springer, New York (2016)
14. Logan, J.D.: Applied Partial Differential Equations. Springer, New York (2015)
15. May, R.M., Oster, G.F.: Bifurcation and dynamic complexity in simple ecological models. Amer. Natur. **110**, 573–599 (1976)
16. Fall, C., Marland, E., Wagner, J., Tyson, J. (eds.): Computational Cell Biology. Interdisciplinary Applied Mathematics, Springer, New York (2002)

17. Michaelis, L., Menten, M.: Die kinetik der invertinwikung. Biochemistry Zeitung **49**, 333–369 (1913)
18. Johnson, K.A., Goody, R.S.: The original Michaelis constant: translation of the 1913 Michaelis-Menten paper. Biochemistry **50**, 8264–8269 (2011)
19. Hill, A.: The possible effects of the aggregation of the molecules of haemoglobin on its dissociation curves. J Physiol Suppl iv–vii (1910)

# Chapter 3
# Mathematical Preliminary–Stochastic Modeling

*Essentially, all models are wrong; some models are useful.*

—George E.P. Box

## 3.1 Introduction

Biological processes are essentially random in terms of both cellular behavior and extracellular environment [1–14]. These stochasticities come from both random molecular interactions (intrinsic noise) and fluctuating microenvironments (extrinsic noise). In particular, gene expression is fundamentally a stochastic process, with randomness in transcription and translation that can lead to cell-to-cell variations in mRNA and protein levels [15–17]. The randomly changing nucleosome state, e.g., DNA methylation or histone modifications, introduces another level of stochastic gene expression [18, 19]. Noise propagation in gene networks has important consequences for cellular functions, being beneficial in some contexts and harmful in others [6, 20]. To describe these stochastic chemical kinetics, stochastic modeling is highlighted [2, 9, 16, 21].

Chemical dynamics have been widely accepted in studying the chemical kinetics of reacting systems with large molecule populations, typically on the order of $10^{23}$. In these systems, the kinetics are nearly deterministic and can be described by a set of ordinary differential equations—the *reaction rate equations*—or partial differential equations if spatial movements are taken into account. Nevertheless, in intracellular molecule kinetics, stochasticity is significant because the numbers of each molecule species are very low. For instance, the promoter of each gene can either be active or inactive during gene expression. In individual bacteria, there are fewer than 20 transcripts of mRNAs from a single gene [22]. The typical molecule numbers of the same protein species in a cell are usually no more than a few thousands. Thus,

© The Author(s), under exclusive license to Springer Nature Switzerland AG 2021
J. Lei, *Systems Biology*, Lecture Notes on Mathematical Modelling in the Life Sciences,
https://doi.org/10.1007/978-3-030-73033-8_3

fluctuations in protein activities are significant due to the effect of low numbers. Reaction rate equations fail to describe these fluctuations.

Essentially, the kinetics of biological molecules in living cells are the consequences of chemically reacting systems. In this chapter, we introduce the main equations for the stochastic modeling of such systems and discuss the numerical methods used in stochastic simulations. While we mainly focus on biochemical reaction systems, the methods introduced here are also applicable to problems with low-number effects, such as birth-death processes in cell population dynamics when the cell counts are small or stochastic epidemic dynamics [23].

## 3.2 Biochemical Reaction Systems

Consider a system of well-stirred mixtures of $N (\geq 1)$ molecular species $\{S_1, \ldots, S_N\}$ inside a fixed volume $\Omega$ and at a constant temperature through $M (\geq 1)$ reaction channels $\{R_1, \ldots, R_M\}$. We can specify the dynamical state of this system by $\mathbf{X}(t) = (X_1(t), \ldots, X_N(t))$, where

$$
\begin{aligned}
X_i(t) = \text{ the number of } S_i \text{ molecule in} \\
\text{the system at time } t, \quad (i = 1, \ldots, N).
\end{aligned}
\tag{3.1}
$$

The system dynamics can be described by the evolution of $\mathbf{X}(t)$ from some given initial state $\mathbf{X}(t_0) = \mathbf{x}_0$. It is obvious that $\mathbf{X}(t)$ is a stochastic process because the time at which a particular reaction occurs is random. Therefore, instead of tracking a single pathway, our goal is to study the evolution of the statistical properties of system states.

Here, we always assume that each reaction, once it occurs, is completed instantaneously. This is to be distinguished from systems involving reactions with a delay or molecular memory [24, 25]. Furthermore, we assume that the system is well stirred such that at any moment, each reaction occurs with equal probability at any position. Under these assumptions, each reaction channel $R_j$ associates with a *propensity function* $a_j$ and a *state-change vector* $\mathbf{v}_j = (v_{j1}, \ldots, v_{jN})$, which are defined such that

$$
\begin{aligned}
a_j(\mathbf{x})dt = \text{ the probability, given } \mathbf{X}(t) = \mathbf{x}, \text{ that one reaction } R_j \\
\text{will occur somewhere inside } \Omega \text{ in the next infinitesimal} \\
\text{time interval } [t, t + dt), \quad (j = 1, \ldots, M).
\end{aligned}
\tag{3.2}
$$

and

$$
\begin{aligned}
v_{ji} = \text{ the change in the number of } S_i \text{ molecule due to} \\
\text{one } R_j \text{ reaction}, \quad (j = 1, \ldots, M; i = 1, \ldots N).
\end{aligned}
\tag{3.3}
$$

**Fig. 3.1** Schematic of
chemical energy for the
reaction $O_2 + 2H_2 \rightarrow 2H_2O$

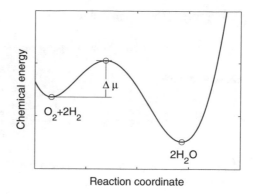

The propensity function and the state-change vector together specify the reaction channel $R_j$. Therefore, the equations given below to describe the evolution of a biochemical reaction system are derived from the propensity functions and state-change vectors connected to the $M$ reaction channels.

The state-change vector of a reaction channel is easy to obtain by counting the numbers of each molecule species that are consumed and produced in the reaction. For instance, if $R_1$ represents the reaction $S_1 + 2S_2 \rightarrow 2S_3$, then $\mathbf{v}_1 = (-1, -2, 2, \ldots)$.

Exact descriptions of the propensity functions are often associated with the *ad hoc* stochasticity of deterministic chemical kinetics [26] and have a solid microphysical basis. In general, the function $a_j$ has the mathematical form [27]

$$a_j(\mathbf{x}) = c_j h_j(\mathbf{x}). \tag{3.4}$$

Here, $c_j$ is the *specific probability rate constant* for the channel $R_j$, which is defined such that $c_j dt$ gives the probability that a randomly chosen combination of $R_j$ reactant molecules will react accordingly in the next infinitesimal time interval $dt$. This probability $c_j dt$ equals the multiple of two parts, the probability that a randomly chosen combination of $R_j$ reactant molecules will collide in the next $dt$, and the probability that colliding reactant molecules will actually react according to $R_j$. The first probability depends on the average relative speeds (which in turn depend on the temperature), the collision sections of reactant molecules, and the system volume $\Omega$. The second probability depends on the chemical energy barrier $\Delta\mu$ of the reaction $R_j$ (Fig. 3.1) and is usually associated with temperature through a Boltzmann factor $e^{-\Delta\mu/k_BT}$, where $k_B$ is the Boltzmann constant.

The function $h_j(\mathbf{x})$ in (3.4) measures the number of distinct combinations of $R_j$ reactant molecules available under state $\mathbf{x}$. It can be easily obtained from the reaction $R_j$. For example, in the above reaction $R_1$, we would have $h_j(\mathbf{x}) = x_1 x_2 (x_2 - 1)/2$, which gives the number of combinations to select one $S_1$ molecule from $x_1$ of them and two $S_2$ molecules from $x_2$ of them.

In general, for a chemical reaction

$$R_j : m_{j1}S_1 + \cdots + m_{jN}S_N \rightarrow n_{j1}S_1 + \cdots + n_{jN}S_N,$$  (3.5)

we would have

$$v_{ji} = n_{ji} - m_{ji}, \quad a_j(\mathbf{x}) \propto \prod_{k=1}^{N} \frac{x_k!}{m_{jk}!(x_k - m_{jk})!}.$$

If $x_k \gg m_{jk}$ for any $k$ and $j$, we have approximately

$$a_j(\mathbf{x}) = c_j \prod_{k=1}^{N} x_k^{m_{jk}}.$$  (3.6)

The *reaction rate constant* $c_j$ can only be obtained from experiments and usually depends on the system volume $\Omega$ and the temperature.

In real systems, most reaction channels are either *monomolecular* or *bimolecular* reactions. For a monomolecular reaction, the reaction rate constant $c_j$ is independent of the system volume $\Omega$. For a bimolecular reaction, the rate constant $c_j$ is inversely proportional to $\Omega$. *Trimolecular* reactions do not physically occur in dilute solutions with appreciable frequency. One can consider a trimolecular reaction as the combined result of two bimolecular reactions that involves an additional short-lived species. For such an "effective trimolecular" reaction, the approximate $c_j$ is proportional to $\Omega^{-2}$ [27].

## 3.3  Mathematical Formulations–Intrinsic Noise

First, we assume that the propensity functions are time independent, i.e., the reaction rates $c_j$ in (3.4) are constants. In this situation, the fluctuations in the system are inherent to the system of interest (*intrinsic noise*). The opposite case is *extrinsic noise*, which arises from the variability in factors that are considered to be external. Mathematical formulations for extrinsic noise are discussed in the next section.

### 3.3.1  Chemical Master Equation

From the propensity function given previously, the state vector $\mathbf{X}(t)$ is a jump-type Markov process on the nonnegative $N$-dimensional integer lattice. In the following analysis of such a system, we mainly focus on the conditional *probability function*

$$P(\mathbf{x}, t|\mathbf{x}_0, t_0) = \text{Prob}\{\mathbf{X}(t) = \mathbf{x}, \text{ given that } \mathbf{X}(t_0) = \mathbf{x}_0\}. \tag{3.7}$$

Hereinafter, we use an upper case letter to denote a random variable and the corresponding lower case letter to denote a possible value of the random variable.

Through the probability function, the *average* or *expectation value* of any quantity $f(\mathbf{X}|\mathbf{x}_0, t_0)$ defined in the system is given by

$$\langle f(\mathbf{X}|\mathbf{x}_0, t_0)\rangle = \sum_{\mathbf{x}} f(\mathbf{x})P(\mathbf{x}, t|\mathbf{x}_0, t_0). \tag{3.8}$$

Physically, in an ensemble of identical systems starting from the same initial state $\mathbf{X}(t_0) = \mathbf{x}_0$, the function $P(\mathbf{x}, t|\mathbf{x}_0, t_0)$ gives the fraction of subsystems with state $\mathbf{X}(t) = \mathbf{x}$ at time $t$.

To derive a time evolution for the probability function $P(\mathbf{x}, t|\mathbf{x}_0, t_0)$, we take a time increment $dt$ and consider the variation between the probability of $\mathbf{X}(t) = \mathbf{x}$ and the probability of $\mathbf{X}(t + dt) = \mathbf{x}$, given that $\mathbf{X}(t_0) = \mathbf{x}_0$. This variation is

$$P(\mathbf{x}, t + dt|\mathbf{x}_0, t_0) - P(\mathbf{x}, t|\mathbf{x}_0, t_0) = \quad \text{Increase of the probability in } dt \atop - \text{ Decrease of the probability in } dt \tag{3.9}$$

We take $dt$ so small such that the probability of having two or more reactions in $dt$ is negligible compared to the probability of having only one reaction. Then, an increase of the probability in $dt$ occurs when a system with state $\mathbf{X}(t) = \mathbf{x} - \mathbf{v}_j$ reacts according to $R_j$ in $(t, t + dt)$, the probability of which is $a_j(\mathbf{x} - \mathbf{v}_j)dt$. Thus,

$$\text{Increase of the probability in } dt = \sum_{j=1}^{M} P(\mathbf{x} - \mathbf{v}_j, t|\mathbf{x}_0, t_0)a_j(\mathbf{x} - \mathbf{v}_j)dt. \tag{3.10}$$

Similarly, when a system with state $\mathbf{X}(t) = \mathbf{x}$ reacts according to any reaction channel $R_j$ in $(t, t + dt)$, the probability $P(\mathbf{x}, t|\mathbf{x}_0, t_0)$ decreases. Thus,

$$\text{Decrease of the probability in } dt = \sum_{j=1}^{M} P(\mathbf{x}, t|\mathbf{x}_0, t_0)a_j(\mathbf{x})dt. \tag{3.11}$$

Substituting (3.10) and (3.11) into (3.9), we obtain

$$P(\mathbf{x}, t + dt|\mathbf{x}_0, t_0) - P(\mathbf{x}, t|\mathbf{x}_0, t_0) = \sum_{j=1}^{M} P(\mathbf{x} - \mathbf{v}_j, t|\mathbf{x}_0, t_0)a_j(\mathbf{x} - \mathbf{v}_j)dt$$
$$- \sum_{j=1}^{M} P(\mathbf{x}, t|\mathbf{x}_0, t_0)a_j(\mathbf{x})dt,$$

which yields, with the limit $dt \to 0$, the *chemical master equation* (CME) [27, 28]:

$$\frac{\partial}{\partial t}P(\mathbf{x}, t|\mathbf{x}_0, t_0) = \sum_{j=1}^{M}[P(\mathbf{x} - \mathbf{v}_j, t|\mathbf{x}_0, t_0)a_j(\mathbf{x} - \mathbf{v}_j) - P(\mathbf{x}, t|\mathbf{x}_0, t_0)a_j(\mathbf{x})]. \quad (3.12)$$

Equation (3.12) is an exact consequence of the reaction channels characterized by the propensity functions and the state-change vectors. If one can solve (3.12) for $P(\mathbf{x}, t|\mathbf{x}_0, t_0)$, we should be able to determine the evolution of the statistical properties of the system. However, such an exact solution of (3.12) can rarely be obtained. We refer the readers to [29–31] for examples of analytically solving the chemical master equation.

In fact, the chemical master equation (3.12) is a set of linear differential equations with constant coefficients. The difficulty in solving the Eq. (3.12) comes from the extremely high dimensionality, which equals the total number of possible states of the system under study. For example, for a system with 100 molecule species, each of which has two possible states ($X_i(t) = 0$ or 1), the system has a total of $2^{100}$ possible states; therefore, the Eq. (3.12) contains $2^{100}$ equations!

### 3.3.2  Fokker-Planck Equation

In a biochemical reaction system, if all components of $\mathbf{X}(t)$ are very large compared to 1, we can regard the components of $\mathbf{X}(t)$ as real numbers. We further assume that the functions $f_j(\mathbf{x}) \equiv a_j(\mathbf{x})P(\mathbf{x}, t|\mathbf{x}_0, t_0)$ are analytic in the variable $\mathbf{x}$. Under these two assumptions, we can use Taylor's expansion to write

$$f_j(\mathbf{x} - \mathbf{v}_j) = f_j(\mathbf{x}) + \sum_{|\mathbf{m}|\geq 1}\prod_{i=1}^{N}\frac{(-1)^{m_i}}{m_i!}\left(v_{ji}\frac{\partial}{\partial x_i}\right)^{m_i}(f_j(\mathbf{x})). \quad (3.13)$$

Here $\mathbf{m} = (m_1, \ldots, m_N) \in \mathbb{Z}^N$, $|\mathbf{m}| = m_1 + \cdots + m_N$. Substituting (3.13) into (3.12) and noting that

$$\prod_{i=1}^{N}\frac{(-1)^{m_i}}{m_i!}\left(v_{ji}\frac{\partial}{\partial x_i}\right)^{m_i} = (-1)^{|\mathbf{m}|}\prod_{i=1}^{N}v_{ji}^{m_i} \times \prod_{i=1}^{N}\left(\frac{1}{m_i!}\frac{\partial^{m_i}}{\partial x_i^{m_i}}\right),$$

we immediately obtain the *chemical Kramers-Moyal equation* (CKME) [27]

$$\frac{\partial}{\partial t}P(\mathbf{x}, t) = \sum_{|\mathbf{m}|\geq 1}(-1)^{|\mathbf{m}|}\prod_{i=1}^{N}\left(\frac{1}{m_i!}\frac{\partial^{m_i}}{\partial x_i^{m_i}}\right)\left(A^{m_1,\ldots,m_N}(\mathbf{x})P(\mathbf{x}, t)\right), \quad (3.14)$$

where

$$A^{m_1,\ldots,m_N}(\mathbf{x}) = \sum_{j=1}^{M} v_{j1}^{m_1} \cdots v_{jN}^{m_N} a_j(\mathbf{x}).$$

Hereinafter, we omit the initial condition $(\mathbf{x}_0, t_0)$. If all functions $f_j(\mathbf{x} - \mathbf{v}_j)$ are smooth, Eq. (3.14) would be equivalent to Eq. (3.12). Therefore, the chemical Kramers-Moyal equation is a "semi-rigorous" consequence of the chemical master equation (3.12).

We often truncate the right-hand side of (3.14) at $|\mathbf{m}| = 2$ and obtain the *chemical Fokker-Planck equation* (CFPE)

$$\frac{\partial}{\partial t} P(\mathbf{x}, t) = -\sum_{i=1}^{N} \frac{\partial}{\partial x_i} A_i(\mathbf{x}) P(\mathbf{x}, t) + \frac{1}{2} \sum_{1 \le i,j \le N} \frac{\partial^2}{\partial x_i \partial x_j} B_{ij}(\mathbf{x}) P(\mathbf{x}, t). \qquad (3.15)$$

where

$$A_i(\mathbf{x}) = \sum_{k=1}^{M} v_{ki} a_k(\mathbf{x}), \quad B_{ij}(\mathbf{x}) = \sum_{k=1}^{M} v_{ki} v_{kj} a_k(\mathbf{x}). \qquad (3.16)$$

The chemical Fokker-Planck equation gives a partial differential equation for the evolution of the probability density of the biochemical reaction system. We will reobtain this chemical Fokker-Planck equation in Sect. 3.4.3 from the chemical Langevin equation.

### 3.3.3 Reaction Rate Equation

If we multiply the chemical master equation (3.12) by $x_i$ and sum over all $\mathbf{x}$, we have

$$\sum_{\mathbf{x}} x_i \frac{\partial}{\partial t} P(\mathbf{x}, t) = \sum_{\mathbf{x}} x_i \sum_{j=1}^{M} [P(\mathbf{x} - \mathbf{v}_j, t) a_j(\mathbf{x} - \mathbf{v}_j) - P(\mathbf{x}, t)].$$

Exchanging the summation over $\mathbf{x}$ and the partial derivative $\partial/\partial t$ on the left-hand side as well as the summations over $\mathbf{x}$ and $j$ on the right-hand side, we have

$$\frac{\partial}{\partial t} \sum_{\mathbf{x}} x_i P(\mathbf{x}, t) = \sum_{j=1}^{M} \sum_{\mathbf{x}} x_i [P(\mathbf{x} - \mathbf{v}_j, t) a_j(\mathbf{x} - \mathbf{v}_j) - P(\mathbf{x}, t)]$$

We note that

$$\langle X_i \rangle = \sum_{\mathbf{x}} x_i P(\mathbf{x}, t)$$

gives the ensemble average of $X_i$, hence

$$\frac{d \langle X_i \rangle}{dt} = \sum_{j=1}^{M} \sum_{\mathbf{x}} x_i [P(\mathbf{x} - \mathbf{v}_j, t) a_j(\mathbf{x} - \mathbf{v}_j) - P(\mathbf{x}, t)]$$

$$= \sum_{j=1}^{M} \left( \sum_{\mathbf{x}} (x_i + v_{ji}) P(\mathbf{x}, t) a_j(\mathbf{x}) - \sum_{\mathbf{x}} P(\mathbf{x}, t) \right)$$

$$= \sum_{j=1}^{M} \sum_{\mathbf{x}} v_{ji} P(\mathbf{x}, t) a_j(\mathbf{x})$$

$$= \sum_{j=1}^{M} v_{ji} \langle a_j(\mathbf{X}) \rangle.$$

Here, we note the average of any quantity defined by (3.8). Now, we have obtained the *chemical ensemble average equation* (CEAE)

$$\frac{d \langle X_i \rangle}{dt} = \sum_{j=1}^{M} v_{ji} \langle a_j(\mathbf{X}) \rangle \quad (i = 1, 2, \ldots, N). \tag{3.17}$$

Equation (3.17) is an exact consequence of (3.12). Nevertheless, since the expectation $\langle a_j(\mathbf{X}) \rangle$ cannot be expressed as a function of $\langle X \rangle$ when $a_i$ is a nonlinear function, (3.17) is usually not a closed-form equation system.

If the approximation

$$\langle a_j(\mathbf{X}) \rangle \approx a_j(\langle \mathbf{X} \rangle) \tag{3.18}$$

is acceptable for all $1 \le i \le N$, the ensemble average Eq. (3.17) can be rewritten as a set of ordinary differential equations

$$\frac{dx_i}{dt} = \sum_{j=1}^{M} v_{ji} a_j(\mathbf{x}) \quad (i = 1, \ldots, N), \tag{3.19}$$

where the components of $\mathbf{x}(t)$ are now considered real variables. Equation (3.19) is referred to as the macroscopic *reaction rate equation* (RRE), or the *chemical rate equation* in some articles.

The condition (3.18) is satisfied when all reaction channels are monomolecular; hence, the propensity functions $a_j(\mathbf{X})$ are linear. Otherwise, the reaction rate Eq. (3.19) is valid only when the fluctuations are not important so that (3.18) holds approximately. In Chap. 5, given an example of a situation in which the reaction rate equation may give different results from the original stochastic model.

The reaction rate Eq. (3.19) is often written in terms of the species concentrations

$$z_i(t) \equiv x_i(t)/\Omega \quad (i = 1, \ldots, N).$$

The "concentration" form of the reaction rate equation has the form

$$\frac{dz_i}{dt} = \sum_{j=1}^{M} v_{ji} \tilde{a}_j(\mathbf{z}) \quad (i = 1, \ldots, N), \tag{3.20}$$

where

$$\tilde{a}_j(\mathbf{z}) = a_j(\Omega \mathbf{z})/\Omega \quad (j = 1, 2, \ldots, M).$$

The function of $\tilde{a}_j$ has the same form as that of $a_j$ but may depend on the volume $\Omega$. For example, when $a_j$ has the form of polynomial (3.6), we have

$$\tilde{a}_j(\mathbf{z}) = c_j \Omega^{-1} \prod_{k=1}^{N} (\Omega z_k)^{m_{jk}} = k_j \prod_{k=1}^{N} z_k^{m_{jk}}, \quad k_j = c_j \Omega^{|\mathbf{m}_j|-1}.$$

Here, we have replaced the rate constant $c_j$ with the *reaction rate constant* $k_j$.

Reaction rate equations are most commonly used in modeling biochemical reaction systems. However, these equations fail to describe the stochastic effects, which may be very important in biological processes. In the next section, we introduce the *chemical Langevin equation*, which can describe the stochasticity and is easy to study, at least numerically.

### 3.3.4 Chemical Langevin Equation

The chemical Langevin equation was derived to yield a Langevin-type approximate equation for the time evolution of a biochemical reaction system. The derivation of the equation, given by Daniel T. Gillespie, was based on the chemical master equation and two explicit dynamical conditions, as detailed below. Most of the following refer to Gillespie's original paper [27].

Suppose that the state of a system at current time $t$ is $\mathbf{X}(t) = \mathbf{x}$. Let $K_j(\mathbf{x}, \tau) \, (\tau > 0)$ be the number of $R_j$ reactions that occur in the subsequent time interval $[t, t + \tau]$ (Fig. 3.2). Since each of these reactions changes the number of $S_i$ populations by $v_{ji}$, the number of $S_i$ molecules in the system at time $t + \tau$ becomes

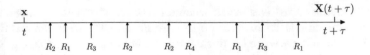

**Fig. 3.2** Illustration of the chemical reactions that occur during the time period from $t$ to $t + \tau$. The system state changes from **x** to $\mathbf{X}(t + \tau)$

$$X_i(t + \tau) = x_i + \sum_{j=1}^{M} K_j(\mathbf{x}, \tau) v_{ji}, \quad (i = 1, \ldots, N). \tag{3.21}$$

We note that $K_j(\mathbf{x}, \tau)$ is a *random variable* and therefore, $X_i(t + \tau)$ is random.

Here, we obtain an approximation of $K_j(\mathbf{x}, \tau)$ by imposing the following conditions:

**Condition (i)**   Require $\tau$ to be *small* enough so that the change in the state during $[t, t + \tau]$ is so slight that none of the propensity functions changes its value "appreciably".

**Condition (ii)**   Require $\tau$ to be *large* enough so that the expected number of the occurrences of each reaction channel $R_j$ in $[t, t + \tau]$ is much larger than 1.

From condition (i), the propensity functions satisfy

$$a_j(\mathbf{X}(t')) \approx a_j(\mathbf{x}), \quad \forall t' \in [t, t + \tau], \; \forall j \in \{1, 2, \ldots, M\}. \tag{3.22}$$

Thus, the probability that the reaction $R_j$ will occur in any infinitesimal interval $d\tau$ within $[t, t + \tau]$ is $a_j(\mathbf{x}) d\tau$. Here, $K_j(\mathbf{x}, \tau)$, the occurrence of the "events" reaction $R_j$ in the time interval $[t, t + \tau]$, is a statistically independent *Poisson* random variable with a mean $a_j(\mathbf{x})\tau$, which is represented by $\mathscr{P}_j(a_j(\mathbf{x})\tau)$. Therefore, (3.21) can be approximated by

$$X_i(t + \tau) = x_i + \sum_{j=1}^{M} v_{ji} \mathscr{P}_j(a_j(\mathbf{x})\tau), \quad (i = 1, \ldots, N) \tag{3.23}$$

according to condition (i).

According to the properties of Poisson random variables, the mean and variance of $\mathscr{P}(a\tau)$ are

$$\langle \mathscr{P}(a\tau) \rangle = \text{var}\{\mathscr{P}(a\tau)\} = a\tau. \tag{3.24}$$

Condition (ii) means

$$a_j(\mathbf{x})\tau \gg 1, \quad \forall j \in \{1, 2, \ldots, M\}. \tag{3.25}$$

The inequality (3.25) allows us to approximate each Poisson random variable by a *normal* random variable with the same mean and variance, i.e., $\mathscr{P}_j(a(\mathbf{x})\tau) \sim \mathscr{N}_j(a(\mathbf{x})\tau, a(\mathbf{x})\tau)$, where $\mathscr{N}_j(m, \sigma^2)$ denotes the normal random variable with mean $m$ and variance $\sigma^2$. This leads to further approximation:

$$X_i(t + \tau) = x_i + \sum_{j=1}^{M} v_{ji} \mathscr{N}_j(a_j(\mathbf{x})\tau, a_j(\mathbf{x})\tau), \quad (i = 1, \ldots, N) \tag{3.26}$$

Here, the $M$ random variables are independent of each other. Note that in the above approximation (3.26), we have converted the molecular population $X_i$ from discretely changing integers to continuously changing real variables.

We note the linear combination theorem for normal random variables, which gives

$$\mathscr{N}(m, \sigma^2) = m + \sigma \mathscr{N}(0, 1); \tag{3.27}$$

equation (3.26) can further be rewritten as

$$X_i(t + \tau) = x_i + \sum_{j=1}^{M} v_{ji} a_j(\mathbf{x})\tau + \sum_{j=1}^{M} v_{ji} \sqrt{a_j(\mathbf{x})\tau} \mathscr{N}_j(0, 1) \quad (i = 1, \ldots, N). \tag{3.28}$$

Now, we are ready to obtain the Langevin-type equations by making some purely notational changes. First, we denote the time interval $\tau$ as $dt$ and write

$$dX_i = X_i(t + dt) - X_i(t).$$

Next, introduce $M$, which is temporally uncorrelated, and independent random process $W_j(t)$, which satisfies

$$dW_j(t) = W_j(t + dt) - W_j(t) = \mathscr{N}_j(0, 1)\sqrt{dt} \quad (j = 1, \ldots, N). \tag{3.29}$$

It is easy to verify that the processes $W_j(t)$ have stationary independent increments with mean 0, i.e.,

$$\langle dW_j(t) \rangle = 0, \quad \langle dW_i(t)dW_j(t') \rangle = \delta_{ij}\delta(t - t')dt, \quad \forall 1 \le i, j \le M, \forall t, t'. \tag{3.30}$$

Here, $\delta_{ij}$ is the Kronecker delta, and $\delta(t - t')$ is the Dirac delta function. Thus, each $W_j$ is a *Wiener process* (or referred to as *Brownian motion*) [28]. Finally, recalling that $\mathbf{x}$ stands for $\mathbf{X}(t)$, the Eq. (3.28) becomes a *Langevin equation* (or *stochastic differential equation*)

$$dX_i = \sum_{j=1}^{M} v_{ji} a_j(\mathbf{X})dt + \sum_{j=1}^{M} v_{ji} \sqrt{a_j(\mathbf{X})} dW_j \quad (i = 1, \ldots, N). \tag{3.31}$$

The stochastic differential equation (3.31) is the desired *chemical Langevin equation* (CLE). The solution of (3.31) with initial condition $\mathbf{X}(0) = \mathbf{X}_0$ is a stochastic process $\mathbf{X}(t)$ that satisfies

$$\mathbf{X}(t) = \mathbf{X}_0 + \sum_{j=1}^{M} \int_0^t v_{ji} a_j(\mathbf{X}(s))ds + \sum_{j=1}^{M} \int_0^t v_{ji} \sqrt{a_j(\mathbf{X}(s))} dW_j(s). \tag{3.32}$$

Here, Itô integrals are used since the stochastic fluctuations are *intrinsic* [28].

In the above arguments, it is obvious that the conditions (i) and (ii) are contradictory to each other. The inequality (3.25) implies that $a_j(\mathbf{x})$ is large when $\tau$ is small enough, as required by condition (i). This is possible only when the system has large molecular populations for each molecular species, since $a_j(\mathbf{x})$ is typically proportional to one or more components of $\mathbf{x}$. In many biological systems, the two conditions cannot be satisfied simultaneously. In this case, the derivation of the chemical Langevin equation may fail. Nevertheless, even when conditions (i) and (ii) are not satisfied simultaneously, the chemical Langevin equation (3.31) can be heuristic for many problems.

### 3.3.5  Discussions of the Chemical Langevin Equation

In many intracellular biochemical systems, such as gene expression and genetic networks, the molecular populations are small, and the reactions are usually slow. In these systems, the conditions to derive the chemical Langevin equation may not be valid. Nevertheless, Eq. (3.31) is still useful for such systems, as it can provide reasonable descriptions of the statistical properties of the kinetic processes. The reasons are given below.

#### 3.3.5.1  Time Evolution of the Mean

If we take the average of both sides of (3.31) and note

$$\langle \sqrt{a_j(\mathbf{X})} dW_j \rangle = 0 \quad (j = 1, \ldots, M)$$

according to the Itô interpretation, we have

$$\frac{d\langle X_i \rangle}{dt} = \sum_{j=1}^{M} v_{ji} \langle a_j(\mathbf{X}) \rangle \quad (i = 1, \ldots, N). \tag{3.33}$$

This gives the same form of the chemical ensemble average equation (3.17) as we obtained from the chemical master equation.

### 3.3.5.2 Time Evolution of the Correlations

The correlations between the molecule numbers are defined as

$$\sigma_{ij}(t) = \langle X_i(t)X_j(t)\rangle - \langle X_i(t)\rangle\langle X_j(t)\rangle, \quad (1 \le i,j \le N). \tag{3.34}$$

Multiplying the chemical master equation (3.12) by $x_i x_j$ and summing over all $\mathbf{x}$,

$$\sum_{\mathbf{x}} x_i x_j \frac{\partial}{\partial t} P(\mathbf{x}, t) = \sum_{\mathbf{x}} x_i x_j \sum_{k=1}^{M} [P(\mathbf{x} - \mathbf{v}, t) a_k(\mathbf{x} - \mathbf{v_j}) - P(\mathbf{x}, t) a_k(\mathbf{x})].$$

Exchanging the summation over $\mathbf{x}$ with the derivative and the summations over $\mathbf{x}$ and $k$, we have

$$\frac{\partial}{\partial t} \sum_{\mathbf{x}} x_i x_j P(\mathbf{x}, t) = \sum_{k=1}^{M} \sum_{\mathbf{x}} x_i x_j (P(\mathbf{x} - \mathbf{v}_k, t) a_k(\mathbf{x} - \mathbf{v}_k) - P(\mathbf{x}, t) a_k(\mathbf{x})).$$

We note that

$$\langle X_i X_j \rangle = \sum_{\mathbf{x}} x_i x_j P(\mathbf{x}, t),$$

and

$$\begin{aligned}
&\sum_{\mathbf{x}} x_i x_j (P(\mathbf{x} - \mathbf{v}, t) a_k(\mathbf{x} - \mathbf{v_j}) - P(\mathbf{x}, t) a_k(\mathbf{x})) \\
&= \sum_{\mathbf{x}} (x_i + v_{ki})(x_j + v_{kj}) P(\mathbf{x}, t) a_k(\mathbf{x}) - \sum_{\mathbf{x}} x_i x_j P(\mathbf{x}, t) \\
&= \sum_{\mathbf{x}} x_i v_{kj} P(\mathbf{x}, t) a_k(\mathbf{x}) + \sum_{\mathbf{x}} v_{ki} x_j P(\mathbf{x}, t) a_k(\mathbf{x}) + \sum_{\mathbf{x}} v_{ki} v_{kj} P(\mathbf{x}, t) a_k(\mathbf{x}) \\
&= \langle v_{kj} X_i a_k(\mathbf{X}) \rangle + \langle v_{ki} X_j a_k(\mathbf{X}) \rangle + \langle v_{ki} v_{kj} a_k(\mathbf{X}) \rangle.
\end{aligned}$$

Hence, we have

$$\frac{d \langle X_i X_j \rangle}{dt} = \sum_{k=1}^{M} \left[ \langle v_{kj} X_i a_k(\mathbf{X}) \rangle + \langle v_{ki} X_j a_k(\mathbf{X}) \rangle + \langle v_{ki} v_{kj} a_k(\mathbf{X}) \rangle \right]. \tag{3.35}$$

Furthermore, the chemical ensemble average equation (3.17) gives

$$\frac{d\langle X_i\rangle\langle X_j\rangle}{dt} = \langle X_i\rangle\frac{d\langle X_j\rangle}{dt} + \langle X_j\rangle\frac{d\langle X_i\rangle}{dt}$$

$$= \langle X_i\rangle\sum_{k=1}^{M}v_{kj}\langle a_k(\mathbf{X})\rangle + \langle X_j\rangle\sum_{k=1}^{M}v_{ki}\langle a_k(\mathbf{X})\rangle$$

$$= \sum_{k=1}^{M}\left[\langle v_{ki}a_k(\mathbf{X})\rangle\langle X_j\rangle + \langle v_{kj}a_k(\mathbf{X})\rangle\langle X_i\rangle\right]. \tag{3.36}$$

Thus, from (3.35) and (3.36), we obtain the *chemical ensemble correlations equation* (CECE)

$$\frac{d\sigma_{ij}}{dt} = \langle A_i(\mathbf{X})(X_j - \langle X_j\rangle)\rangle + \langle (X_i - \langle X_i\rangle)A_j(\mathbf{X})\rangle + \langle B_{ij}(\mathbf{X})\rangle, \tag{3.37}$$

where $A_i(\mathbf{X})$ and $B_{ij}(\mathbf{X})$ have been defined in (3.16) as

$$A_i(\mathbf{X}) = \sum_{k=1}^{M}v_{ki}a_k(\mathbf{X}), \quad B_{ij}(\mathbf{X}) = \sum_{k=1}^{M}v_{ki}v_{kj}a_k(\mathbf{X}).$$

Equation (3.37) gives the *exact* form of the chemical master equation.

Now, we derive (3.37) from the chemical Langevin equation (3.31). Applying the Itô formula (2.43), we have

$$d(X_iX_j) = X_i(dX_j) + X_j(dX_i) + (dX_i)(dX_j)$$

$$= \sum_{k=1}^{M}(X_jv_{ki}a_k(\mathbf{X}) + X_iv_{kj}a_k(\mathbf{X}) + v_{ki}v_{kj}a_k(\mathbf{X}))dt$$

$$+ \sum_{k=1}^{M}(X_iv_{kj}\sqrt{a_k(\mathbf{X})} + X_jv_{kj}\sqrt{a_k(\mathbf{X})})dW_k + o(dt)$$

Taking the average of both sides and noting that

$$\langle (X_iv_{kj}\sqrt{a_k(\mathbf{X})} + X_jv_{kj}\sqrt{a_k(\mathbf{X})})dW_k\rangle = 0,$$

we reobtain (3.35), which yields the same form of the chemical ensemble correlations equation (3.37).

At the states of near equilibrium, if the random fluctuations are not important, we have approximately

$$A_i(\mathbf{X}) \approx A_i(\langle\mathbf{X}\rangle) + \sum_{l=1}^{N}\frac{\partial A_i(\langle\mathbf{X}\rangle)}{\partial X_l}(X_l - \langle X_l\rangle), \quad \langle B_{ij}(\mathbf{X})\rangle \approx B_{ij}(\langle\mathbf{X}\rangle),$$

where $\langle \mathbf{X} \rangle$ represents the equilibrium state so that

$$\frac{d\langle \mathbf{X} \rangle}{dt} = \mathbf{0}.$$

Substituting the above approximations into (3.37) and defining the matrices

$$\sigma = (\sigma_{ij}), \quad A = \left( \frac{\partial A_i(\langle \mathbf{X} \rangle)}{\partial X_l} \right), \quad B = (B_{ij}(\langle \mathbf{X} \rangle)), \tag{3.38}$$

we have

$$\frac{d\sigma}{dt} = (A\sigma + \sigma A^T) + B. \tag{3.39}$$

Equation (3.39) gives the linear approximation of the time evolution of the correlation functions near the equilibrium state. This gives an example of the well-known *fluctuation-dissipation theorem* [28].

### 3.3.5.3   Fokker-Planck Equation

From the chemical Langevin equation (3.31) and the discussions in Sect. 2.4.3, note that the flux $J_i(\mathbf{x}, t)$ is now given by

$$J_i(\mathbf{x}, t) = A_i(\mathbf{x})P(\mathbf{x}, t) - \frac{1}{2} \sum_{j=1}^{N} \frac{\partial}{\partial x_j} (B_{ij}(\mathbf{x})P(\mathbf{x}, t))$$

to reobtain the chemical Fokker-Planck equation (3.15). Thus, both the chemical Langevin equation and the chemical master equation yield the same form of the chemical Fokker-Planck equation.

### 3.3.5.4   Remarks

From the above discussions of the chemical Langevin equation and the chemical master equation, we should note the following:

1. The chemical Langevin equation and chemical master equation yield the same forms of the chemical ensemble average equation (3.17), chemical ensemble correlation equation (3.37), and chemical Fokker-Planck equation (3.15).
2. Despite the same forms of Eqs. (3.17) and (3.37), since the averages are taken with respect to the probability functions $P(\mathbf{x}, t)$, which are obtained from the solutions of either the chemical master equation or the chemical Langevin equation, the two equations (CLE and CME) do not always yield the same dynamics of the ensemble average and correlation. When all reaction channels are

monomolecular, the two Eqs. (3.17) and (3.37) are closed, as we discussed before.

3. Both the chemical Langevin equation and the chemical master equation yield the same form of the chemical Fokker-Planck equation, which is a closed equation. Thus, up to the second-order approximation, the two equations give the same time evolutions of the probability function $P(\mathbf{x}, t)$.

## 3.4  Mathematical Formulations–Fluctuations in Kinetic Parameters

In the above chemical master equations, we assume that the reaction rates $c_j$ in (3.4) are constants. This is only a rough approximation of the real world in which the cell environments are stochastic and therefore, the reaction rates are often random. Here, we introduce mathematical formulations for situations in which there are fluctuations in kinetic parameters.

### 3.4.1  Reaction Rate as a Random Process

When the reaction rate $c_j$ is random, the previous propensity function $a_j(\mathbf{x})$ in (3.4) should be rewritten as

$$a_j(\mathbf{x}, t) = c_j(t) h_j(\mathbf{x}). \tag{3.40}$$

Here, $c_j(t)$, as previously described, is the specific probability reaction rate for reaction channel $R_j$ at time $t$. This reaction rate is usually random, depending on the fluctuating environment whose explicit time dependence is not known. A reasonable approximation is given below.

In previous discussions, we obtained $c_j \propto e^{-\Delta\mu_j/k_B T}$. If there are noise perturbations to the energy barrier, we can replace $\Delta\mu_j$ with $\Delta\bar{\mu}_j - k_B T \eta_j(t)$, where $\eta_j(t)$ is a stochastic process. Consequently, we replace the reaction rate $c_j(t)$ with

$$c_j(t) = \bar{c}_j e^{\eta_j(t)} / \langle e^{\eta_j(t)} \rangle, \tag{3.41}$$

where $\bar{c}_j$ is a constant that measures the mean of the reaction rate.

In many cases, the noise perturbation $\eta(t)$ (here, we omit the subscript $j$) can be described by an *Ornsterin-Uhlenbeck process*, which is given by the solution of the following stochastic differential equation:

$$d\eta = -(\eta/\tau) dt + (\sigma/\tau) dW. \tag{3.42}$$

Here, $W$ is a Wiener process, and $\tau$ and $\sigma$ are positive constants measuring the auto-correlation time and variance, respectively. It is easy to verify that $\eta(t)$ is normally distributed and has an exponentially decaying stationary autocorrelation function [28, 32–34]

$$\langle \eta(t)\eta(t') \rangle = \frac{\sigma^2}{2\tau} e^{-|t-t'|/\tau}. \tag{3.43}$$

The Ornstein-Uhlenbeck process is an example of *color noise*. When $\tau \to 0$, $\eta(t)$ approaches *Gaussian white noise*.

When $\eta(t)$ is an Ornstein-Uhlenbeck process, the stationary distribution of the reaction rate $c_j(t)$ is a *log-normal* random variable. The log-normal distribution has been seen in many applications [35]. For instance, log-normal distributions rather than normal distributions have been measured for gene expression rates [36, 37] and translation kinetics [38].

When $\eta(t)$ is an Ornstein-Uhlenbeck process, we have [39]

$$\langle e^{\eta(t)} \rangle = e^{\sigma^2/(4\tau)}. \tag{3.44}$$

Therefore, the reaction rate (3.41) can be rewritten as

$$c_j(t) = \bar{c}_j e^{\eta_j(t)-\sigma_j^2/(4\tau_j)}. \tag{3.45}$$

Thus, the propensity function (3.40) now becomes

$$a_j(\mathbf{x}, t) = \bar{c}_j e^{\eta_j(t)-\sigma_j^2/(4\tau_j)} h_j(\mathbf{x}), \tag{3.46}$$

where $\eta_j(t)$ satisfies an equation of form (3.42).

Next, we introduce generalizations of the reaction rate equations and the chemical Langevin equations to describe reaction systems with fluctuations in reaction rates. We note that in this situation, the state variable $\mathbf{X}$ alone is not enough to describe the state of a system, and the state of the environment has to be taken into account as well. Thus, to generalize the chemical master equation or the Fokker-Planck equation, we should replace the previous probability function with $P(\mathbf{x}, \mathbf{c}, t | \mathbf{x}_0, \mathbf{c}_0, t_0)$, where $\mathbf{c} = \{c_1, \ldots, c_M\}$ represents the state of the environment. This consideration is omitted here.

### 3.4.2  Reaction Rate Equation

Substituting the propensity functions of forms (3.46) and (3.42) into the reaction rate equation (3.19), we obtain the following stochastic differential equations:

$$\begin{cases} dX_i = \left( \sum_{j=1}^{M} v_{ji}\bar{c}_j e^{\eta_j - \sigma_j^2/(4\tau_j)} h_j(\mathbf{X}) \right) dt, \quad (i = 1, \dots, N) \\[4mm] d\eta_j = -(\eta_j/\tau_j)dt + (\sigma_j/\tau_j)dW_j, \quad (j = 1, \dots, M) \end{cases} \tag{3.47}$$

Here $\bar{c}_j$, $\tau_j$, and $\sigma_j$ are positive constants. Equation (3.47) generalizes the reaction rate equation to describe a reaction system with fluctuations in kinetic parameters.

We note that the above equations are different from the additive noise

$$d\mathbf{x} = \sum_{j=1}^{M} v_{ji}c_j h_j(\mathbf{x})dt + \sigma dW \tag{3.48}$$

or the multiplicative noise

$$d\mathbf{x} = \sum_{j=1}^{M} v_{ji}c_j h_j(\mathbf{x})dt + \sigma g(\mathbf{x})dW \tag{3.49}$$

in some studies [40–42]. Equations (3.48) and (3.49) are standard stochastic differential equations and hence mathematically simpler; however, the physical bases of the source of noise are not clearly defined.

### 3.4.3  Chemical Langevin Equation

Similar to the above strategy, substituting the propensity functions of the forms (3.46) and (3.42) into the chemical Langevin equation (3.31), we obtain the following stochastic differential equations:

$$\begin{cases} dX_i = \left( \sum_{j=1}^{M} v_{ji}\bar{c}_j e^{\eta_j - \sigma_j^2/(4\tau_j)} h_j(\mathbf{X}) \right) dt \\[2mm] \qquad + \sum_{j=1}^{M} v_{ji} \left( \bar{c}_j e^{\eta_j - \sigma_j^2/(4\tau_j)} h_j(\mathbf{X}) \right)^{1/2} dW_j, \quad (i = 1, \dots, N) \\[4mm] d\eta_j = -(\eta_j/\tau_j)dt + (\sigma_j/\tau_j)d\hat{W}_j, \quad (j = 1, \dots, M) \end{cases}$$

Here, $\hat{W}_j$ are an independent Wiener processes, and $\bar{c}_j$, $\tau_j$, and $\sigma_j$ are positive constants, as previously described. This set of generalized chemical Langevin equations provides a way to describe the stochasticity of chemical systems with both intrinsic noise and fluctuations in kinetic parameters.

Equation (3.50) is not a standard stochastic differential equation and is usually not easy to analyze. When the external noise $\sigma_j$ is small and the autocorrelation time $\tau \to 0$, we can replace the perturbed coefficient (3.41) with

$$c_j(t) = \bar{c}_j(1 + \sigma_j \xi_j(t)), \tag{3.50}$$

with $\xi_j(t)$ being Gaussian white noise. Therewith, the first equation in (3.50) can be rewritten as

$$dX_i = \left( \sum_{j=1}^{M} v_{ji} \bar{c}_j (1 + \sigma_j \xi_j(t)) h_j(\mathbf{X}) \right) dt + \sum_{j=1}^{M} v_{ji} \left( \bar{c}_j (1 + \xi_j(t)) h_j(\mathbf{X}) \right)^{1/2} dW_j.$$
$$\tag{3.51}$$

If we further adopt the approximation

$$\xi_j(t)dt \to dV_j, \quad (1 + \xi_j(t))^{1/2} dW_j \to dW_j,$$

where $dV_j$ represents a Gaussian noise in the sense of Stratonovich [28], and equation (3.51) is rewritten as

$$dX_i = \sum_{j=1}^{M} v_{ji} \bar{c}_j h_j(\mathbf{X}) dt + \sum_{j=1}^{M} v_{ji} \bar{c}_j \sigma_j h_j(\mathbf{X}) \circ dV_j + \sum_{j=1}^{M} v_{ji} \sqrt{\bar{c}_j h_j(\mathbf{X})} dW_j,$$

which gives a set of stochastic differential equations

$$dX_i = \left( \sum_{j=1}^{M} v_{ji} a_j(\mathbf{X}) \right) dt + \sum_{j=1}^{M} v_{ji} \left( \sqrt{a_j(\mathbf{X})} dW_j + \sigma_j a_j(\mathbf{X}) \circ dV_j \right), \quad (i = 1, \ldots, N). \tag{3.52}$$

Here we note $a_j(\mathbf{X}) = \bar{c}_j h_j(\mathbf{X})$. We have applied the notation $\circ$ for the Stratonovich interpretation, which is often used for external noise, while the Itô interpretation is more suitable for intrinsic noise [28]. Mathematically, we write the Stratonovich interpretation $\circ$ in (3.52) with the Itô interpretation through a transformation [43]

$$v_{ji} \sigma_j a_j(\mathbf{X}) \circ dV_j \mapsto v_{ji} \sigma_j a_j(\mathbf{X}) dV_j + \frac{1}{2} \sigma_j^2 \sum_k v_{ji} v_{jk} \frac{\partial a_j(\mathbf{X})}{\partial X_k} a_j(\mathbf{X}) dt.$$

Therefore, (3.52) can be written as the following stochastic differential equation under the Itô interpretation:

$$dX_i = \sum_{j=1}^{M} \left( v_{ji} a_j(\mathbf{X}) + \frac{1}{2} \sigma_j^2 \sum_{k=1}^{N} v_{ji} v_{jk} \frac{\partial a_j(\mathbf{X})}{\partial X_k} a_j(\mathbf{X}) \right) dt \tag{3.53}$$

$$+ \sum_{j=1}^{M} v_{ji} \left( \sqrt{a_j(\mathbf{X})} dW_j + \sigma_j a_j(\mathbf{X}) dV_j \right) \quad (i = 1, \ldots, N).$$

Equations (3.52) or (3.53) describe the stochastic processes of biochemical reactions with both intrinsic noise and fluctuations in the kinetic rates.

## 3.5   Stochastic Simulations

We have introduced several equations for modeling systems of biochemical reactions. Here, we introduce stochastic simulation methods that intend to mimic a random process $\mathbf{X}(t)$. Once we have enough sampling pathways of the random process, we are able to calculate the probability density function $P(\mathbf{x}, t)$ and other statistical behaviors, such as the mean trajectory and correlations.

### 3.5.1   Stochastic Simulation Algorithm

To simulate the sampling pathways, assuming that the system has state $\mathbf{x}$ at time $t$, we need to determine when the next reaction may occur and which reaction will occur.

Assume that the system is in state $\mathbf{x}$ at time $t$. We define the *next-reaction density function* $p(\tau, \mu | \mathbf{x}, t)$ as (Fig. 3.3)

$$
\begin{aligned}
p(\tau, \mu | \mathbf{x}, t)d\tau = \ &\text{the probability that, given } \mathbf{X}(t) = \mathbf{x}, \text{ the next} \\
&\text{reaction in } \Omega \text{ will occur in the infinitesimal time} \\
&\text{interval } [t + \tau, t + \tau + d\tau), \text{ and will be the} \\
&\text{reaction } R_\mu.
\end{aligned}
\tag{3.54}
$$

To obtain the probability $p(\tau, \mu | \mathbf{x}, t)$, we note that

$$
p(\tau, \mu | \mathbf{x}, t) = p_0(\tau | \mathbf{x}, t) \times a_\mu(\mathbf{x}),
\tag{3.55}
$$

where $p_0(\tau | \mathbf{x}, t)$ is the probability of having no reaction occur during the interval $(t, t + \tau)$. According to the definition of the propensity function, $a_\mu(\mathbf{x})d\tau$ gives the probability of the reaction $R_\mu$ occurring at the infinitesimal time interval $[t + \tau, t + \tau + d\tau)$.

Now, to derive the formulation for $p_0(\tau | \mathbf{x}, t)$, we note that $p_0(0 | \mathbf{x}, t) = 1$. From the multiplicity principle of the probability, when $\tau' < \tau$, we have

$$
p_0(\tau' + d\tau' | \mathbf{x}, t) = p_0(\tau' | \mathbf{x}, t) \times [1 - \sum_{v=1}^{M} a_v(\mathbf{x})d\tau'].
$$

Here, the term inside the bracket means the probability of having no reaction occur within a small time frame $d\tau'$. From the above relation, we have the differential equation for $p_0(\tau' | \mathbf{x}, t)$:

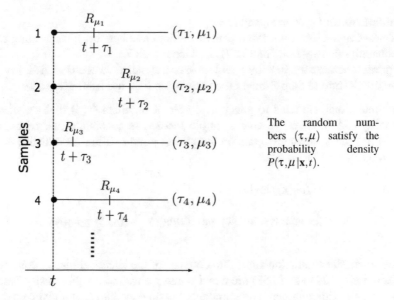

The random numbers $(\tau, \mu)$ satisfy the probability density $P(\tau, \mu \,|\mathbf{x}, t)$.

**Fig. 3.3** Illustration of the definition of the probability $p(\tau, \mu\,|\mathbf{x}, t)$

$$\frac{dp_0(\tau'\,|\mathbf{x}, t)}{d\tau'} = -\sum_{v=1}^{M} a_v(\mathbf{x})p_0(\tau'\,|\mathbf{x}, t), \quad p_0(0|\mathbf{x}, t) = 1. \tag{3.56}$$

Thus, it is easy to obtain

$$p_0(\tau\,|\mathbf{x}, t) = \exp[-\sum_{v=1}^{M} a_v(\mathbf{x})\tau].$$

Hence, from (3.55), letting

$$a_0(\mathbf{x}) = \sum_{v=1}^{M} a_v(\mathbf{x}),$$

we have the probability

$$p(\tau, \mu\,|\mathbf{x}, t) = a_\mu(\mathbf{x}) \exp(-a_0(\mathbf{x})\tau), \quad (0 \le \tau < \infty, \mu = 1, \ldots, M). \tag{3.57}$$

This formula provides the basis for the *stochastic simulation algorithm*(SSA) (also known as the *Gillespie algorithm*) [44, 45].

The main steps of the stochastic simulation algorithm are as follows (for details, refer to [45]):

1. **Initialization**: Let $\mathbf{X} = \mathbf{x}_0$ and $t = t_0$.
2. **Monte Carlo step**: Generate a pair of random numbers $(\tau, \mu)$ according to the probability density function (3.57) (replace $\mathbf{x}$ with $\mathbf{X}$).
3. **Update**: Increase the time by $\tau$ and replace the molecule count with $\mathbf{X} + \mathbf{v}_\mu$.
4. **Iterate**: Return to Step 2 unless the simulation time has been exceeded.

In simulations, we need to generate a pair of numbers $(\tau, j)$ that satisfies the probability (3.57). Here, we give a simple method to generate a pair of numbers $(\tau, j)$ [45]: First, generate two random numbers $r_1$ and $r_2$ of uniform distribution in the unit interval and then take

$$\tau = (1/a_0(\mathbf{x})) \ln(1/r_1) \tag{3.58}$$

$$\mu = \text{the smallest integer satisfying} \sum_{\nu=1}^{\mu} a_\nu(\mathbf{x}) > r_2 a_0(\mathbf{x}). \tag{3.59}$$

The stochastic simulation algorithm consists of the chemical master equation in the sense that (3.12) and (3.57) are exact consequences of the propensity function (3.2). This algorithm numerically simulates the time evolution of a given chemical system and gives a sample trajectory of the real system.

### 3.5.2 Tau-Leaping Algorithm

In the derivation of the chemical Langevin equation, we can always select $\tau$ small enough such that the following *leap condition* is satisfied:

*Leap condition:* Require $\tau$ to be small enough that the change in the state during $[t, t + \tau)$ will be so slight that no propensity function will suffer an appreciable change in its value.

Consequently, the previous arguments indicate that we can approximately leap the system with a time $\tau$ by taking

$$\mathbf{X}(t + \tau) \approx \mathbf{x} + \sum_{j=1}^{M} \mathscr{P}_j(a_j(\mathbf{x})\tau)\mathbf{v}_j, \tag{3.60}$$

where $\mathbf{x} = \mathbf{X}(t)$.

Equation (3.60) is the base of the *tau-leaping algorithm* [46]: Starting from the current state $\mathbf{x}$, we first choose a value $\tau$ that satisfies the leap condition. Next, we generate for each $j$ a random number $k_j$ according to a Poisson distribution with a mean $a_j(\mathbf{x})\tau$. Finally, we update the state from $\mathbf{x}$ to $\mathbf{x} + \sum_j k_j \mathbf{v}_j$ and increase the time by $\tau$.

Two practical issues need to be resolved to effectively apply the tau-leaping algorithm. First, how can we estimate the largest value of $\tau$ that satisfies the leap condition? Second, how can we ensure that the generated $k_j$ values do not result in negative populations?

The original estimation of the largest value of $\tau$ was given by Gillespie and is sketched below [46]. If $\tau$ satisfies the leap condition from (3.60), the average state changes over a time $\tau$ is

$$\bar{\boldsymbol{\lambda}} = \sum_{j=1}^{M} \langle \mathscr{P}_j(a_j(\mathbf{x})\tau) \rangle \mathbf{v}_j = \tau \boldsymbol{\xi}(\mathbf{x}),$$

where

$$\boldsymbol{\xi}(\mathbf{x}) = \sum_{j=1}^{M} a_j(\mathbf{x})\mathbf{v}_j \tag{3.61}$$

is the state change per unit time. Therefore, the average change in the propensity function $a_j(\mathbf{x})$ during $[t, t+\tau)$ is given by $|a_j(\mathbf{x}+\bar{\boldsymbol{\lambda}}) - a_j(\mathbf{x})|$. According to the leap condition, we should have

$$|a_j(\mathbf{x}+\bar{\boldsymbol{\lambda}}) - a_j(\mathbf{x})| \le \varepsilon a_0(\mathbf{x}), \quad (j = 1, \ldots, M), \tag{3.62}$$

where $0 < \varepsilon \ll 1$, and

$$a_0(\mathbf{x}) = \sum_{j=1}^{M} a_j(\mathbf{x}). \tag{3.63}$$

Since

$$|a_j(\mathbf{x}+\bar{\boldsymbol{\lambda}}) - a_j(\mathbf{x})| \approx \bar{\boldsymbol{\lambda}} \cdot \nabla a_j(\mathbf{x}) = \sum_{i=1}^{N} \tau \xi_i(\mathbf{x}) \frac{\partial a_j(\mathbf{x})}{\partial x_i},$$

let

$$b_{ji}(\mathbf{x}) = \frac{\partial a_j(\mathbf{x})}{\partial x_i}, \tag{3.64}$$

and the condition (3.62) can be approximated by

$$\tau \left| \sum_{i=1}^{N} \xi_i(\mathbf{x}) b_{ji}(\mathbf{x}) \right| \le \varepsilon a_0(\mathbf{x}), \quad (j = 1, \ldots, M),$$

which yields

$$\tau \le \varepsilon a_0(\mathbf{x}) / |\sum_{i=1}^{N} \xi_i(\mathbf{x}) b_{ji}(\mathbf{x})|, \quad (j = 1, \ldots, M).$$

Thus, the largest value of $\tau$ is given by

$$\tau = \min_{1 \leq j \leq M} \left\{ \varepsilon a_0(\mathbf{x}) / | \sum_{i=1}^{N} \boldsymbol{\xi}_i(\mathbf{x}) b_{ji}(\mathbf{x}) | \right\}. \tag{3.65}$$

In [47] and [48], two successive refinements were made. The latest $\tau$ selection procedure given in [48] is more accurate, easier to code, and faster to execute than the earlier procedures but is logically more complicated.

If the $\tau$ value generated above is much larger than the time required for the stochastic simulation algorithm, this approximate procedure is faster than the exact stochastic simulation algorithm. However, if $\tau$ turns out to be less than a few multiples of the time required for the stochastic simulation algorithm to make an exact time step (on the order of $1/a_0(\mathbf{x})$), it would be better to use the stochastic simulation algorithm instead.

To avoid negative populations in tau-leaping, several strategies have been proposed, in which the unbounded Poisson random numbers $k_j$ are replaced by bonded binomial random numbers [49, 50]. In 2005, a new Poisson tau-leaping procedure was proposed to resolve this issue [51]. In this new Poisson tau-leaping procedure, all reaction channels are separated into two classes: *critical reactions* that may exhaust one of its reactants after some firings and *noncritical reactions* otherwise. Next, the noncritical reactions are handled by the regular tau-leaping method to obtain a leap time $\tau'$. The exact stochastic simulation algorithm is then applied to the critical reactions, which gives the time $\tau''$ and the index $j_c$ for the next critical reaction. The actual time step $\tau$ is then taken to be the smaller value of $\tau'$ and $\tau''$. If it is the former, no critical reactions fire, and if it is the latter, only one critical reaction $R_{j_c}$ fires.

The numerical scheme of the modified tau-leaping procedure that can avoid negative populations is given below (refer to [51] for details):

1. **Initialization**: Let $\mathbf{X} = \mathbf{x}_0$ and $t = t_0$.
2. **Identify the critical reactions**: Identify the reaction channels $R_j$ for which $a_j(\mathbf{X}) > 0$ and may exhaust one of its reactants after some firings.
3. **Calculate the leap time**: Compute the largest leap time $\tau'$ for the noncritical reactions.
4. **Monte Carlo step**: Generate $(\tau'', j_c)$ for the next critical reaction according to the modified density function given by (3.57).
5. **Determine the next step**:

   a. If $\tau' < \tau''$: Take $\tau = \tau'$. For all critical reactions $R_j$, set $k_j = 0$. For all noncritical reactions $R_j$, generate $k_j$ as a Poisson random variable with mean $a_j(\mathbf{X})\tau$.
   b. If $\tau'' \leq \tau'$: Take $\tau = \tau''$. Set $k_{j_c} = 1$, and for all other critical reactions, set $k_j = 0$. For all noncritical reactions $R_j$, generate $k_j$ as a Poisson random variable with mean $a_j(\mathbf{X})\tau$.

6. **Update**: Increase the time by $\tau$ and replace the molecule count by $\mathbf{X} + \sum_{j=1}^{M} k_j \mathbf{v}_j$.
7. **Iterate**: Return to Step 2 unless the simulation time has been exceeded.

### 3.5.3 Other Simulation Methods

In addition to the prominent approximate acceleration procedure tau-leaping, there are some other strategies that tend to speed up the stochastic simulation algorithm. Here, we briefly outline two of the most promising methods.

Many real systems in biological processes involve chemical reactions with different time scales, and "fast" reactions fire much more frequently than "slow" reactions. Procedures to handle such systems often involve a stochastic generalization of the quasi-steady-state assumption or partial (rapid) equilibrium methods of deterministic chemical kinetics [52–56]. The *slow-scale stochastic simulation algorithm* (ssSSA) (or *multiscale stochastic simulation algorithm*) is a systematic procedure for partitioning the system into fast and slow reactions and simulating slow reactions by specially modified propensity functions [52, 53, 57].

Another approach to simulate multiscale chemical reaction systems includes different kinds of *hybrid methods* [58–61]. Hybrid methods combine the deterministic reaction rate equation with the stochastic simulation algorithm. The idea is to split the system into two regimes: the continuous regimes of large molecule population species and the discrete regimes of small molecule population species. The continuous regimes are treated by ordinary differential equations, while the discrete regimes are simulated by the stochastic simulation algorithm. Hybrid methods efficiently utilize the multiscale properties of the system. However, many problems remain unsolved because of the lack of rigorous theoretical foundations [3].

There are many approaches attempting to find the numerical solution of the chemical master equation [29, 62–68]. In addition, numerical methods for the Langevin equation have been well documented [69]. We will not get into these two subjects here.

## 3.6  Summary

In modeling well-stirred biochemical reaction systems, the chemical master equation provides an 'exact' description of the time evolution of the states. However, it is difficult to directly study the chemical master equation because of the dimension problem. Several approximations are therefore developed, including the Fokker-Planck equation, reaction rate equation, and chemical Langevin equation. The reaction rate equation is widely used when fluctuations are not important. When the fluctuations are significant, the chemical Langevin equation can provide a reasonable description of the statistical properties of the kinetics, although the conditions in deriving the chemical Langevin equation may not hold.

When there are noise perturbations to the kinetic parameters, there is no simple way to model the system dynamics because the time dependence of the environmental variables can be very complicated. In a particular case, we can replace the reaction rates with log-normal random variables and generalize the reaction rate equation or the chemical Langevin equation to describe the dynamics of a chemical system with extrinsic noise.

The stochastic simulation algorithm (SSA) is an 'exact' numerical simulation that shares the same fundamental basis as the chemical master equation. The approximate explicit tau-leaping procedure, on the other hand, is closely related to the chemical Langevin equation. The robustness and efficiency of the two methods have been considerably improved, and these procedures seem to be nearing maturity [2]. Other strategies have been developed for simulating systems that are dynamically stiff [2, 3].

The techniques introduced here can be generalized to many scale problems, where the fluxes of molecules, cells or individuals can be visualized as small number fluctuations due to random birth-death processes.

We conclude this chapter with Fig. 3.4, which summarizes the theoretical structure of stochastic modeling for chemical kinetics (also refer to [2]).

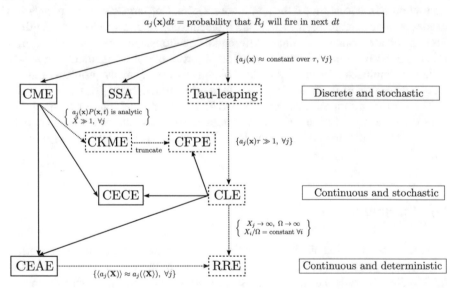

**Fig. 3.4**  Theoretical structure of stochastic chemical kinetics. Everything originates from the fundamental premise of the propensity function given at the top box. The solid arrows show the exact inference routes, and the dotted arrows show the approximate routes, with justified conditions in braces. The solid boxes are exact results: the chemical master equation (CME), the stochastic simulation algorithm (SSA), the chemical ensemble coefficient equation (CECE), and the chemical ensemble average equation (CEAE). The dashed boxes are approximate results: the Tau-leaping algorithm, the chemical Langevin equation (CLE), the chemical Kramers-Moyal equation (CKME), the chemical Fokker-Planck equation (CFPE), and the reaction rate equation (RRE). Replotted from [2]

# References

1. Fiering, S., Whitelaw, E., Martin, D.: To be or not to be active: the stochastic nature of enhancer action. Bioessays **22**, 381–387 (2000)
2. Gillespie, D.T.: Stochastic simulation of chemical kinetics. Annu. Rev. Phys. Chem. **58**, 35–55 (2007)
3. Li, D., Li, C.: Noise-induced dynamics in the mixed-feedback-loop network motif. Phys. Rev. E **77**, 011903 (2008)
4. McAdams, H.H., Arkin, A.: Stochastic mechanisms in gene expression. Proc. Natl. Acad. Sci. USA **94**, 814–819 (1997)
5. McAdams, H.H., Arkin, A.: It's a noisy business! Genetic regulation at the nanomolar scale. Trends Genet. **15**, 65–69 (1999)
6. Raj, A., van Oudenaarden, A.: Nature, nurture, or chance: stochastic gene expression and its consequences. Cell **135**, 216–226 (2008)
7. Rao, C.V., Wolf, D.M., Arkin, A.P.: Control, exploitation and tolerance of intracellular noise. Nature **420**, 231–237 (2002)
8. Samoilov, M.S., Price, G., Arkin, A.P.: From fluctuations to phenotypes: the physiology of noise. Sci STKE **2006**, re17 (2006)
9. Shahrezaei, V., Swain, P.S.: The stochastic nature of biochemical networks. Curr. Opin. Biotechnol. **19**, 369–374 (2008b)
10. Lin, J., Amir, A.: The effects of Stochasticity at the single-cell level and cell size control on the population growth. Cell Syst. **5**, 358-367.e4 (2017)
11. Brennan, M.D., Cheong, R., Levchenko, A.: How information theory handles cell signaling and uncertainty. Science **338**, 334–335 (2012)
12. Pelkmans, L.L.: Using cell-to-cell variability-a new era in molecular biology. Science **336**, 425–426 (2012)
13. Munsky, B., Neuert, G., van Oudenaarden, A.: Using gene expression noise to understand gene regulation. Science **336**, 183–187 (2012)
14. Losick, R., Desplan, C.: Stochasticity and cell fate. Science **320**, 65–68 (2008)
15. Elowitz, M.B., Elowitz, M.B., Levine, A.J., Siggia, E.D., Swain, P.S.: Stochastic gene expression in a single cell. Science **297**, 1183–1186 (2002)
16. Kærn, M., Elston, T.C., Blake, W.J., Collins, J.J.: Stochasticity in gene expression: from theories to phenotypes. Nat. Rev. Genet. **6**, 451–464 (2005)
17. Paulsson, J.: Summing up the noise in gene networks. Nature **427**, 415–418 (2004)
18. Bintu, L., Yong, J., Antebi, Y.E., McCue, K., Kazuki, Y., Uno, N., Oshimura, M., Elowitz, M.B.: Dynamics of epigenetic regulation at the single-cell level. Science **351**, 720–724 (2016)
19. Probst, A.V., Dunleavy, E., Almouzni, G.: Epigenetic inheritance during the cell cycle. Nat. Rev. Mol. Cell Biol. **10**, 192–206 (2009)
20. Pedraza, J.M.: Noise propagation in gene networks. Science **307**, 1965–1969 (2005)
21. Higham, D.J.: Modeling and simulating chemical reactions. Siam Rev. **50**, 347–368 (2008)
22. Golding, I., Paulsson, J., Zawilski, S.M., Cox, E.C.: Real-time kinetics of gene activity in individual bacteria. Cell **123**, 1025–1036 (2005)
23. Keeling, M.J., Rohani, P.: Modeling Infectious Diseases. Princeton University Press, Princeton (2015)
24. Schlicht, R., Winkler, G.: A delay stochastic process with applications in molecular biology. J. Math. Biol. **57**, 613–648 (2008)
25. Zhang, J., Zhou, T.: Markovian approaches to modeling intracellular reaction processes with molecular memory. Proc. Natl. Acad. Sci. USA **116**, 23542–23550 (2019)
26. Oppenheim, I., Shuler, K., Weiss, G.: Stochastic and determinsitic formulation of chemical rate ewquations. J. Chem. Phs. **50**, 460–466 (1969)
27. Gillespie, D.T.: The chemical Langevin equation. J. Chem. Phys. **113**, 297–306 (2000)
28. van Kampen, N.: Stochastic Processes in Physics and Chemistry. North-Holland, Amsterdam (2007)

29. Jahnke, T., Huisinga, W.: Solving the chemical master equation for monomolecular reaction systems analytically. J. Math. Biol. **54**, 1–26 (2006)
30. Shahrezaei, V., Swain, P.S.: Analytical distributions for stochastic gene expression. Proc. Natl. Acad. Sci. USA **105**, 17256–17261 (2008a)
31. Zhou, S., Zhou, S.S., Lo, W.-C.W., Lo, W.-C., Suhalim, J.L.J., Suhalim, J.L., Digman, M.A.M., Digman, M.A., Gratton, E.E., Gratton, E., Nie, Q.Q., Nie, Q., Lander, A.D.A., Lander, A.D.: Free extracellular diffusion creates the Dpp Morphogen gradient of the drosophila wing disc. Curr. Biol. **22**, 668–675 (2012)
32. Gillespie, D.: Markov Processes: An Introduction for Physical Scientists. Academic Press, San Diego, CA (1992)
33. Uhlenbeck, G., Ornstein, L.: On the theory of the Brownian motion. Phys. Rev. **36**, 823–841 (1930)
34. van Kampen, N.: Lagenvin-like equation with colored noise. J. Stat. Phys. **54**, 1289–1308 (1989)
35. Limpert, E., Stahel, W., Abbt, M.: Log-normal distributions across the sciences: keys and clues. BioScience **51**, 341–352 (2001)
36. Rosenfeld, N., Young, J.W., Alon, U., Swain, P.S., Elowitz, M.B.: Gene regulation at the single-cell level. Science **307**, 1962–1965 (2005)
37. Sommer, S.S., Rin, N.A.: The lognormal distribution fits the decay profile of eukaryotic mRNA. Biochem. Biophys. Res. Commun. **90**, 135–141 (1979)
38. Xia, W., Lei, J.: Formulation of the protein synthesis rate with sequence information. MBE **15**, 507–522 (2018)
39. Crow, E.L., Shimizu, K.: eds. Lognormal Distributions, Theory and Applications. New York (1988)
40. Hasty, J., Pradines, J., Dolnik, M., Collins, J.J.: Noise-based switches and amplifiers for gene expression. Proc. Natl. Acad. Sci. USA **97**, 2075–2080 (2000)
41. Jin, Y., Lindsey, M.: Stability analysis of genetic regulatory network with additive noises. BMC Genomics **9**(Suppl 1), S21 (2008)
42. Wang, J., Zhang, J., Yuan, Z., Zhou, T.: Noise-induced switches in network systems of the genetic toggle switch. BMC Syst. Biol. **1**, 50 (2007)
43. Øksendal, B.: Stochastic Differential Equations, 6th edn. Springer, Berlin (2005)
44. Gillespie, D.T.: A general method for numerically simulating the stochastic time evolution of coupled chemical reactions. J. Comput. Phys. **22**, 403–434 (1976)
45. Gillespie, D.: Exact stochastic simulation of coupled chemical reactions. J. Phys. Chem. **81**, 2340–2361 (1977)
46. Gillespie, D.: Approximate accelerated stochastic simulation of chemically reacting systems. J. Chem. Phys. **115**, 1716 (2001)
47. Gillespie, D.T., Petzold, L.R.: Improved leap-size selection for accelerated stochastic simulation. J. Chem. Phys. **119**, 8229–8234 (2003)
48. Cao, Y., Gillespie, D.T., Petzold, L.R.: Efficient step size selection for the tau-leaping simulation method. J. Chem. Phys. **124**, 044109 (2006)
49. Chatterjee, A., Vlachos, D.G., Katsoulakis, M.A.: Binomial distribution based $\tau$-leap accelerated stochastic simulation. J. Chem. Phys. **122**, 024112–024112 (2005)
50. Tian, T., Burrage, K.: Binomial leap methods for simulating stochastic chemical kinetics. J. Chem. Phys. **121**, 10356–10364 (2004)
51. Cao, Y., Gillespie, D.T., Petzold, L.R.: Avoiding negative populations in explicit Poisson tau-leaping. J. Chem. Phys. **123**, 054104 (2005c)
52. Cao, Y., Gillespie, D., Petzold, L.: Multiscale stochastic simulation algorithm with stochastic partial equilibrium assumption for chemically reacting systems. J. Comput. Phys. **206**, 395–411 (2005a)
53. Cao, Y., Gillespie, D.T., Petzold, L.R.: The slow-scale stochastic simulation algorithm. J. Chem. Phys. **122**, 14116 (2005d)
54. Goutsias, J.: Quasiequilibrium approximation of fast reaction kinetics in stochastic biochemical systems. J. Chem. Phys. **122**, 184102–184102 (2005)

55. Rao, C., Arkin, A.: Stochastic chemical kinetics and the quasi-steady-state assumption: application to the Gillespie algorithm. J. Chem. Phys. **118**, 4999 (2003)
56. Samant, A., Vlachos, D.G.: Overcoming stiffness in stochastic simulation stemming from partial equilibrium: a multiscale Monte Carlo algorithm. J. Chem. Phys. **123**, 144114–144114 (2005)
57. Cao, Y., Gillespie, D.T., Petzold, L.R.: Accelerated stochastic simulation of the stiff enzyme-substrate reaction. J. Chem. Phys. **123**, 144917 (2005b)
58. Alfonsi, A., Cancès, E., Turinici, G., Di Ventura, B., Huisinga, W.: Adaptive simulation of hybrid stochastic and deterministic models for biochemical systems. ESAIM: Proc. *14*, 1–13 (2005)
59. Erban, R., Kevrekidis, I.G., Adalsteinsson, D., Elston, T.C.: Gene regulatory networks: a coarse-grained, equation-free approach to multiscale computation. J. Chem. Phys. **124**, 084106 (2006)
60. Haseltine, E.L., Rawlings, J.B.: On the origins of approximations for stochastic chemical kinetics. J. Chem. Phys. **123**, 164115–164115 (2005)
61. Salis, H., Kaznessis, Y.: Accurate hybrid stochastic simulation of a system of coupled chemical or biochemical reactions. J. Chem. Phys. **122**, 054103 (2005)
62. Deuflhard, P., Huisinga, W., Jahnke, T., Wulkow, M.: Adaptive discrete Galerkin methods applied to the chemical master equation. SIAM J. Sci. Comput. **30**, 2990–3011 (2008)
63. Engblom, S.: Galerkin spectral method applied to the chemical master equation. Commun. Comput. Phys. **5**, 871–896 (2009b)
64. Engblom, S.: Spectral approximation of solutions to the chemical master equation. J. Comput. Appl. Math. **229**, 208–221 (2009a)
65. Hegland, M., Hellander, A., Lötstedt, P.: Sparse grids and hybrid methods for the chemical master equation. BIT **48**, 265–283 (2008)
66. Jahnke, T., Huisinga, W.: A dynamical low-rank approach to the chemical master equation. Bull. Math. Biol. **70**, 2283–2302 (2008)
67. Khoo, C.F., Hegland, M.: The total quasi-steady state assumption: its justification by singular perturbation and its application to the chemical master equation. ANZIAM J. **50**, C429–C443 (2008)
68. Munsky, B., Khammash, M.: A multiple time interval finite state projection algorithm for the solution to the chemical master equation. J. Comput. Phys. **226**, 818–835 (2007)
69. Kloeden, P.E., Platen, E.: Numerical Solutions of Stochastic Differential Equation. Springer, New York (1992)

# Chapter 4
# Stochastic Modeling of Gene Expression

*What I cannot create, I do not understand.*

—Richard Feynman

## 4.1 Introduction

Gene expression is the most fundamental biological process by which the genetic information encoded in the DNA sequence is "interpreted" through the assembly of functional molecules. Gene expression connects the information flux from genotype to phenotype, which is associated with the transcriptome of all genes encoded in the genome. There are two common cell types: prokaryotic and eukaryotic. Prokaryotic gene expression and eukaryotic gene expression are the two cellular processes responsible for the expression of genes in the genome to produce a functional gene product. In general, both processes proceed through two steps: transcription and translation. In prokaryotes, closely related genes are clustered to form operons and thus produce a polycistronic mRNA. Meanwhile, functional genes undergo expression individually and produce monocistronic mRNAs. Furthermore, transcription and translation simultaneously occur in the cytoplasm of prokaryotes. In eukaryotes, they occur separately; the former occurs inside the nucleus, and the latter occurs in the cytoplasm. Moreover, eukaryotes undergo both posttranscriptional and translational modifications.

The biochemical reactions involved in the gene expression process are essentially stochastic. The importance of stochasticity in gene expression has long been recognized [1–3]. Recent advances in single-cell sequencing technologies have provided opportunities for direct measurements of transcription at the individual cell level [4, 5]. For a particular gene of interest, the transcription level varies from cell to cell and over time in a single cell [6]. The cell-to-cell variance in gene expression may come from different structures of the gene regulation networks or random fluctuations in the gene expression process. The cellular components interact with one another in

J. Lei, *Systems Biology*, Lecture Notes on Mathematical Modelling in the Life Sciences, https://doi.org/10.1007/978-3-030-73033-8_4

complex regulatory networks, and a fluctuation in the amount of even a single component may affect the performance of the entire system. The quantitative modeling of the sources of stochasticity in gene expression is essential for understanding the cell-to-cell variance that leads to heterogeneity in many biological processes [7].

In this chapter, we introduce mathematical models for different aspects of the gene expression process in both prokaryotes and eukaryotes, including transcription, translation, and epigenetic changes of histone modification and DNA methylation in eukaryotic cells. The models mainly focus on the molecular kinetics and the associated stochastic effects on gene expression.

## 4.2  Gene Expression in Prokaryotes

Prokaryotes are small, single-celled living organisms. The two domains of prokaryotes are bacteria and archaebacteria, both of which lack a defined nucleus and have circular pieces of DNA located in the nucleoid. Prokaryotic cells do not have a nucleus or organelles, so gene expression happens in the open cytoplasm, and all stages can occur simultaneously.

Gene expression includes two main stages, transcription and translation (Fig. 4.1). During transcription, the cell translates DNA into a messenger RNA (mRNA) molecule. During translation, the cell makes the proteins from the mRNA. Both transcription and translation happen in the cytoplasm of prokaryotes.

The *transcription* process includes three parts: initiation, chain elongation, and termination. Transcription starts when DNA unwinds and *RNA polymerase* (RNAP) binds to the promoter. A *promoter* is a special DNA sequence that exists at the beginning of a specific gene. DNA polymerase relies on one DNA strand as the template to build a new strand of RNA called the RNA transcript.

During the chain elongation phase, RNA polymerase moves along the DNA template strand and makes an mRNA molecule. The RNA strand becomes longer as more nucleotides are added. Essentially, RNA polymerase moves along the DNA strand in the 3' to 5' direction to accomplish the elongation process. Bacteria can create *polycistronic* mRNAs, each of which codes multiple proteins.

During the termination phase, the transcription process stops. There are two types of termination phases in prokaryotes: Rho-dependent termination and Rho-independent termination. In *Rho-dependent termination*, a special protein factor called Rho interrupts transcription and terminates it. The Rho protein factor attaches to the RNA strand at a specific binding site and then moves along the strand to reach the RNA polymerase in the transcription bubble. In *Rho-independent termination*, the RNA molecule makes a loop and detaches. The RNA polymerase reaches a DNA sequence on the template strand that is the terminator and has many cytosine (C) and guanine (G) nucleotides. The new RNA strand starts to fold into a hairpin shape through the binding of C and G nucleotides. This process stops the RNA polymerase from moving.

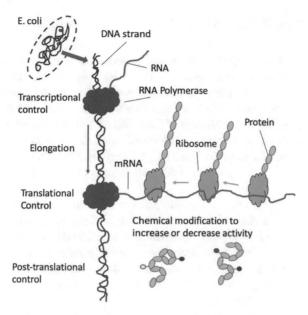

**Fig. 4.1** Illustration of prokaryotic gene expression. Replotted from `https://courses.lumenlearning.com/`

Translation creates a protein molecule or polypeptide based on the RNA template created during transcription. In bacteria, the new RNA strand is immediately ready for *translation*, which can occur immediately and sometimes starts during transcription.

During translation, the mRNA strand acts as a template for making amino acids that become proteins. The cell decodes the mRNA to accomplish this. The initiation of translation requires transfer RNAs (tRNAs), ribosomes, and mRNAs. In bacteria, the translation process starts when a small ribosomal unit attaches to the mRNA at a *Shine-Dalgarno sequence*. The Shine-Dalgarno sequence is a special ribosomal binding area in both bacteria and archaebacteria, approximately eight nucleotides from the start codon AUG.

During the elongation step of the translation process, the amino acid chain becomes longer when the tRNAs add amino acids to the polypeptide chain. Each tRNA molecule has an anticodon for an amino acid. A tRNA starts working in the P site of the ribosome. A tRNA that matches the codon can go to the A site next to the P site. Then, a peptide bond can form between the amino acids. The ribosome moves along the mRNA so that the amino acids form a polypeptide chain.

Termination of the translation process occurs when a stop codon is reached. When a stop codon enters the A site, the translation process stops because the stop codon does not have a complementary tRNA. Proteins called release factors that fit into the P site can recognize stop codons and prevent peptide bonds from forming.

## 4.3  Stochasticity in Prokaryotic Gene Expression

Gene expression is essentially a stochastic process, so clonal populations of cells exhibit substantial phenotypic variation. There are two sources of stochasticity in gene expression: intrinsic noise and extrinsic noise. First, gene expression is essentially a sequence of biochemical reactions that are inherently stochastic due to the random births, deaths, and collisions of the molecules. The inherent stochasticity is often referred to as intrinsic noise [1, 3]. Second, the cell environment is complicated and the expression of a particular gene may be regulated by other molecular species that are, themselves, gene products with populations that vary over time and from cell to cell [8].

Experimentally, it is not trivial to distinguish intrinsic noise from extrinsic noise in vivo [9, 10]. The variation in environmental conditions may produce complicated effects on the fluctuations in gene expression and is referred to as extrinsic noise. The extrinsic sources of noise arise independently of the gene and are usually related to the variability in the rate constants that are associated with the biochemical reactions in gene expression. In general, the total variation in gene expression is the joint effect of both intrinsic and extrinsic noise.

In this section, starting from processes of transcription and translation in a single cell, we introduce methods to model both the intrinsic and extrinsic noise of gene expression. The intrinsic noise comes from fluctuations generated by stochastic promoter activation, promoter inactivation, and mRNA and protein production and degradation. The extrinsic noise usually originates from the total noise of the protein concentrations from upstream gene expression and transcription factors and can be very complicated.

### 4.3.1  Intrinsic Noise

Essentially, prokaryotic gene expression is a sequence of biochemical reactions that are associated with mRNA and protein production, transitions between promoter states (binding and unbinding with RNA polymerase), and the degradation of mRNAs and proteins. Typical steps in gene expression are illustrated in Fig. 4.2 (also refer to [8, 11]). Note that transcription is a process of mRNA synthesis as specified by the gene, in which only the information is read out and the gene is not consumed. Similarly, proteins are translated from an mRNA sequence according to the genetic code, and the mRNA is not consumed in this step.

In Fig. 4.2, $\lambda_1^+$ and $\lambda_1^-$ are the activation and inactivation rates of the promoter, respectively. The transcription and translation rates are $\lambda_2$ and $\lambda_3$, and the degradation (dilution) rates of mRNAs and proteins are $\delta_2$ and $\delta_3$, respectively. Let $X_1$, $X_2$, and $X_3$ be the amounts of active promoters (binding with RNAP to initiate the transcription

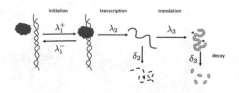

| $j$ | $R_j$ | $a_j(\mathbf{X})$ | $\mathbf{v}_j$ |
|---|---|---|---|
| 1 | $X_0 \to X_1$ | $\lambda_1^+ X_0$ | $(-1,1,0,0)$ |
| 2 | $X_1 \to X_0$ | $\lambda_1^- X_1$ | $(1,-1,0,0)$ |
| 3 | $X_1 \to X_2$ | $\lambda_2 X_1$ | $(0,0,1,0)$ |
| 4 | $X_2 \to \emptyset$ | $\delta_2 X_2$ | $(0,0,-1,0)$ |
| 5 | $X_2 \to X_3$ | $\lambda_3 X_2$ | $(0,0,0,1)$ |
| 6 | $X_3 \to \emptyset$ | $\delta_3 X_3$ | $(0,0,0,-1)$ |

**Fig. 4.2** (From [8]) Illustration of the model of prokaryotic gene expression. Each step represents the biochemical reactions that are associated with the binding and unbinding of promoters with RNAP and the production and degradation of mRNAs and proteins. The reaction channels, propensity functions, and state-change vectors are shown in the left table. Here, $\mathbf{X} = (X_0, X_1, X_2, X_3)$ represents the amounts of free promoters, active promoters (binding with RNAP), mRNAs, and proteins, respectively. In general, when we consider a gene with $n$ copies, then $X_0 = n - X_1$

process), mRNAs, and proteins, respectively. When the copy number of the gene is $n$, the number of inactive promoters is $X_0 = n - X_1$. Most genes have a single copy with $n = 1$. The propensity functions and state-change vectors of each reaction channel are listed in Fig. 4.2. We omit extrinsic noise due to fluctuations in the reaction rates, and the kinetics of gene expression can be described by the chemical master equation (refer to (3.12))

$$\frac{\partial P(X_1, X_2, X_3, t)}{\partial t} = \lambda_1^+(n - X_1 + 1)P(X_1 - 1, X_2, X_3, t) - \lambda_1^+(n - X_1)P(X_1, X_2, X_3, t)$$
$$+ \lambda_1^-(X_1 + 1)P(X_1 + 1, X_2, X_3, t) - \lambda_1^- X_1 P(X_1, X_2, X_3, t)$$
$$+ \lambda_2 X_1 P(X_1, X_2 - 1, X_3, t) - \lambda_2 X_1 P(X_1, X_2, X_3, t)$$
$$+ \delta_2(X_2 + 1)P(X_1, X_2 + 1, X_3, t) - \delta_2 X_2 P(X_1, X_2, X_3, t)$$
$$+ \lambda_3 X_2 P(X_1, X_2, X_3 - 1, t) - \lambda_3 X_2 P(X_1, X_2, X_3, t)$$
$$+ \delta_3(X_3 + 1)P(X_1, X_2, X_3 + 1, t) - \delta_3 X_3 P(X_1, X_2, X_3, t) \quad (4.1)$$
$$(0 \le X_1 \le n, X_2, X_3 \ge 0)$$

Despite the simplicity, it is not easy to exactly solve the master equation (4.1) because the numbers of $X_2$ and $X_3$ are unbounded and hence, the equation is infinite dimensional. Approximation solutions can be obtained by the generating function method [12]. Alternatively, the stochastic process can be obtained through the stochastic simulation algorithm in Chap. 3. Figure 4.3 shows the random kinetics of protein numbers.

Here, we introduce the method for obtaining the stationary fluctuation from the chemical master equation and the generation function method for obtaining the analytical distribution.

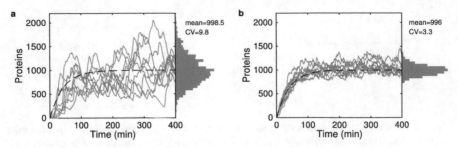

**Fig. 4.3** Kinetics of gene expression in prokaryotes. Sample dynamics of protein numbers based on stochastic simulation are shown in the figure panels together with the histograms of the protein numbers. The dashed lines show the average levels of protein numbers obtained by solving the reaction rate equation (4.2). The parameters are as follows: **a** $n = 1$, $\lambda_1^+ = 0.005$ s$^{-1}$, $\lambda_1^- = 0.03$ s$^{-1}$, $\lambda_2 = 0.07$ s$^{-1}$, $\delta_2 = 0.005$ s$^{-1}$, $\lambda_3 = 0.2$ s$^{-1}$, $\delta_3 = 0.0004$ s$^{-1}$; **b** $\lambda_1^+ = 0.05$ s$^{-1}$, $\lambda_1^- = 0.3$ s$^{-1}$, $\lambda_2 = 0.7$ s$^{-1}$, $\delta_2 = 0.05$ s$^{-1}$, $\lambda_3 = 0.2$ s$^{-1}$, and $\delta_3 = 0.0004$ s$^{-1}$. The mean and coefficient of variance (CV) of the protein numbers at the stationary state are shown

### 4.3.1.1   Intrinsic Stationary Fluctuations

Let

$$\langle X_i(t) \rangle = \sum_{X_1=0}^{n} \sum_{X_2=0}^{+\infty} \sum_{X_3=0}^{+\infty} X_i P(X_1, X_2, X_3, t), \quad (i = 1, 2, 3)$$

be the average numbers of active promoters, mRNAs, and proteins. The reaction rate Eq. (3.17) gives the dynamical equation for the average numbers

$$\begin{cases} \dfrac{d\langle X_1 \rangle}{dt} = \lambda_1^+ (n - \langle X_1 \rangle) - \lambda_1^- \langle X_1 \rangle, \\[2mm] \dfrac{d\langle X_2 \rangle}{dt} = \lambda_2 \langle X_1 \rangle - \delta_2 \langle X_2 \rangle, \\[2mm] \dfrac{d\langle X_3 \rangle}{dt} = \lambda_3 \langle X_2 \rangle - \delta_3 \langle X_3 \rangle. \end{cases} \qquad (4.2)$$

The Eq. (4.2) is a linear differential equation. Given the initial conditions $\langle X_i \rangle$ ($i = 1, 2, 3$) at time $t = 0$, (4.2) has a unique solution, and the solution approaches a unique steady state that is independent of the initial condition. The solutions of (4.2) are shown by dashed curves in Fig. 4.3.

The steady state of (4.2) is obtained by solving

$$\frac{d\langle X_1 \rangle}{dt} = \frac{d\langle X_2 \rangle}{dt} = \frac{d\langle X_3 \rangle}{dt} = 0,$$

which gives

$$\langle X_1 \rangle = g_1 n, \quad \langle X_2 \rangle = g_2 \langle X_1 \rangle, \quad \langle X_3 \rangle = g_3 \langle X_2 \rangle, \tag{4.3}$$

where

$$g_1 = \frac{\lambda_1^+}{\lambda_1^+ + \lambda_1^-}, \quad g_2 = \frac{\lambda_2}{\delta_2}, \quad g_3 = \frac{\lambda_3}{\delta_3}. \tag{4.4}$$

Here, $g_1$ is the activation rate of the promoter, $g_2$ is the transcriptional efficiency for the average number of mRNAs products when the promoter is active, and $g_3$ is the *translational efficiency* for the number of proteins produced by one mRNA.

The stationary fluctuation of intrinsic noise is often measured by the strength of intrinsic fluctuation, which is defined as the relative deviation from the average, i.e., the ratio $\eta$ of the standard deviation to the mean. This ratio $\eta$ is typically referred to as the *coefficient of variation* (CV). We also refer to $\eta^2$ as the *stationary fluctuation*. The coefficient of variation is nondimensional and hence can be used to compare different scale systems.

To calculate the coefficient of variation of protein products, we define the variance matrix $\sigma = (\sigma_{ij})$ as

$$\sigma_{ij} = \langle (X_i - \langle X_i \rangle)(X_j - \langle X_j \rangle) \rangle$$

at the linear approximation. The coefficient of variation $\eta = \eta_{ij}$ is defined by the variances as

$$\eta_{ij} = \frac{\sigma_{ij}}{\langle X_i \rangle \langle X_j \rangle}, \quad (i = 1, 2, 3). \tag{4.5}$$

The variance matrix $\sigma$ satisfies the equation (refer to (3.39))

$$\frac{d\sigma}{dt} = (A\sigma + \sigma A^T) + B. \tag{4.6}$$

Here, the coefficient matrices $A$ and $B$ are

$$A = \begin{bmatrix} -\delta_1 & 0 & 0 \\ \lambda_2 & -\delta_2 & 0 \\ 0 & \lambda_3 & -\delta_3 \end{bmatrix}, \quad B = \begin{bmatrix} 2\lambda_1^- \langle X_1 \rangle & 0 & 0 \\ 0 & 2\delta_2 \langle X_2 \rangle & 0 \\ 0 & 0 & 2\delta_2 \langle X_3 \rangle \end{bmatrix},$$

where $\delta_1 = \lambda_1^+ + \lambda_1^-$.

At the stationary state, the variance matrix satisfies

$$(A\sigma + \sigma A^T) + B = 0. \tag{4.7}$$

Expanding equation (4.7), we obtain the linear equations for $\sigma_{ij}$

$$-2\delta_1\sigma_{11} + 2\lambda_1^-\langle X_1\rangle = 0,$$
$$\lambda_2\sigma_{11} - (\delta_1 + \delta_2)\sigma_{12} = 0,$$
$$\lambda_3\sigma_{12} - (\delta_1 + \delta_3)\sigma_{13} = 0,$$
$$2\lambda_2\sigma_{12} - 2\delta_2\sigma_{22} + 2\delta_2\langle X_2\rangle = 0,$$
$$\lambda_2\sigma_{13} + \lambda_3\sigma_{22} - (\delta_2 + \delta_3)\sigma_{23} = 0,$$
$$2\lambda_3\delta_{23} - 2\delta_3\sigma_{33} + 2\delta_3\langle X_3\rangle = 0.$$

The variances $\sigma_{ij}$ are obtained by solving the above equation, which gives

$$\sigma_{11} = \frac{\lambda_1^-}{\delta_1}\langle X_1\rangle,$$
$$\sigma_{12} = \frac{\lambda_2}{\delta_1 + \delta_2}\sigma_{11},$$
$$\sigma_{13} = \frac{\lambda_3}{\delta_1 + \delta_3}\sigma_{12},$$
$$\sigma_{22} = \langle X_2\rangle + g_2\sigma_{12},$$
$$\sigma_{23} = \frac{\lambda_2}{\delta_2 + \delta_3}\sigma_{13} + \frac{\lambda_3}{\delta_2 + \delta_3}\sigma_{22},$$
$$\sigma_{33} = \langle X_3\rangle + g_3\sigma_{23}.$$

From the obtained variances, the coefficients of variation for active promoters, mRNAs, and proteins are given by (here, $\eta_i^2 = \eta_{ii}$)

$$\eta_1^2 = \frac{1 - g_1}{\langle X_1\rangle}, \tag{4.8}$$

$$\eta_2^2 = \frac{1}{\langle X_2\rangle} + \frac{\tau_1}{\tau_1 + \tau_2}\frac{1 - g_1}{\langle X_1\rangle}, \tag{4.9}$$

$$\eta_3^2 = \frac{1}{\langle X_3\rangle} + \frac{\tau_2}{\tau_3 + \tau_2}\left(\frac{1}{\langle X_2\rangle} + \frac{\tau_1}{\tau_1 + \tau_2}\frac{1 - g_1}{\langle X_1\rangle}\right)$$
$$+ \frac{1 - g_1}{\langle X_1\rangle}\frac{\tau_1}{\tau_1 + \tau_2}\frac{\tau_2}{\tau_2 + \tau_3}\frac{\tau_1\tau_3/\tau_2}{\tau_1 + \tau_3}, \tag{4.10}$$

where $\tau_i$ $(i = 1, 2, 3)$ are the average lifetimes of the active gene, mRNA, and protein, respectively. Here, $\eta_3^2$ gives the stationary fluctuation of protein numbers [2, 3].

From (4.8), the stationary fluctuation of the active promoter fraction is inversely proportional to the average active promoter number.

From (4.9), the stationary fluctuation of the mRNA number consists of two parts: the intrinsic noise of translation and mRNA degradation, which is inversely proportional to the average mRNA number, and the random promoter activity that multiples the relative lifetime of the active promoter.

From (4.10), the stationary fluctuation of the protein number consists of three parts: the intrinsic noise of translation and protein degradation, which is inversely proportional to the average protein number, the fluctuations in mRNA number, which is multiplied by the average lifetime of mRNAs, and the fluctuations due to random promoter activity, which is multiplied by a factor associated with the average lifetime of the active promoter and mRNAs. In Eq. (4.10), the fluctuation of the protein number is given by the average molecule (active promoter, mRNA, and protein) numbers and the average lifetimes, and is independent of the transcriptional efficiency ($g_2$) and the translational efficiency ($g_3$). Moreover, the fluctuation is inversely proportional to the molecule numbers; hence, lower expression yields large stationary fluctuations.

When the kinetics of promoter activation/inactivation are slow compared with the average lifetimes of mRNAs and proteins, i.e., $\tau_1 \gg \tau_2, \tau_3$, (4.10) yields

$$\eta_3^2 \approx \frac{1}{\langle X_3 \rangle} + \frac{\tau_2}{\tau_2 + \tau_3} \frac{1}{\langle X_2 \rangle} + \frac{1 - g_1}{\langle X_1 \rangle}. \tag{4.11}$$

For a single-copy gene ($n = 1$), if the promoter activation rate is slow ($\langle X_1 \rangle = g_1 \ll 1$), the fluctuation $\eta_3^2$ becomes large (Fig. 4.4a). Biologically, the promoter stays at either the active state or inactive state for a long time, which gives the mechanism of transcriptional bursting [1].

When the kinetics of promoter activation/inactivation are fast compared with the average lifetimes of mRNAs and proteins, i.e., $\tau_1 \ll \tau_2, \tau_3$, the protein fluctuation due to fluctuations in promoter activity can be neglected, and hence, (4.10) yields

$$\eta_3^2 \approx \frac{1}{\langle X_3 \rangle} + \frac{\tau_2}{\tau_2 + \tau_3} \frac{1}{\langle X_2 \rangle}. \tag{4.12}$$

In this case, if both the numbers of mRNAs and proteins are large (highly expressed genes), the stationary fluctuation is small and the reaction rate equation (4.2) can describe the dynamics of protein numbers well (Fig. 4.4b). Alternatively, since $\langle X_2 \rangle = \langle X_3 \rangle / g_3$, we can rewrite $\eta_3^2$ based on the translational efficiency as

$$\eta_3^2 \approx \frac{1}{\langle X_3 \rangle} + \frac{\tau_2}{\tau_2 + \tau_3} \frac{g_3}{\langle X_3 \rangle}. \tag{4.13}$$

Hence, for genes with the same production (average protein level), a larger translation efficiency (lower transcription level) yields higher fluctuation. This is the basis of the mechanism of translational bursting [1].

### 4.3.1.2  Analytical Distributions

Now, we assume that the promoter is always active ($X_1 \equiv 1$) and introduce the method of *generating function* to obtain the analytical distribution of protein numbers [12].

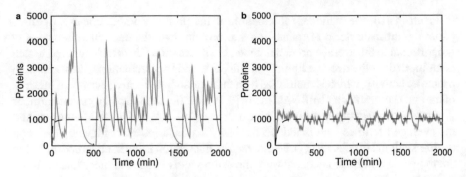

**Fig. 4.4** Promoter activity and transcriptional bursting. **a** Slow promoter activity with $\lambda_1^+ = 0.0005\ \text{s}^{-1}$ and $\lambda_1^- = 0.003\text{s}^{-1}$. **b** Fast promoter activity with $\lambda_1^+ = 0.05\ \text{s}^{-1}$ and $\lambda_1^- = 0.3\ \text{s}^{-1}$. The other parameters are the same as those in Fig. 4.3. The dashed black lines show the average protein number obtained from the reaction rate equation (4.2)

Since the promoter is always active, we can only consider the mRNA number and the protein number, and the chemical master equation (4.1) becomes

$$\frac{\partial P_{m,n}(t)}{\partial t} = \lambda_2(P_{m-1,n}(t) - P_{m,n}(t)) + \delta_2((m+1)P_{m+1,n}(t) - mP_{m,n}(t))$$
$$+ \lambda_3(mP_{m,n-1}(t) - mP_{m,n}(t))$$
$$+ \delta_3((n+1)P_{m,n+1}(t) - nP_{m,n}(t)). \tag{4.14}$$

Here, we denote $m$ as the mRNA number and $n$ as the protein number for convenience, and $P_{m,n}(t)$ is the probability of values $(m, n)$ at time $t$.

The generating function $F(z', z, t)$ is defined from $P_{m,n}(t)$ as

$$F(z', z, t) = \sum_{m,n} P_{m,n}(t) z'^m z^n. \tag{4.15}$$

Here, the summation is taken over $m, n \geq 0$. From (4.15), we have the partial derivatives

$$\frac{\partial F}{\partial z'} = \sum_{m,n} mP_{m,n}(t) z'^{m-1} z^n,$$

$$\frac{\partial F}{\partial z} = \sum_{m,n} nP_{m,n}(t) z'^m z^{n-1}.$$

Hence, from (4.14)

$$\frac{\partial F}{\partial t} = \sum_{m,n} \frac{\partial P_{m,n}(t)}{\partial t} z'^m z^n$$

$$= \sum_{m,n} z'^m z^n [\lambda_2(P_{m-1,n} - P_{m,n}) + \lambda_3(m P_{m,n-1} - m P_{m,n})$$

$$+ \delta_2((m+1)P_{m+1,n} - m P_{m,n}) + \delta_3((n+1)P_{m,n+1}) - n P_{m,n})]$$

$$= \lambda_2 \left( \sum_{m,n} P_{m-1,n} z'^m z^n - \sum_{m,n} P_{m,n} z'^m z^n \right)$$

$$+ \lambda_3 \left( \sum_{m,n} m P_{m,n-1} z'^m z^n - \sum_{m,n} m P_{m,n} z'^m z^n \right)$$

$$+ \delta_2 \left( \sum_{m,n} (m+1) P_{m+1,n} z'^m z^n - \sum_{m,n} m P_{m,n} z'^m z^n \right)$$

$$+ \delta_3 \left( \sum_{m,n} (n+1) P_{m,n+1} z'^m z^n - \sum_{m,n} n P_{m,n} z'^m z^n \right)$$

$$= \lambda_2(z'-1) \sum_{m,n} P_{m,n} z'^m z^n + \lambda_3(z-1)z' \sum_{m,n} m P_{m,n} z'^{m-1} z^n$$

$$+ \delta_2(1-z') \sum_{m,n} m P_{m,n} z'^{m-1} z^n + \delta_3(1-z) \sum_{m,n} n P_{m,n} z'^m z^{n-1}$$

$$= \lambda_2(z'-1)F + \lambda_3 z'(z-1)\frac{\partial F}{\partial z'} + \delta_2(1-z')\frac{\partial F}{\partial z'} + \delta_3(1-z)\frac{\partial F}{\partial z}.$$

Thus, we have

$$\frac{\partial F}{\partial t} - (\lambda_3 z'(z-1) + \delta_2(1-z'))\frac{\partial F}{\partial z'} - \delta_3(1-z)\frac{\partial F}{\partial z} = \lambda_2(z'-1)F.$$

Let $v = z - 1$, $u = z' - 1$, and

$$\tau = \delta_3 t, a = \frac{\lambda_2}{\delta_3}, b = \frac{\lambda_3}{\delta_2}, \gamma = \frac{\delta_2}{\delta_3},$$

we obtain a first-order partial differentiation equation for the generating function $F(u, v, \tau)$

$$\frac{\partial F}{\partial \tau} - \gamma [bv(1+u) - u]\frac{\partial F}{\partial u} + v\frac{\partial F}{\partial v} = auF. \tag{4.16}$$

The boundary condition of (4.16) is

$$F(0, 0, \tau) = \sum_{m,n} P_{m,n}(\tau) \equiv 1, \quad \forall \tau \geq 0. \tag{4.17}$$

Thus, to obtain the analytic distribution $P_{m,n}$, we first solve equation (4.16) with boundary condition (4.17) and then expand the generating function $F(z', z, t)$ as a Taylor series. The distribution $P_{m,n}$ is then given by the coefficients of the series.

Now, we solve equation (4.16) through the method of characteristic line.

The characteristic curves of (4.16) are expressed invariantly by the Lagrange-Charpit equations [13].

$$\frac{d\tau}{1} = \frac{du}{-\gamma(bv(1+u) - u)} = \frac{dv}{v} = \frac{dF}{auF}. \tag{4.18}$$

If we take $\tau$ as an independent parameter of the curves, these equations can be written as a system of ordinary differential equations for $u(\tau)$, $v(\tau)$, and $F(\tau)$:

$$\begin{cases} \dfrac{dv}{d\tau} = v \\ \dfrac{du}{d\tau} = -\gamma(bv(1+u) - u) \\ \dfrac{dF}{d\tau} = auF. \end{cases} \tag{4.19}$$

In general, assuming the initial condition

$$u = u_0, v = v_0, F = F_0, \quad \text{when } t = 0 \ (\tau = 0), \tag{4.20}$$

we can solve the Eq. (4.19) for

$$u = u(\tau, u_0, v_0, F_0), v = v(\tau, u_0, v_0, F_0), F = F(\tau, u_0, v_0, F_0).$$

The generating function $F(u, v, \tau)$ is obtained by eliminating $u_0$, $v_0$, and $F_0$ from the above solution and the boundary condition (4.17).

Given the initial condition (4.20), we have the solutions

$$\begin{cases} v = v_0 e^\tau, \\ u(v) = e^{-\gamma bv} v^\gamma [e^{\gamma bv_0} v_0^{-\gamma} u_0 - b\gamma \displaystyle\int_{v_0}^v e^{\gamma bv'} v'^{-\gamma} dv'], \\ F(v) = F_0 e^{a \int_{v_0}^v \frac{u(v')}{v'} dv'}. \end{cases} \tag{4.21}$$

It is in general difficult to obtain the explicit expression of $F(u, v, \tau)$ from (4.21). Nevertheless, we are able to obtain a simple form when the proteins are much more stable than the mRNAs, i.e., $\gamma \gg 1$ ($\delta_2 \gg \delta_3$).

When $\gamma \gg 1$, we have

$$\frac{1}{\gamma}\frac{du}{dv} = -((b - \frac{1}{v})u + b),$$

which yields

$$u(v) \approx \frac{bv}{1 - bv}$$

by the quasi-steady-state assumption. Hence, we have

$$\frac{dF}{dv} = \frac{ab}{1 - bv} F, \quad F(v_0) = F_0, \tag{4.22}$$

which gives

$$F = F_0 \left[ \frac{1 - bv_0}{1 - bv} \right]^a.$$

Finally, we eliminate $v_0$ and $F_0$ from the conditions

$$v = v_0 e^{\tau}, \quad F|_{v=0} = 1,$$

which gives $F_0 = 1$ and

$$F = \left[ \frac{1 - bve^{-\tau}}{1 - bv} \right]^a$$

or

$$F(z', z, \tau) = \left[ \frac{1 + be^{-\tau} - bze^{-\tau}}{1 + b - bz} \right]^a. \tag{4.23}$$

From (4.23), when $\gamma \gg 1$, the generating function $F$ is independent of $z'$. A large $\gamma$ implies that most of the mass of the joint probability distribution of mRNA and protein is peaked at $m = 0 : P_{m,n} \approx P_{0,n}$, and hence, $F$ is merely a function of $z$ and $\tau$. The distribution function $P_n(\tau)$ $(= P_{0,n}(\tau))$ is given by the Taylor expansion of $F$ in $z$ [12]

$$P_n(\tau) = \frac{1}{n!} \frac{\partial^n F}{\partial z^n} \Big|_{z=0}$$

$$= \frac{\Gamma(a + n)}{\Gamma(n + 1)\Gamma(a)} \left( \frac{b}{1+b} \right)^n \left( \frac{1+be^{-\tau}}{1 + b} \right)^a {}_2F_1(-n, -a, 1 - a - n; \frac{1+b}{e^{\tau} + b}).$$

Here, ${}_2F_1(a, b, c; z)$ is a hypergeometric function, and $\Gamma$ denotes the gamma function.

At the steady state when $\tau \gg 1$, we obtain the stationary distribution

$$P_n = \frac{\Gamma(a + n)}{\Gamma(n + 1)\Gamma(a)} \left( \frac{b}{1+b} \right)^n \left( \frac{1}{1+b} \right)^a, \tag{4.24}$$

which is a negative binomial distribution. The distribution (4.24) approaches to a gamma distribution

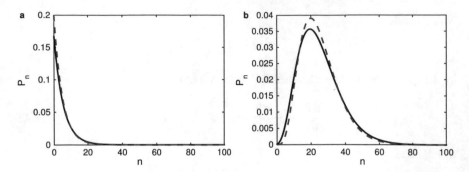

**Fig. 4.5** Analytic distribution of protein numbers. Distributions calculated from the negative bino-mial distribution (4.24) (black solid lines) and the gamma distribution (4.25) (red dashed lines). **a** $a = 1, b = 5$. **b** $a = 5, b = 5$

$$P_n \rightarrow \frac{n^{a-1} e^{-n/b}}{b^a \Gamma(a)} \tag{4.25}$$

when $n$ is large [12]. The obtained distribution establishes a direct correspondence between the steady-state distribution and the kinetic parameters that describe protein expression. The parameters $a$ and $b$ determine the shape of the distribution $P_n$. The distribution peaks at $n = 0$ when $a$ is small and at a value near the average protein level when $a$ is large (Fig. 4.5).

The gamma distribution of protein numbers can also be derived more intuitively from a continuous approach [14]. When $\gamma = \delta_2/\delta_3 \gg 1$, fluctuations in the mRNA number can be integrated out, and the proteins can be considered to be produced in random uncorrelated events, where each event results in an exponentially distributed number of proteins [15, 16]. Thus, considering the distribution of a protein in a population of cells, $p(x)$, we have a *continuous master equation* [14]

$$\frac{\partial p(x)}{\partial t} = \frac{\partial}{\partial x} [\delta_3 x p(x)] + \lambda_2 \int_0^x dx' w(x, x') p(x'). \tag{4.26}$$

Here, $x = n/V$ is the concentration of a protein in a cell. The first term on the right-hand side of (4.26) describes decreases in the concentration of proteins. The second term describes protein production in bursts. The function $w(x, x') = w(x|x') - \delta(x - x')$, with $w(x|x')$ being the conditional probability that the protein concentration in a cell will be $x$ after a protein production burst, given that the concentration was $x'$ before the burst. The $\delta$ function conserves the total probability density by accounting for the loss of density at the existing protein concentration as a result of a burst away from $x$. Usually, the bursting size $x - x'$ is independent of the current protein concentration and follows some characteristic distribution $v(x - x')$. Under these assumptions, $w(x, x') = w(x - x') = v(x - x') - \delta(x - x')$. The steady state is given by $\partial p(x)/\partial t = 0$, which gives

$$-\frac{\partial}{\partial x}[xp(x)] = aw * p(x), \qquad (4.27)$$

with the convolution integral denoted by $w * p = \int_0^x w(x - x')p(x')dx'$, and $a = \lambda_2/\delta_3$. The Laplace transform of (4.27) gives a first-order differential equation

$$s\frac{\partial \hat{p}}{\partial s} = a\hat{w}\hat{p}. \qquad (4.28)$$

Here, $s$ is the Laplace space variable, while $\hat{w}$ and $\hat{p}$ denote the Laplace transformations of $w(x)$ and $p(x)$, respectively. When the burst size distribution is given by an exponential distribution $v(x) = (1/b)\exp(-x/b)$, with an average burst size $b = \lambda_3/\delta_2$, the corresponding Laplace transform of $w(x)$ is $\hat{w}(s) = -s/(s + 1/b)$. Accordingly, we solve (4.28) and have

$$\hat{p}(s) = (s + (1/b))^{-a}. \qquad (4.29)$$

Transforming back to the real space, we obtain the solution for the steady-state distribution

$$p(x) = \frac{1}{b^a \Gamma(a)} x^{a-1} e^{-x/b}. \qquad (4.30)$$

The gamma distribution gives the stationary distribution of protein products based on the two-stage ON-OFF model (Fig. 4.2) of gene expression. In mammalian cells, there are more patterns of protein distributions due to the complex interactions in transcript variability [17].

## 4.3.2 Extrinsic Noise

### 4.3.2.1 Stochastic Differential Equation with both Intrinsic Noise and Fluctuations in Kinetic Rates

To consider the effects of extrinsic noise on gene expression, following the discussions in Sect. 3.4, we introduce external perturbations to the kinetic rates so that

$$\lambda_1^+ \to \lambda_1^+ + f_{\lambda_1^+}\zeta_{\lambda_1^+}(t),$$
$$\lambda_1^- \to \lambda_1^- + f_{\lambda_1^-}\zeta_{\lambda_1^-}(t),$$
$$\lambda_2 \to \lambda_2 + f_{\lambda_2}\zeta_{\lambda_2}(t),$$
$$\delta_2 \to \delta_2 + f_{\delta_2}\zeta_{\delta_2}(t),$$
$$\lambda_3 \to \lambda_3 + f_{\lambda_3}\zeta_{\lambda_3}(t),$$
$$\delta_3 \to \delta_3 + f_{\delta_3}\zeta_{\delta_3}(t).$$

Here, $f_i$ $(i = \lambda_1^+, \lambda_1^-, \lambda_2, \delta_2, \lambda_3, \delta_3)$ are noise strengths, and $\zeta_i$ are Gaussian white noises. Thus, from the chemical Langevin equation (3.52), we obtain the following stochastic differential equations:

$$
\begin{aligned}
dX_1 = {}& (\lambda_1^+(n - X_1) - \lambda_1^- X_1)dt + \sqrt{\lambda_1^+(n - X_1)}dW_1 - \sqrt{\lambda_1^- X_1}dW_2 \quad (4.31) \\
& + f_{\lambda_1^+}(n - X_1) \circ dV_1 - f_{\lambda_1^-}X_1 \circ dV_2,
\end{aligned}
$$

$$
\begin{aligned}
dX_2 = {}& (\lambda_2 X_1 - \delta_2 X_2)dt + \sqrt{\lambda_2 X_1}dW_3 - \sqrt{\delta_2 X_2}dW_4 \quad\quad\quad (4.32) \\
& + f_{\lambda_2}X_1 \circ dV_3 - f_{\delta_2}X_2 \circ dV_4,
\end{aligned}
$$

$$
\begin{aligned}
dX_3 = {}& (\lambda_3 X_2 - \delta_3 X_3)dt + \sqrt{\lambda_3 X_2}dW_5 - \sqrt{\delta_3 X_3}dW_6 \quad\quad\quad (4.33) \\
& + f_{\lambda_3}X_2 \circ dV_5 - f_{\delta_3}X_3 \circ dV_6.
\end{aligned}
$$

Here, $\{W_i(t), t \geq 0\}$ and $\{V_i(t), t \geq 0\}$ are Wiener processes. We apply the Itô interpretation to the intrinsic noise and the Stratonovich interpretation to the extrinsic noise.

Following the transformation in (3.53), we rewrite the above equations into Itô interpretation stochastic differential equations

$$
\begin{aligned}
dX_1 = {}& (\lambda_{1,\mathrm{obs}}^+(n - X_1) - \lambda_{1,\mathrm{obs}}^- X_1)dt + \sqrt{\lambda_1^+(n - X_1)}dW_1 \\
& - \sqrt{\lambda_1^- X_1}dW_2 + f_{\lambda_1^+}(n - X_1)dV_1 - f_{\lambda_1^-}X_1 dV_2, \quad (4.34)
\end{aligned}
$$

$$
\begin{aligned}
dX_2 = {}& (\lambda_2 X_1 - \delta_{2,\mathrm{obs}}X_2)dt + \sqrt{\lambda_2 X_1}dW_3 - \sqrt{\delta_2 X_2}dW_4 \\
& + f_{\lambda_2}X_1 dV_3 - f_{\delta_2}X_2 dV_4, \quad\quad\quad\quad\quad\quad\quad\quad\quad (4.35)
\end{aligned}
$$

$$
\begin{aligned}
dX_3 = {}& (\lambda_3 X_2 - \delta_{3,\mathrm{obs}}X_3)dt + \sqrt{\lambda_3 X_2}dW_5 - \sqrt{\delta_3 X_3}dW_6 \\
& + f_{\lambda_3}X_2 dV_5 - f_{\delta_3}X_3 dV_6, \quad\quad\quad\quad\quad\quad\quad\quad\quad (4.36)
\end{aligned}
$$

where

$$
\lambda_{1,\mathrm{obs}}^+ = \lambda_1^+ - \frac{1}{2}f_{\lambda_1^+}^2, \quad \lambda_{1,\mathrm{obs}}^- = \lambda_1^- - \frac{1}{2}f_{\lambda_1^-}^2,
$$

$$
\delta_{2,\mathrm{obs}} = \delta_2 - \frac{1}{2}f_{\delta_2}^2, \quad \delta_{3,\mathrm{obs}} = \delta_3 - \frac{1}{2}f_{\delta_3}^2
$$

are observable reaction rates under extrinsic noise perturbations. Please note that the above discussions are valid only when the extrinsic noises are weak enough so that the observed rates are positive, i.e.,

$$
f_c < \sqrt{2c}, \quad (c = \lambda_1^+, \lambda_1^-, \delta_2, \delta_3).
$$

Equations (4.34)–(4.36) constitute a highly simplified representation of the problem. In biological systems, the time course of reaction rates depends on the environmental conditions and the amount of other proteins produced from upstream gene expression, which are usually random and time-correlated in complex ways. Experimental observations have shown that external noise can have a long correlation time (colored noise), which is at a time scale of approximately one cell cycle (from 40 min in bacteria to 200 min in humans) [18–20]. Here, we neglect the time correlation and adopt the simplest assumption that the perturbations are independent white noises. For a discussion of colored external noise, please refer to [21].

#### 4.3.2.2 Stationary Fluctuations in Single Gene Expression

From the Itô stochastic differential equations (4.34)–(4.36), the equations for the averages $\langle X_i \rangle$ are given by

$$\frac{d\langle X_1 \rangle}{dt} = \lambda_{1,\text{obs}}^+ (n - \langle X_1 \rangle) - \lambda_{1,\text{obs}}^- \langle X_1 \rangle, \tag{4.37}$$

$$\frac{d\langle X_2 \rangle}{dt} = \lambda_2 \langle X_1 \rangle - \delta_{2,\text{obs}} \langle X_2 \rangle, \tag{4.38}$$

$$\frac{d\langle X_3 \rangle}{dt} = \lambda_3 \langle X_2 \rangle - \delta_{3,\text{obs}} \langle X_3 \rangle. \tag{4.39}$$

Introducing the observed degradation rate

$$\delta_{1,\text{obs}} = \lambda_{1,\text{obs}}^+ + \lambda_{1,\text{obs}}^-,$$

and the net production rates

$$g_1 = \lambda_{1,\text{obs}}^+ / \lambda_{1,\text{obs}}^-,$$
$$g_2 = \lambda_2 / \delta_{2,\text{obs}}, \quad g_3 = \lambda_3 / \delta_{3,\text{obs}}, \tag{4.40}$$

we obtain the average numbers of molecules in the stationary state

$$\langle X_1 \rangle = \frac{g_1}{1 + g_1} n, \quad \langle X_2 \rangle = g_2 \langle X_1 \rangle, \quad \langle X_3 \rangle = g_3 \langle X_2 \rangle. \tag{4.41}$$

From (4.41), the averages of the molecule numbers depend on extrinsic noise strengths. Fluctuations in $\lambda_1^-$, $\delta_2$ or $\delta_3$ may increase the mean protein number, while fluctuations in $\lambda_1^+$ may decrease the mean protein number. Mathematically, this effect originates from the Stratonovich interpretation for extrinsic noises. The same effect can be found by stochastic simulation [21]. This consistency supports the justification of applying the Stratonovich interpretation for extrinsic noises.

It is easy to gather from (4.41) that the stationary solution exists only when

$$f_{\lambda_1^+}^2 + f_{\lambda_1^-}^2 < 2(\lambda_1^+ + \lambda_1^-), \quad f_{\delta_2}^2 < 2\delta_2, \quad f_{\delta_3}^2 < 2\delta_3, \tag{4.42}$$

which means that the noise perturbations to the degradation rates are weak.

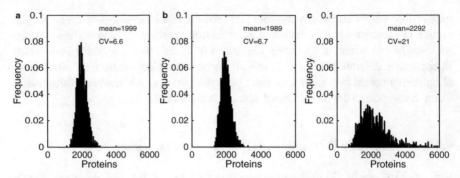

**Fig. 4.6** Protein number distributions under perturbations with extrinsic noise. **a** Distribution of protein numbers when intrinsic fluctuations are included. **b** Distribution of protein numbers when there are extrinsic noise perturbations to the translational rate $\lambda_3$. **c** Distribution of protein numbers when there are extrinsic noise perturbations to the protein degradation rate $\delta_3$. The mean and CV of the protein numbers for each case are also shown. In all panels, the results are obtained from 1000 sample trajectories obtained by solving the stochastic differential equations (4.34)–(4.36) numerically. The parameters are $n = 1$, $\lambda_1^+ = 0.005$ s$^{-1}$, $\lambda_1^- = 0.03$ s$^{-1}$, $\lambda_2 = 0.07$ s$^{-1}$, $\delta_2 = 0.005$ s$^{-1}$, $\lambda_3 = 0.2$ s$^{-1}$, and $\delta_3 = 0.0004$ s$^{-1}$. The Fano factors for the extrinsic noises are 0.25 in (**b**) and (**c**). Replotted from [8]

Similar to the discussions in Sect. 4.3.1.1, the stationary variances $\sigma = (\sigma_{ij})$ are obtained from the equation (also refer to [8])

$$(A\sigma + \sigma A^T + B)_{ij} + \frac{1}{2}\text{trace}(F^{ij}\sigma) = 0, \tag{4.43}$$

where the matrices $A$, $B$ and $F^{ij}$ are calculated from (4.34) and (4.36). The simulation results for the distribution and stationary fluctuations are given in Fig. 4.6.

Explicitly, it is easy to obtain

$$A = \begin{bmatrix} -\delta_{1,\text{obs}} & 0 & 0 \\ \lambda_2 & -\delta_{2,\text{obs}} & 0 \\ 0 & \lambda_3 & -\delta_{3,\text{obs}} \end{bmatrix},$$

and $B$ is a diagonalization matrix with

$$B_{11} = 2\lambda_{1,\text{obs}}^-(1 + \frac{\zeta_{\lambda_1^-} + \zeta_{\lambda_1^+}}{4})\langle X_1 \rangle + \left[ f_{\lambda_1^+}^2 (1 - g_1)^2/g_1^2 + f_{\lambda_1^-}^2 \right] \langle X_1 \rangle^2,$$

$$B_{22} = 2\delta_{2,\text{obs}}(1 + \frac{\zeta_{\delta_2}}{4})\langle X_2 \rangle + (f_{\lambda_2}^2/g_2^2 + f_{\delta_2}^2)\langle X_2 \rangle^2,$$

$$B_{33} = 2\delta_{3,\text{obs}}(1 + \frac{\zeta_{\delta_3}}{4})\langle X_3 \rangle + (f_{\lambda_3}^2/g_3^2 + f_{\delta_3}^2)\langle X_3 \rangle^2,$$

where

$$\zeta_{\lambda_1^+} = f_{\lambda_1^+}^2/\lambda_{1,\text{obs}}^+, \quad \zeta_{\lambda_1^-} = f_{\lambda_1^-}^2/\lambda_{1,\text{obs}}^-,$$
$$\zeta_{\lambda_2} = f_{\lambda_2}^2/\lambda_2, \quad \zeta_{\delta_2} = f_{\delta_2}^2/\delta_{2,\text{obs}},$$
$$\zeta_{\lambda_3} = f_{\lambda_3}^2/\lambda_3, \quad \zeta_{\delta_3} = f_{\delta_3}^2/\delta_{3,\text{obs}}, \tag{4.44}$$

are noise strengths. The matrices $F^{ij} = 0$ when $i \neq j$, and

$$F^{11} = \text{diag}\{2(f_{\lambda_1^+}^2 + f_{\lambda_1^-}^2), 0, 0\},$$
$$F^{22} = \text{diag}\{2f_{\lambda_2}^2, 2f_{\delta_2}^2, 0\},$$
$$F^{33} = \text{diag}\{0, 2f_{\lambda_3}^2, 2f_{\delta_3}^2\}.$$

The noise strengths are related to the total fluctuation in protein numbers. It is more straightforward to associate them with experimental observations by the *Fano factors*, which are defined as

$$F_c = \frac{f_c^2}{c}, \quad (c = \lambda_1^+, \lambda_1^-, \lambda_2, \delta_2, \lambda_3, \delta_3). \tag{4.45}$$

The noise strengths and the Fano factors for extrinsic noises are connected by

$$\zeta_{\lambda_1^+} = \frac{F_{\lambda_1^+}}{1 - \frac{1}{2}F_{\lambda_1^+}}, \quad \zeta_{\lambda_1^-} = \frac{F_{\lambda_1^-}}{1 - \frac{1}{2}F_{\lambda_1^-}},$$
$$\zeta_{\lambda_2} = F_{\lambda_2}, \quad \zeta_{\delta_2} = \frac{F_{\delta_2}}{1 - \frac{1}{2}F_{\delta_2}}, \tag{4.46}$$
$$\zeta_{\lambda_3} = F_{\lambda_3}, \quad \zeta_{\delta_3} = \frac{F_{\delta_3}}{1 - \frac{1}{2}F_{\delta_3}}.$$

Moreover, we define

$$\zeta_{\delta_1} = \frac{f_{\lambda_1^+}^2 + f_{\lambda_1^-}^2}{\lambda_1^+ + \lambda_1^-} = \frac{g_1}{1 + g_1}\zeta_{\lambda_1^+} + \frac{1}{1 + g_1}\zeta_{\lambda_1^-}$$

as the average noise strength of the transition process.

Substituting $A$, $B$, and $F$ into (4.43) and solving for $\sigma$, we have

$$\sigma_{11} = \frac{B_{11}}{2\delta_{1,\text{obs}} - (f_{\lambda_1^+}^2 + f_{\lambda_1^-}^2)}, \tag{4.47}$$

$$\sigma_{12} = \frac{\lambda_2}{\delta_{1,\text{obs}} + \delta_{2,\text{obs}}}\sigma_{11}, \tag{4.48}$$

$$\sigma_{22} = \frac{1}{2\delta_{2,\text{obs}} - f_{\delta_2}^2}\left(B_{22} + 2\lambda_2\sigma_{12} + f_{\lambda_2}^2\sigma_{11}\right), \tag{4.49}$$

$$\sigma_{23} = \frac{\lambda_2 \lambda_3}{(\delta_{1,obs} + \delta_{3,obs})(\delta_{2,obs} + \delta_{3,obs})} \sigma_{12} + \frac{\lambda_3}{\delta_{2,obs} + \delta_{3,obs}} \sigma_{22}, \tag{4.50}$$

$$\sigma_{33} = \frac{1}{2\delta_{3,obs} - f_{\delta_3}^2} \left( B_{33} + 2\lambda_3 \sigma_{23} + f_{\lambda_3}^2 \sigma_{22} \right). \tag{4.51}$$

From the obtained variances, the stationary fluctuations $\eta_{ij}$ are

$$\eta_{11} = k_1 \frac{\lambda_{1,obs}^-}{\delta_{1,obs}} \left[ \frac{1}{\langle X_1 \rangle} + \frac{1}{2}(\frac{\zeta_{\lambda_1^+}}{g_1} + \zeta_{\lambda_1^-}) + \frac{1}{4}\frac{\zeta_{\lambda_1^+} + \zeta_{\lambda_1^-}}{\langle X_1 \rangle} \right], \tag{4.52}$$

$$\eta_{22} = k_2 \left[ \frac{1}{\langle X_2 \rangle} + \frac{1}{2}(\frac{\zeta_{\lambda_2}}{g_2}(1 + \eta_{11}) + \zeta_{\delta_2}) + \frac{1}{4}\frac{\zeta_{\delta_2}}{\langle X_2 \rangle} + \eta_{12} \right], \tag{4.53}$$

$$\eta_{33} = k_3 \left[ \frac{1}{\langle X_3 \rangle} + \frac{1}{2}(\frac{\zeta_{\lambda_3}}{g_3}(1 + \eta_{22}) + \zeta_{\delta_3}) + \frac{1}{4}\frac{\zeta_{\delta_3}}{\langle X_3 \rangle} + \eta_{23} \right], \tag{4.54}$$

and

$$\eta_{12} = \frac{\tau_1}{\tau_1 + \tau_2} \eta_{11}, \tag{4.55}$$

$$\eta_{23} = \frac{\tau_1 \tau_3 / \tau_2}{\tau_1 + \tau_3} \frac{\tau_1}{\tau_1 + \tau_2} \frac{\tau_2}{\tau_2 + \tau_3} \eta_{11} + \frac{\tau_2}{\tau_2 + \tau_3} \eta_{22}, \tag{4.56}$$

where

$$k_i = \frac{1}{1 - \frac{1}{2}\zeta_{\delta_i}}, \quad (i = 1, 2, 3) \tag{4.57}$$

are impact factors, and

$$\tau_i = 1/\delta_{i,obs}, \quad (i = 1, 2, 3)$$

are average lifetimes. Here, we note (4.57), and the results of (4.52)–(4.54) are valid only when

$$\zeta_{\delta_i} < 2 \quad (i = 1, 2, 3). \tag{4.58}$$

The condition (4.58) automatically implies (4.42). The discussions below would be limited to the restriction (4.58).

Equations (4.52)–(4.56) give complete formulations for the stationary fluctuations in gene expression with both intrinsic and extrinsic noises. We often use $\eta_i^2 = \eta_{ii} (i = 1, 2, 3)$ for stationary fluctuations in the number of active genes, mRNAs and proteins molecules.

From (4.52)–(4.54), the total fluctuation in protein numbers can be decomposed into different components. In (4.54), the stationary fluctuation consists of contributions from the intrinsic fluctuation $k_3/\langle X_3 \rangle$ and the extrinsic fluctuation $\frac{k_3}{2}(\frac{\zeta_{\lambda_3}}{g_2}(1 + \eta_1^2) + \zeta_{\delta_3})$, from the correlation between intrinsic and extrinsic noises

$\frac{k_3}{4} \frac{\zeta_{\delta_3}}{\langle X_3 \rangle}$, and from the fluctuations in the numbers of upstream molecules $k_3 \eta_{23}$. The transmission fluctuation depends on the average time of both transcription and translation, which are altered by extrinsic noises. Mathematically, this effect again originates from the Stratonovich interpretation for extrinsic noise. Both contributions from the extrinsic fluctuation and from the correlation are proportional to the extrinsic noise strengths. All components are proportional to an impact factor $k_3$, which increases rapidly when $\zeta_{\delta_3}$ approaches 2. Similar results are also held for the stationary fluctuations in the numbers of active genes and mRNA molecules.

## 4.4 Gene Expression in Eukaryotes

Eukaryotes are organisms made up of cells that possess a membrane-bound nucleus as well as membrane-bound organelles. The nucleus holds DNA in the form of chromosomes. Eukaryotic organisms may be multicellular or single-celled organisms. All animals are eukaryotes, and other eukaryotes include plants, fungi, and protists.

Eukaryotic gene expression involves the remodeling of nucleosome structures, DNA packing, the activation of promoters, transcription, RNA splicing, and translation [22]. Eukaryotic DNA is stored inside the nucleus; hence, transcription also occurs inside the nucleus. After transcription and posttranscriptional modifications, the mature mRNA leaves the nucleus and travels to the cytoplasm, where it undergoes translation in the ribosome (Fig. 4.7).

In eukaryotes, most gene expression is regulated by *transcription factors* that bind to DNA regulatory sequences situated upstream where transcription is initiated. Transcription factors are proteins that bind to DNA-regulatory sequences (enhancers and silencers), usually localized in the 5'-upstream region of target genes, to modulate the rate of gene transcription. More than 5% of human genes are predicted to encode

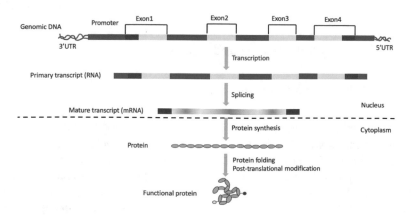

**Fig. 4.7** The basic process of gene expression in eukaryotes

transcription factors [23]. The activities of these proteins are controlled by a diverse array of regulatory pathways. Once activated, transcription factors will bind to gene regulatory elements, interact with other components of the transcription machinery, promote access to DNA and facilitate the recruitment of RNA polymerase (RNAP) enzymes to the transcriptional start site. Usually, three RNA polymerases (RNAP I, II, and III) are responsible for the transcription of different types of RNAs: RNAP I transcribes rRNA, RNAP II transcribes mRNA, and RNAP III transcribes tRNA.

After RNAP II initiates transcription, the nascent RNA is modified by the addition of a "cap" structure at its 5' end. This cap initially serves to protect the new transcript from attack by nucleases and later serves as a binding site for proteins involved in exporting the mature mRNA into the cytoplasm and its translation into protein [24]. RNAP II begins RNA synthesis with transcription "elongation", in which the polymerase moves from 5' to 3' along the gene sequence and extends the transcript. Coding sequences in the gene (exons) are often interrupted by long noncoding sequences (introns), which are removed by pre-mRNA splicing during elongation (Fig. 4.7) [25, 26]. Upon reaching the end of a gene, RNAP II stops transcription ("termination"), the newly synthesized RNA is cleaved ("cleavage") and a polyadenosine (poly(A)) tail is added to the 3' end of the transcript ("polyadenylation") [24].

In eukaryotes, translation is a cytoplasmic event. Therefore, processed mRNAs must be transported from the nucleus to the cytoplasm before translation can occur. The export of mRNA is mediated by factors that bind to the mRNA molecules in the nucleus and direct them into the cytoplasm through interactions with proteins that line the nuclear pores [27]. The translation of mRNA into protein takes place on large ribonucleoprotein complexes called ribosomes and is mechanistically analogous to transcription [28]. It begins with the location of the start codon by translational initiation factors in conjunction with the subunits of the ribosome and involves elongation and termination phases [29]. The nascent polypeptide chain then undergoes folding [30] and often undergoes posttranslational chemical modifications [31, 32] to have a biologically functional active protein.

## 4.5  Noise in Eukaryotic Gene Expression

### 4.5.1  Stochastic Simulation

Eukaryotes contain a large set of promoter elements, including the TATA box. During eukaryotic gene expression, initial transcription factors assemble with the initiation complex to induce the modification of promoter configuration to initiate transcript elongation. Figure 4.8 illustrates a schematic model of gene expression that incorporates the main features specific to eukaryotic transcription, including the sequential assembly of the core transcription apparatus, slow chromatin remodeling, the rate-limiting binding of TATA box-binding protein (TBP) and pulsatile mRNA produc-

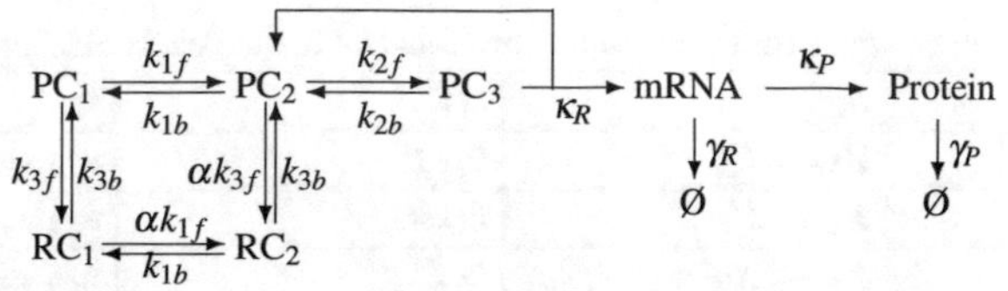

**Fig. 4.8** A model of gene expression in eukaryotes. The states $PC_1$, $PC_2$, and $PC_3$ represent the inactive (or silenced) promoter, an intermediate complex with TBP and various transcription factors bound, and the preinitiation complex where all components required for transcription are assembled on the promoter, respectively. Reinitiation is modeled as a return to the intermediate complex $PC_2$ (and the subsequent transition back to $PC_3$) after transcription initiation has occurred from $PC_3$. The states $RC_1$ and $RC_2$ represent different repressed promoter configurations. The factor $\alpha$ characterizes the masking of unoccupied DNA-binding sites in the repressed complex $RC_1$ and the intermediate complex $PC_2$. Replotted from [33]

**Table 4.1** Description of the dynamic variables in the stochastic model

| Variable | Symbol | Description |
|---|---|---|
| $X_1$ | $PC_1$ | Inactive or silenced promoter (unable to bind RNAP) |
| $X_2$ | $PC_2$ | Intermediate promoter complex including transcription factor IIA (TFIIA) TBP and associated factors (capable of binding RNAP) |
| $X_3$ | $PC_3$ | Preinitiation complex including RNAP |
| $X_4$ | $RC_1$ | Inactive promoter complex with repressor bound (unable to bind RNAP) |
| $X_5$ | $RC_2$ | Intermediate promoter complex with repressor bound (unable to bind RNAP) |
| $X_6$ | mRNA | Number of mRNA transcripts |
| $X_7$ | Protein | Number of proteins |

tion due to reinitiation [33]. The model assumes random transitions among different promoter states in transcription initiation. Here, there are three promoter complex states ($PC_1$, $PC_2$, and $PC_3$) and two repressed states ($RC_1$ and $RC_2$). Transcription elongation and translation are modeled as single-step processes with the stochastic degradation of mRNA and protein.

The model can be simulated with Gillespie's algorithm [33]. The state of the system is represented by a vector $\mathbf{X} = (X_1, X_2, \ldots, X_7)$ with 7 entries corresponding to each dynamical variable (Table 4.1). There are 14 reaction channels, and the corresponding propensity functions and state-change vectors are shown in Table 4.2.

Figure 4.9 shows the model construction and stochastic simulation results of gene expression based on the reaction channels in Table 4.2. In this model, anhydrotetracycline (ATc) and galactose (GAL) are required to induce the gene expression reporter (*yEGFP*) (Fig. 4.9a). Galactose activates transcription through chromatin remodeling and/or the recruitment of TBP so that both rates $k_{1f}$ and $k_{1b}$ are dependent on GAL

**Table 4.2**  Reaction channels for the transcriptional control in the yeast *GAL1* promoter

| Reaction | $R_i$ | $a_i(\mathbf{X})$ | $\mathbf{v}_{i,j}$ |
|---|---|---|---|
| 1 | $X_1 \rightarrow X_2$ | $k_{1f} X_1$ | $v_{1,1} = -1, v_{1,2} = 1$ |
| 2 | $X_2 \rightarrow X_1$ | $k_{1b} X_2$ | $v_{2,1} = 1, v_{2,2} = -1$ |
| 3 | $X_2 \rightarrow X_3$ | $k_{2f} X_2$ | $v_{3,2} = -1, v_{3,3} = 1$ |
| 4 | $X_3 \rightarrow X_2$ | $k_{2b} X_3$ | $v_{4,2} = 1, v_{4,3} = -1$ |
| 5 | $X_1 \rightarrow X_4$ | $k_{3f} X_1$ | $v_{5,1} = -1, v_{5,4} = 1$ |
| 6 | $X_4 \rightarrow X_1$ | $k_{3b} X_4$ | $v_{6,1} = 1, v_{6,4} = -1$ |
| 7 | $X_4 \rightarrow X_5$ | $\alpha k_{1f} X_4$ | $v_{7,4} = -1, v_{7,5} = 1$ |
| 8 | $X_5 \rightarrow X_4$ | $k_{1b} X_5$ | $v_{8,4} = 1, v_{8,5} = -1$ |
| 9 | $X_2 \rightarrow X_5$ | $\alpha k_{3f} X_2$ | $v_{9,2} = -1, v_{9,5} = 1$ |
| 10 | $X_5 \rightarrow X_2$ | $k_{3b} X_5$ | $v_{10,2} = 1, v_{10,5} = -1$ |
| 11 | $X_3 \rightarrow X_6 + X_2$ | $\kappa_R X_3$ | $v_{11,2} = 1, v_{11,3} = -1, v_{11,6} = 1$ |
| 12 | $X_6 \rightarrow \emptyset$ | $\gamma_R X_6$ | $v_{12,6} = -1$ |
| 13 | $\rightarrow X_7$ | $\kappa_P X_6$ | $v_{13,7} = 1$ |
| 14 | $X_7 \rightarrow \emptyset$ | $\gamma_P X_7$ | $v_{14,7} = -1$ |

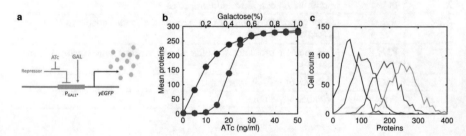

**Fig. 4.9** Transcriptional control in the yeast *GAL1* promoter. **a** Anhydrotetracycline (ATc) and galactose (GAL) are required to induce the gene expression reporter (*yEGFP*). Galactose activates transcription through chromatin remodeling and/or the recruitment of TBP. ATc binds to the repressor (TetR) to mediate the transcription process. **b** Simulated dose-response curve of $P_{GAL1*}$ expression to ATc at galactose induction (0.8%, red points) and to galactose at ATc induction (40 ng/ml, blue points). **c**. Transient response of cells induced with 0.8% galactose and 40 ng/ml ATc. Histograms correspond to time points from preinduction (black) to late induction (green). The parameters used in simulations are as follows (referred to [33]): $\kappa_R = 1$, $\kappa_P = 5$, $\gamma_R = 1$, $\gamma_P = 0.0125$, $k_{2f} = 50$, $k_{2b} = 10$, $k_{3b} = 10$, $\alpha = 0.025$, and $k_{1f} = 0.02 + 0.2 \times GAL$, $k_{1b} = 0.07/GAL + 0.007/GAL + 0.1 \times GAL + 0.01$, $k_{3f} = 200 \times (n_{\text{repressor}})^2/(1 + c_I^4 \times ATc^4)^2$, $n_{\text{repressor}} = 1$, $c_I = 0.1 \text{ng/ml}$

concentration. ATc binds to the repressor (TetR) to mediate the transcription process, and $k_{3f}$ decreases with ATc concentration. The dose-response curves of gene expression to different values of ATc and galactose induction are shown in Fig. 4.9b. The simulated time evolution of the distribution of protein levels from preinduction to late induction is given in Fig. 4.9c.

## 4.5.2  Analytic mRNA Distribution

Now, we obtain the stationary distribution of mRNA numbers through the method of generating functions. We refer the readers to [34, 35] for more details on the analytical distribution of a multistate stochastic model of gene expression.

In model Fig. 4.8, the promoter can only be in one of the 5 states, $PC_i$ ($i = 1, 2, 3$) or $RC_i$ ($i = 1, 2$), and there is no feedback from protein to transcription. Hence, we only consider the mRNA distribution. Let $P_k(m, t)$ represent the probability of having $m$ mRNAs when the promoter is at the $k$-th state ($k = 1, 2, \ldots, 5$, for states $X_k$, respectively) at time $t$. According to the propensity functions in Table 4.2, the chemical master equation reads

$$
\begin{aligned}
\frac{dP_1(m, t)}{dt} &= -(k_{1f} + k_{3f})P_1(m, t) + k_{1b}P_2(m, t) + k_{3b}P_4(m, t) \\
&\quad + \gamma_R(\mathbb{E} - \mathbb{I})[mP_1(m, t)], \\
\frac{dP_2(m, t)}{dt} &= -(k_{2f} + k_{1b} + \alpha k_{3f})P_2(m, t) \\
&\quad + k_{2b}P_3(m, t) + k_{1f}P_1(m, t) + k_{3b}P_5(m, t) \\
&\quad + \kappa_R\mathbb{E}^{-1}[P_3(m, t)] + \gamma_R(\mathbb{E} - \mathbb{I})[mP_2(m, t)], \\
\frac{dP_3(m, t)}{dt} &= -k_{2b}P_3(m, t) + k_{2f}P_2(m, t) - \kappa_R P_3(m, t) \qquad\qquad (4.59) \\
&\quad + \gamma_R(\mathbb{E} - \mathbb{I})[mP_3(m, t)], \\
\frac{dP_4(m, t)}{dt} &= -(k_{3b} + \alpha k_{1f})P_4(m, t) + k_{3f}P_1(m, t) + k_{1b}P_5(m, t) \\
&\quad + \gamma_R(\mathbb{E} - \mathbb{I})[mP_4(m, t)], \\
\frac{dP_5(m, t)}{dt} &= -(k_{3b} + k_{1b})P_5(m, t) + \alpha k_{3f}P_2(m, t) + \alpha k_{1f}P_4(m, t) \\
&\quad + \gamma_R(\mathbb{E} - \mathbb{I})[mP_5(m, t)],
\end{aligned}
$$

where $\mathbb{I}$ is the unit operator and $\mathbb{E}$ and $\mathbb{E}^{-1}$ are step operators, i.e., for any function $f$ and any integer $n$, we have $\mathbb{E}[f(n)] = f(n + 1)$, $\mathbb{E}^{-1}[f(n)] = f(n - 1)$.

Introducing the generating functions

$$
G_k(z, t) = \sum_{m=0}^{\infty} z^m P_k(m, t), \quad (k = 1, 2, \ldots, 5),
$$

the master equation (4.59) can be transformed into the following partial differential equations:

$$\frac{\partial G_1(z, t)}{\partial t} = -(k_{1f} + k_{3f})G_1(z, t) + k_{1b}G_2(z, t) + k_{3b}G_4(z, t)$$
$$+ \gamma_R(1 - z)\frac{\partial G_1(z, t)}{\partial z},$$

$$\frac{\partial G_2(z, t)}{\partial t} = -(k_{2f} + k_{1b} + \alpha k_{3f})G_2(z, t) + k_{2b}G_3(z, t) + k_{1f}G_1(z, t)$$
$$+ k_{3b}G_5(z, t) + \kappa_R z G_3(z, t) + \gamma_R(1 - z)\frac{\partial G_2}{\partial z},$$

$$\frac{\partial G_3(z, t)}{\partial t} = -k_{2b}G_3(z, t) + k_{2f}G_2(z, t)$$
$$- \kappa_R G_3(z, t) + \gamma_R(1 - z)\frac{\partial G_3(z, t)}{\partial z}, \tag{4.60}$$

$$\frac{\partial G_4(z, t)}{\partial t} = -(k_{3b} + \alpha k_{1f})G_4(z, t) + k_{3f}G_1(z, t) + k_{1b}G_5(z, t)$$
$$+ \gamma_R(1 - z)\frac{\partial G_4(z, t)}{\partial z},$$

$$\frac{\partial G_5(z, t)}{\partial t} = -(k_{3b} + k_{1b})G_5(z, t) + \alpha k_{3f}G_2(z, t) + \alpha k_{1f}G_4(z, t)$$
$$+ \gamma_R(1 - z)\frac{\partial G_5(z, t)}{\partial z}.$$

We are looking for the steady states with $\partial G_k(z, t)/\partial t = 0$. Thus, we consider the following ordinary differential equations:

$$\begin{aligned}
sG'_1(s) &= -(f_1 + f_3)G_1(s) + b_1G_2(s) + b_3G_4(s), \\
sG'_2(s) &= f_1G_1(s) - (f_2 + b_1 + \alpha f_3)G_2(s) + b_2G_3(s) \\
&\quad + b_3G_5(s) + (s + \kappa)G_3(s) \\
sG'_3(s) &= f_2G_2(s) - b_2G_3(s) - G_3(s), \\
sG'_4(s) &= f_3G_1(s) - (b_3 + \alpha f_1)G_4(s) + b_1G_5(s), \\
sG'_5(s) &= \alpha f_3G_2(s) + \alpha f_1G_4(s) - (b_1 + b_3)G_5(s),
\end{aligned} \tag{4.61}$$

where $s = \kappa(z - 1)$ is now taken as the variable of $G_k$, where $\kappa = \frac{\kappa_R}{\gamma_R}$ and $G'_k(s) = \frac{dG_k(s)}{ds}$

$$b_i = \frac{k_{ib}}{\gamma_R}, \quad f_i = \frac{k_{if}}{\gamma_R}, \quad i = 1, 2, 3.$$

Let

$$G(s) = \sum_{k=1}^{5} G_k(s)$$

be the generating function at the steady state. If we can solve the function $G_k(s)$ or the generating function $G(s)$ from (4.61), the steady-state mRNA distribution $P(m)$ is given by

$$P(m) = \frac{\kappa^m}{m!} \frac{d^m G(s)}{ds^m}\bigg|_{s=0} = \frac{1}{m!} \frac{d^m G(z)}{dz^m}\bigg|_{z=1}. \tag{4.62}$$

Let $\mathbf{G}(s) = (G_1(s), \ldots, G_5(s))^T$ and

$$A = \begin{bmatrix} -(f_1 + f_3) & b_1 & 0 & b_3 & 0 \\ f_1 & -(f_2 + b_1 + \alpha f_3) & b_2 + \kappa & 0 & b_3 \\ 0 & f_2 & -b_2 & 0 & 0 \\ f_3 & 0 & 0 & -(b_3 + \alpha f_1) & b_1 \\ 0 & \alpha f_3 & 0 & \alpha f_1 & -(b_1 + b_3) \end{bmatrix},$$

$$J = \begin{bmatrix} 0 & 0 & 0 & 0 & 0 \\ 0 & 0 & 1 & 0 & 0 \\ 0 & 0 & -1 & 0 & 0 \\ 0 & 0 & 0 & 0 & 0 \\ 0 & 0 & 0 & 0 & 0 \end{bmatrix},$$

equation (4.61) reads

$$s\mathbf{G}'(s) = A\mathbf{G}(s) + sJ\mathbf{G}(s). \tag{4.63}$$

Write the Taylor expansion of $\mathbf{G}(s)$ as

$$\mathbf{G}(s) = \sum_{m=0}^{\infty} s^m \mathbf{u}_m.$$

Substituting the expansion into (4.63), we obtain

$$A\mathbf{u}_0 = 0, \quad \text{and} \quad (A - mI)\mathbf{u}_m = -J\mathbf{u}_{m-1}, \quad (m \geq 1),$$

where $I$ is the unit matrix. Hence, $(A - mI)$ is invertible for any $m \geq 1$ and given $\mathbf{u}_0$, the coefficients $\mathbf{u}_m$ are obtained iteratively from

$$\mathbf{u}_m = -(A - mI)^{-1} J\mathbf{u}_{m-1},$$
$$= (-1)^m \prod_{j=1}^{m} (A - jI)^{-1} J\mathbf{u}_0 \quad m = 1, 2, \ldots. \tag{4.64}$$

From (4.62), the mRNA distribution is given by

$$P(m) = \kappa^m \langle e, \mathbf{u}_m \rangle, \quad \mathbf{e} = (1, 1, 1, 1, 1). \tag{4.65}$$

The coefficient $\mathbf{u}_0$ satisfies the normalization condition

$$1 = \sum_{m=0}^{\infty} P(m) = \sum_{m=0}^{\infty} \kappa^m \langle \mathbf{e}, \mathbf{u}_m \rangle = \langle \mathbf{e}, \mathbf{u}_0 \rangle + \sum_{m=1}^{\infty} \kappa^m \langle \mathbf{e}, (-1)^m \prod_{j=1}^{m} (A - jI)^{-1} J \mathbf{u}_0 \rangle.$$

Thus, writing

$$Q = I + \sum_{m=1}^{\infty} (-\kappa)^m \prod_{j=1}^{m} (A - jI)^{-1} J,$$

the coefficient $\mathbf{u}_0$ is obtained by solving

$$\begin{cases} A\mathbf{u}_0 = 0 \\ \langle \mathbf{e}, Q\mathbf{u}_0 \rangle = 1. \end{cases} \tag{4.66}$$

Finally, the analytic steady-state distribution of mRNA numbers is obtained by solving (4.66) and calculating $P(m)$ iteratively from (4.64) and (4.65).

## 4.6 Epigenetic Regulation of Transcription

In eukaryotic gene expression, slow chromatin remodeling is associated with epigenetic regulation, such as histone modifications and DNA methylation [36–38]. Hence, stochastic inheritance in epigenetic states may introduce further sources for the random dynamics in gene expression.

The DNA in eukaryotic cells is not naked but packaged into a highly organized and compact nucleoprotein structure known as chromatin. The basic organizational unit of chromatin is the nucleosome, which consists of 146 bp of DNA wrapped almost twice around a protein core containing two copies each of four *histone* proteins: H2A, H2B, H3, and H4 [39]. The histone N-termini (tails) can undergo various post-translational covalent modifications, including acetylation, phosphorylation, methylation, ubiquitination and ADP-ribosylation [40], and the modifications can lead to either active or repressive gene expression states [36, 37]. Chromatin remodeling is a dynamic modification of the chromatin architecture that allows access of condensed genomic DNA to regulatory transcription machinery proteins and thereby controls gene expression. Such remodeling is carried out by: (1) covalent histone modifications by specific enzymes [41], e.g., histone acetyltransferases (HATs), deacetylases, methyltransferases, and kinases; and (2) ATP-dependent chromatin remodeling complexes with either move, eject, or restructure nucleosomes.

DNA methylation is a process by which methyl groups are added to DNA segments, mostly the fifth position of cytosine (5-methylcytosine, 5mC), and is primarily restricted to palindromic CpG (CG/GC) dinucleotides [42, 43]. The methylation of DNA sequences can inhibit protein-DNA interactions and thereby block the transcription process. DNA methylation can also inhibit expression in other ways: some DNA sequences are recognized only when specific repressors are methylated and then switch off nearby genes, often by recruiting histone deacetylase. DNA methylation patterns are dynamically regulated during development and are important for the stable silencing of gene expression, maintenance of genome stability, and establishment of genomic imprinting [44, 45]. DNA methylation patterns are largely erased and then re-established between generations in mammals. Almost all of the methylation from the parents is erased, first during gametogenesis and again in early embryogenesis, with demethylation and remethylation occurring each time. These processes of DNA methylation reprogramming are essential for mammalian embryo development [43, 46]. Abnormal *de novo* DNA methylation is also associated with aging and tumorigenesis, whereby widespread methylation changes in the subpopulation of normal cells serve as a "driver" to prevent the activation of key genes required to induce terminal differentiation and its accompanying inhibition of cell proliferation [47, 48].

Both patterns of histone modification and DNA methylation can be transmitted over a number of cell cycles. Hence, when modeling gene expression in eukaryotes over many cell cycles, we need to consider the random inheritance of epigenetic information during cell cycling.

## 4.6.1 Stochastic Modeling of Histone Modification and Transcriptional Regulation

### 4.6.1.1 Stochastic Model of Histone Modification

Here, we introduce a stochastic model of histone modification dynamics referred to [49, 50].

Consider a DNA segment with $N$ nucleosomes. Each nucleosome contains 2 copies of the core histone H3, and the tail can undergo various types of covalent modifications, such as methylation and acetylation. Two types of methylation, the trimethylation of H3 lysine 4 (H3K4me3) and the trimethylation of H3 lysine 27 (H3K27me3), are particularly interested in interfering with the transcription process. H3K4me3 (active mark) is associated with active transcription, while H3K27me3 (repressive mark) is associated with transcriptional repression. Each H3 histone can be in one of the following three states: unmodified (U), modified by an active mark (A), or modified by a repressive mark (R) [50]. Each nucleosome can be in one of the six physically distinct nucleosome states (Fig. 4.10a):

- UU: The two H3 histones are unmodified.
- AU: One H3 histone is modified with an active mark, and the other one is unmodified.
- UR: One H3 histone is modified with a repressive mark, and the other one is unmodified.
- AA: Both H3 histones are modified with active marks.
- AR: One H3 histone is modified with an active mark, and the other one is unmodified with a repressive mark.

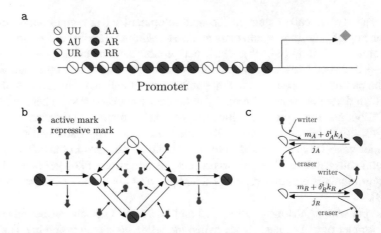

**Fig. 4.10** A stochastic dynamic model of histone modification in a DNA segment. **a** A DNA segment with multiple nucleosomes with the H3 histone proteins in each nucleosome, which can be in one of the 6 possible states: UU, AU, UR, AA, AR, and RR. **b** Dynamics of nucleosome state transition. The 6 nucleosome states transition between each other through the chemical reactions of methylation/demethylation according to the left panel. **c** The reactions of adding/removing an active marker (upper panel) or repressive marker (bottom panel), which are regulated by the corresponding enzymes. Replotted from [49]

- RR: Both H3 histones are modified with repressive marks.

The nucleosome states are dynamically changing due to the chemical reactions of methylation/demethylation, which are regulated by the corresponding enzymes (Fig. 4.10B). The involved biochemical reactions are classified as the methylation of unmodified sites (U) with either active (A) or repressive (R) markers or the demethylation of the modified sites (Fig. 4.10B, right panel). Upon methylation, the modified site is recognized by a chromatin-binding protein, also known as a "reader" protein, which in turn recruits other proteins that lead to additional chromatin-modifying activities, including "writers" and "erasers" that add or remove specific histone post-translational modifications, respectively [51–53]. The eraser removes the methyl group from the methylated nucleosome, and the writer induces the methylation of the neighboring nucleosome [52]. These processes of nucleosome state transition can be described by a stochastic simulation algorithm.

The methylation rate is dependent on the state of the neighboring nucleosome through the reader protein. To specify the neighboring state-mediated methylation process, we introduce a prefactor $\delta_X^i (X = A, R)$ for the $i$'th nucleosome to represent the state of its neighboring nucleosome:

$$\delta_X^i = \begin{cases} 1, & \text{if the } (i-1)'\text{th nucleosome is marked with } X, \\ 0, & \text{otherwise.} \end{cases} \tag{4.67}$$

Here, a nucleosome is marked with A (or R) when one H3 contains the mark A (or R). Let $m_X (X = A, R)$ be the basal methylation rate of an unmodified nucleosome

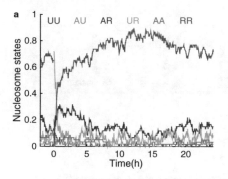
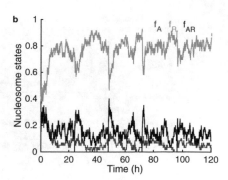

**Fig. 4.11** Nucleosome state dynamics. **a** Sample dynamics of nucleosome states in one cell cycle (starting at 0 h). Curves show the fraction of nucleosome states at each state. **b** The nucleosome states $f_A$, $f_R$, and $f_{AR}$ over a simulation of 10 cell cycles. Here, $f_A = f_{AU} + f_{AA}$, $f_R = f_{UR} + f_{RR}$. In simulations, parameters were taken as $N = 60$, $m_A = 0.96 \, \text{h}^{-1}$, $m_R = 0.31 \, \text{h}^{-1}$, $k_A = 1.39 \, \text{h}^{-1}$, $k_R = 2.34 \, \text{h}^{-1}$, $j_A = 0.65 \, \text{h}^{-1}$, and $j_R = 0.21 \, \text{h}^{-1}$

and $k_X (X = A, R)$ be the neighboring state-mediated methylation rate; then, the effective methylation rate of the $i$'th nucleosome is given by $m_X + \delta_X^i k_X$.

Demethylation is independent of the neighboring nucleosome and is denoted as $j_X (X = A, R)$.

Both rates $k_X$ and $j_X$ are dependent on the corresponding enzyme activities.

During DNA replication, each copy of the H3 histone in the parental nucleosomes is randomly assigned to one of the two daughter strands so that the nucleosome state of the daughter strand inherits the state from the parental strand, while the corresponding site on the newly synthesized strand is assigned an unmodified state [52]. Thus, we have the following transition probabilities of nucleosome states from mother to daughter cells:

$$p_{UU,UU} = 1,$$
$$p_{AU,UU} = 1/2, \quad p_{AU,AU} = 1/2,$$
$$p_{UR,UU} = 1/2, \quad p_{UR,UR} = 1/2,$$
$$p_{AA,AU} = 1,$$
$$p_{AR,AU} = 1/2, \quad p_{AR,UR} = 1/2,$$
$$p_{RR,UR} = 1.$$

The above formulations provide a numerical scheme for chromatin dynamics in long-run cell cycling at the individual cell level. Stochastic simulations based on the model give the dynamics of the nucleosome states of a single cell (Fig. 4.11). Here, the nucleosome states are represented by the fraction of nucleosomes at one of the six possible states ($f_X$, $X = UU, AU, UR, AA, AR, RR$), and $f_A = f_{AU} + f_{AA}$, $f_R = f_{UR} + f_{RR}$. There are sudden changes in the epigenetic state at each cell cycle along with DNA replication, after which the epigenetic states quickly return to their stationary states (Fig. 4.11).

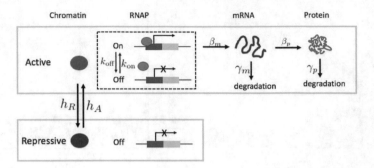

**Fig. 4.12** A two-state chromatin model of gene expression under epigenetic regulation. The chromatin state stochastically transitions between active (A) and repressive (R) states. In the active chromatin state, RNAP repetitively binds to the promoter so that the promoter switches between ON and OFF states, and the transcription process starts when RNAP binds to the promoter

### 4.6.1.2  Chromatin State Transition Regulated by Nucleosome Modifications

The molecular mechanisms of how histone modifications modulate the process of transcription are highly complex. Here, we refer to a simple model of gene expression dynamics under epigenetic regulation [36] and introduce a two-state chromatin model of gene expression (Fig. 4.12). In this model, the promoter configuration stochastically transitions between active (A) and repressive (R) states, and the kinetics of chromatin state transition are mediated by the nucleosome states.

The molecular details of how nucleosome states may affect the chromatin state transition remain unclear. Here, we assume that each nucleosome affects the rates of chromatin state transition by changing the chemical potential barrier between different states. In particular, we assume that (1) the repressive markers (UR or RR) tend to induce the repressive state, (2) the active markers (AU or AA) induce the activation of chromatin, and (3) the bivalent state (AR) further inhibits the activation of the chromatin state. Hence, let $f_A = f_{AU} + f_{AA}$ and $f_R = f_{UR} + f_{RR}$, we formulate the dependence of the chromatin state transition rates on the nucleosome states through exponential functions

$$h_R = \bar{h}_R e^{-\mu_R(1-f_R)}, \tag{4.68}$$

$$h_A = \bar{h}_A e^{-\mu_A(1-f_A)-\mu_{AR}f_{AR}}. \tag{4.69}$$

When the promoter is active, we refer to the previous standard model of transcription and translation. RNAP repeatedly binds to the promoter and triggers the transcription process. The promoter switches between ON and OFF states with rates $k_{on}$ and $k_{off}$, respectively. mRNA is synthesized at a rate $\beta_m$ when the promoter is ON and degraded or diluted at a rate $\gamma_m$. The translation and degradation/dilution rates of the proteins are $\beta_p$ and $\gamma_p$, respectively. Here, we note that mRNA degradation and protein production/degradation are independent of the chromatin state.

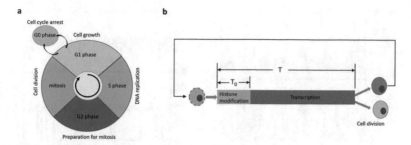

**Fig. 4.13** Modeling gene expression over cell cycles. **a** The eukaryotic cell cycle. The cell cycle is a series of events that take place in a cell, leading to its division and the duplication of its DNA (DNA replication) to produce two daughter cells. In eukaryotes, most of the cell cycle is spent in the interphase, the period during which is called the S phase (for DNA synthesis). The cell is in the G1 phase before the S phase and in the G2 phase after the S phase. The actual process of cell division occurs during the mitotic phase (M phase), which includes mitosis (the division of the cell's nucleus) and cytokinesis (the division of the cytoplasm). Cells that have temporarily or reversibly stopped dividing are said to have entered a state of quiescence, the G0 phase (cell cycle arrest). **b** The simplified scheme of modeling a cell cycle, which is separated into three components: histone modification, transcription, and cell division. During histone modification, the nucleosome states are randomly changed according to the previous kinetic model of histone modification; in the transcription stage, the dynamics of chromatin state transition and transcription evolve according to the two-state model in Fig. 4.12; in cell division, each cell divides into two daughter cells, and the mRNA and protein numbers of daughter cells are reassigned based on their numbers in the mother cells

To model the long-term dynamics of gene expression over the cell cycle, we need to consider the series of events taking place in a cell that lead to cell division and DNA replication (Fig. 4.13a). In eukaryotes, the cell cycle includes interphase and the mitotic (M) phase. During interphase, the cell grows, accumulates nutrients needed for mitosis, prepares for cell division and duplicates its DNA. During the mitotic phase, the chromosomes separate, and the chromosomes and cytoplasm are partitioned into two new daughter cells. Moreover, to ensure the proper division of the cell, there are control mechanisms known as cell cycle checkpoints [54]. The dynamics of cell cycle regulation have been extensively studied in many studies [55, 56].

Here, we do not try to couple the dynamics of cell cycle regulation with details on chromatin state transition and gene expression. Instead, we simplified the process and focused on the effects of the random inheritance of histone modification on the regulation of gene expression. To this end, there are different ways to combine the processes of histone modification, chromatin state transition, and transcription considered in the model (Fig. 4.12). In simulations, we can either put all biochemical reactions together within a cell cycle or separate the events into orderly sequencing components. Here, we apply a simplified numerical scheme, as shown in Fig. 4.13b. We assume that the kinetics of histone modification mainly happen at the stage of DNA replication, during which transcription can be omitted; transcription follows the proposed two-state model over the rest of the cell cycle. At the end of a cell cycle,

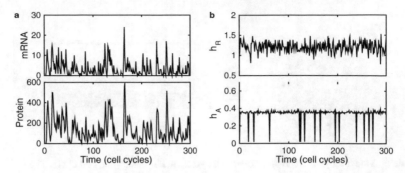

**Fig. 4.14** Chromatin state dynamics and stochasticity in gene expression. **a** Sample dynamics of mRNA and protein numbers over 300 cell cycles. **b** Stochastic rates of chromatin state transition regulated by histone modifications. The parameters are: $\bar{h}_R = 1.69$ h$^{-1}$, $\bar{h}_A = 0.93$ h$^{-1}$, $\mu_R = 2.0$, $\mu_A = 4.0$, $\mu_{AR} = 1.0$, $k_{on} = 8.0$ h$^{-1}$, $k_{off} = 1.0$ h$^{-1}$, $\beta_m = 2.0$ h$^{-1}$, $\gamma_m = 0.03$ h$^{-1}$, $\beta_p = 2.0$ h$^{-1}$, and $\gamma_p = 0.05$ h$^{-1}$; the other parameter values are the same as those in Fig. 4.11; one cycle corresponds to $T = 24$ $h$, and $T_0 = 3$ $h$. In simulations, for simplicity, we assumed that daughter cells have the same numbers of mRNAs and proteins as their mother cells

one cell is divided into two cells so that the histone modification marks are randomly assigned to daughter cells, and the mRNA and protein numbers in the daughter cells are reset. Figure 4.14 shows the sample dynamics of mRNA and protein numbers over a simulation of 300 cycles, as well as the fluctuating chromatin state transition rates due to random histone modifications. In this simulation, for simplicity, we omitted cell growth and assumed that the daughter cells have the same mRNA and protein numbers as their mother cells. The simulations showed that gene expression can automatically switch between high- and low-level expression in different cycles (Fig. 4.14).

### 4.6.1.3    Autoactivation Through Feedback on Enzyme Activity to Regulate Nucleosome Modifications

Histone modifications are regulated by various enzymes that are involved in the inheritance of modified markers during DNA replication [52]; the activities of these enzymes can be modulated by gene expression products, either directly or indirectly, and hence form a feedback loop. To model the feedback regulation, we assume that the protein products promote the writer activity for the active marker and inhibit the writer activity for the repressive marker, so the neighboring state-mediated methylation rates $k_A$ and $k_R$ can be given by Hill type functions of the protein number $P$, i.e.,

$$k_A = \bar{k}_A \frac{P^{m_0}}{K_A^{m_0} + P^{m_0}}, \quad k_R = \bar{k}_R \frac{K_R^{m_1}}{K_R^{m_1} + P^{m_1}}. \tag{4.70}$$

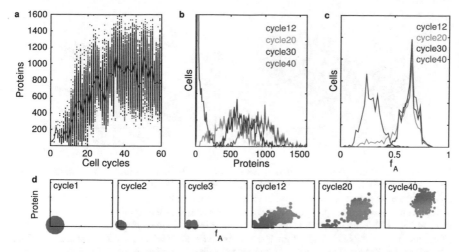

**Fig. 4.15** Evolution dynamics of cell population expansion. **a** Evolution dynamics of the protein numbers in a population of cells originating from one cell following cell divisions over 100 cell cycles. Here, each dot represents the protein number of a cell at the given cell cycle. In simulations, we began from one cell. For each cell, we performed the stochastic simulation for gene expression for 24 h (1 cycle), followed by cell division in which one cell divided into two cells and the nucleosome states of daughter cells randomly changed from those of mother cells following the rule as previously described. **b** Densities of protein levels in all cells at cycles 12, 20, 30, and 40. **c** Densities of the fraction of active nucleosome markers ($f_A$) in all cells at cycles 12, 20, 30, and 40. **d** Active nucleosome markers ($f_A$) and protein levels of individual cells at different cycles. In simulations, the parameters for the feedback regulations in (4.70) were as follows: $\bar{k}_A = 0.93$, $K_A = 100$, $m_0 = 4$, $\bar{k}_R = 1.69$, $K_R = 200$, and $m_1 = 6$. The other parameter values are the same as those in Fig. 4.14

Similarly, other types of feedback regulation can be formulated through various dependencies between the enzyme activities and the protein numbers.

A positive feedback loop in gene expression regulation may yield the coexistence of two phenotypes due to bistability, i.e., either a high or low level of expression can be seen in a cell (refer to Chap. 5). External stimuli or control is often introduced to regulate the switches between different phenotypes [57]. Nevertheless, the random inheritance of the epigenetic state during cell cycling may introduce an extra level of perturbations to cell-to-cell variance and hence can automatically induce switches in cell types. Figure 4.15 shows a simulation result of clone expansion for 60 cycles starting from a single cell with a low level of gene expression. The results show obvious switches in the cells from low to high levels of gene expression. At cycle 20, most cells have low protein numbers and active nucleosome marker states and switch to high protein numbers at cycle 40. The cell-to-cell variance during cell cycling shows the process of spontaneous cell type switches (Fig. 4.15d). This simple model shows that the process of stem cell differentiation can be modeled through the combination of transcriptional regulation with stochastic inheritance of epigenetic regulation during cell cycling.

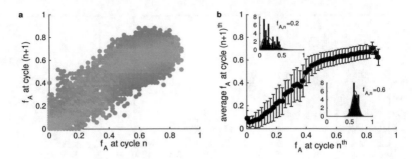

**Fig. 4.16** Nucleosome state transitions across cell cycles. **a** Diagram of the iterative map for the nucleosome activities $f_A$ across sequential cell cycles (from the $n$th to the $(n + 1)$th cycles). The colors of each dot represent the probability at the $(n + 1)$th cycle given the $f_A$ value at the $n$th cycle; a bright color means higher probability. **b** The mean (black dots) and standard deviation (error bars) of $f_A$ at $(n + 1)$th given the value at the $n$th cycle. The insets are the probability densities of the two situations, $f_{A,n} = 0.2$ and $f_{A,n} = 0.6$. The bars from simulated data and red solid lines are fitted with a beta distribution. The data were obtained from the simulation in Fig. 4.14

Histone modifications can be inherited over the cell cycles; hence, the transition of nucleosome states during cell cycling is crucial for the study of tissue development and cellular heterogeneity. To quantify the nucleosome transition across cell cycles, based on the proposed model, we define the nucleosome state at each cycle as the value of nucleosome activity $f_A$ in the previous cell division and find the iterative map of $f_A$ from one cycle to the next (Fig. 4.16a). This map clearly shows that in each cell division, the nucleosome state post-cell division (the daughter cell) is obviously correlated with the state prior to cell division (the mother cell). Moreover, we calculate the mean and variance of $f_A$ at the $(n + 1)$'th cycle versus those of $f_A$ at the $n$'th cycle (Fig. 4.16b). The probability density of $f_A$ at the $(n + 1)$'th cycle is given by a conditional probability through the state at the $n$'th cycle, which can be formulated as conditional beta distribution

$$P(f_{A,n+1} = x | f_{A,n} = y) = \frac{x^{a(y)-1}(1 - x)^{b(y)-1}}{B(a(y), b(y))}, \quad B(a, b) = \frac{\Gamma(a)\Gamma(b)}{\Gamma(a + b)},$$
(4.71)

where $a(y)$, and $b(y)$ are shape parameters depending on the nucleosome state $y$ of the mother cell, and $\Gamma(z)$ is the gamma function. Here, $f_{A,n}$ means the value $f_A$ at the $n$'th cycle. Figure 4.16b shows that the beta distribution fits well with the simulation data.

### 4.6.2 Stochastic Modeling of DNA Methylation

In this section, we introduce a stochastic model of DNA methylation. The molecular basis of DNA methylation patterns has been studied at the *single* CpG site level, and

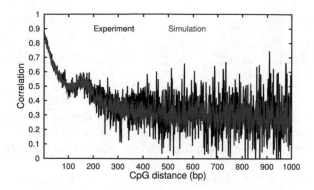

**Fig. 4.17** Pearson correlations obtained from experimental data (black) and from model simulation (red). Data were obtained from the parental strain chromosome 1 at the 2-cell stage of a mouse embryo (data source: GSM1386021). The parameters used in numerical simulations were as follows: $\alpha = 12$, $p = 1$, $\beta = 0.195$, $\gamma = 0.036$, and $c = 0.13$. Refer to Table 4.3 for the kinetic parameters. Replotted from [58]

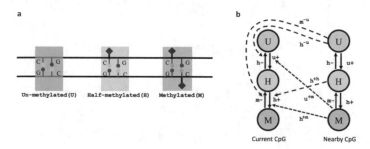

**Fig. 4.18** Schematic illustration of the model of DNA methylation. **a** The three possible states of a CpG are as follows: unmethylated (U), half-methylated (H), and methylated (M). **b** Transitions between the methylation states of a CpG site. Solid arrows represent the transition between the states, and dashed arrows show the correlations with the nearby CpGs. Replotted from [58]

CpG sites are regulated by various types of enzymes, such as DNMT1/UHRF1 for the maintenance of methylation marks, TET families for active demethylation, and DNMT3a/DNMT3b for *de novo* methylation [45]. Further analysis of various data sets suggested a clear correlation between the methylation levels of adjacent CpG sites [58, 59]. There is a strong correlation with the Pearson correlation $r > 0.8$ when the CpG distance is 2 bp; the correlation displays a local maximum at a distance of approximately 160 bp and then decreases with the CpG distance (Fig. 4.17).

To model the kinetics of DNA methylation with the long-distance correlation in adjacent CpG sides, we establish a stochastic simulation scheme for the chemical reactions of the methylation/demethylation of each CpG site (Fig. 4.18a). In the model, each CpG site is in one of the following three possible states: unmethylated (U), half-methylated (H), or methylated (M). Each CpG site can transition between the three states U, H, and M, and the transition rates are affected by the status of nearby CpGs (Fig. 4.18). We further assume that there is a correlation between the

**Table 4.3** Default kinetic parameter values used in the simulations (refer to [60])

| $u^+$ | $h^+$ | $m^-$ | $h^-$ | $h^{+h}$ | $h^{+m}$ | $u^{+m}$ | $h^{-u}$ | $m^{-u}$ |
|-------|-------|-------|-------|----------|----------|----------|----------|----------|
| 0.008 | 0.008 | 0.04  | 0.04  | 0.24     | 0.24     | 0.24     | 0.05     | 0.05     |

methylation and the demethylation of CpG sites and that the correlation is dependent on the pairwise CpG distances (Fig. 4.18b). Hence, we introduce a correlation strength function $\phi(d)$ that is a function of the distance (in bp) between the two CpGs.

Mathematically, the kinetic rates of the $i$'th CpG in a DNA sequence are given by

$$
\begin{cases}
u_i^+ = u^+ + \sum_{j=1}^{N} p_{i,j} \cdot \chi_j(m) \cdot \phi(d_{i,j}) \cdot (u^{+m} - u^+), \\
h_i^+ = h^+ + \sum_{j=1}^{N} p_{i,j} \cdot \chi_j(m) \cdot \phi(d_{i,j}) \cdot (h^{+m} - h^+) \\
\qquad\quad + \sum_{j=1}^{N} p_{i,j} \cdot \chi_j(h) \cdot \phi(d_{i,j}) \cdot (h^{+h} - h^+), \\
m_i^- = m^- + \sum_{j=1}^{N} p_{i,j} \cdot \chi_j(u) \cdot \phi(d_{i,j}) \cdot (m^{-u} - m^-), \\
h_i^- = h^- + \sum_{j=1}^{N} p_{i,j} \cdot \chi_j(u) \cdot \phi(d_{i,j}) \cdot (h^{-u} - h^-).
\end{cases}
\tag{4.72}
$$

Here, $p_{i,j} = 1$ when the $i$'th CpG is affected by the $j$'th CpG; otherwise, $p_{i,j} = 0$. The index $\chi_j(s) = 1$ $(s = u, h, m)$ if the $j$'th CpG has the status $s$; otherwise $\chi_j(s) = 0$. We simply consider the nearby correlation so that $p_{i,j} = 1$ only when $j = i \pm 1$. An example of the kinetic parameters is listed in Table 4.3. In model (4.72), $d_{i,j}$ represents the distance (along the DNA sequence) between CpG sites $i$ and $j$, and the location of each CpG site can be obtained from the DNA sequence of a chromosome.

Based on a study that fits model simulation with data analysis, the correlation function $\phi(d)$ is taken as [58]

$$
\phi(d) = \frac{1}{(\beta + \gamma d)^p} + \frac{\alpha}{30\sqrt{2\pi}} \exp\left[ -\frac{(d - 160)^2}{2 \times 30^2} \right] + c. \tag{4.73}
$$

Here, the function $\phi(d)$ consists of three parts: the power law decay, a bell-shaped bulb, and a constant tail. The power law decay represents the local correlation between nearby CpG sites. We note that the bell-shaped bulb is central at 160 bp, right at the distance of approximately one nucleosome. This suggests a possible correlation between histone modification and DNA methylation. The constant tail coefficient $c$

is related to the global methylation level, which is dependent on the enzyme activities that regulate DNA methylation.

Biologically, H3K9me2 often shows nucleation sites overlapping with the CpG islands (CGIs) [61] and is involved in the regulation of DNA methylation [62–64]. Moreover, the loss of DNA methylation enhances the removal of H3K9me3 under transcriptional stimulus [65]. This evidence supports the correlation between nucleosome modifications and DNA methylation.

The kinetic rates in (4.72) define the dynamics within each cell cycle. To include the effect of reallocation at DNA replication, we need to reset the status $m$ to $h$ and $h$ to either $h$ or $u$, at the beginning of each cell cycle. For a given DNA sequence and a well-defined function $\phi(d)$, stochastic simulation based on the above kinetic rates for multiple cycles can result in a stationary DNA methylation pattern, from which the model-predicted correlation profile can be calculated and compared with the experimental data (Fig. 4.17).

## 4.7 A Unified Model of Gene Expression with Epigenetic Modification

Now, we revisit the basic model of gene expression in Fig. 4.2 and include the process of cell division. We further assume that the promoter activation and inactivation (or chromatin state transition) rates $\lambda_1^+$ and $\lambda_1^-$ are subjected to the regulation of epigenetic modifications, e.g., histone modification or DNA methylation. Let $\beta_n$ ($0 \leq \beta_n \leq 1$) represent the epigenetic state of the gene promoter at cycle $n$. Similar to the above argument in Sect. 4.6.1.2, we write

$$\lambda_1^+(\beta_n) = h_0 e^{-\mu_0 \beta_n}, \tag{4.74}$$

$$\lambda_1^-(\beta_n) = h_1 e^{-\mu_1(1-\beta_n)} \tag{4.75}$$

at cycle $n$. Here, we assume that an epigenetic state with a higher $\beta$ value implies lower levels of gene expression.

To describe the changes in $\beta_n$ over cell cycling, we refer to the beta distribution (4.71). According to the probability density function (4.71), the mean and variance of $\beta_{n+1}$, given $\beta_n = y$, are

$$E(\beta_{n+1}|\beta_n = y) = \frac{a(y)}{a(y) + b(y)} := \phi(y), \tag{4.76}$$

$$var(\beta_{n+1}|\beta_n = y) = \frac{a(y)b(y)}{(a(y) + b(y))^2(a(y) + b(y) + 1)} := \psi(y). \tag{4.77}$$

We note that

$$\psi(y) = \frac{1}{a(y) + b(y) + 1} \phi(y)(1 - \phi(y)).$$

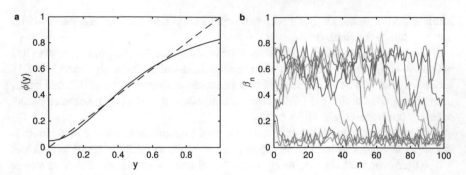

**Fig. 4.19** An example of epigenetic state dynamics. **a** The function $\phi(y)$ (solid line) in (4.79). **b** Plots of 10 independent trajectories of $\beta_n$. Here $\phi_0 = 0.05$, $\phi_1 = 1.0$, $m = 2.0$, $y_0 = 0.53$, $\eta_0 = 60$

Hence, if we assume the functions $0 < \phi(y) < 1$ $(0 < y < 1)$ and $\eta(y) > 0$ so that

$$\phi(y) = E(\beta_{n+1}|\beta_n = y), \quad \psi(y) = \frac{1}{\eta(y) + 1}\phi(y)(1 - \phi(y)),$$

the shape coefficients are given by

$$a(y) = \eta(y)\phi(y), \quad b(y) = \eta(y)(1 - \phi(y)). \tag{4.78}$$

Consequently, while we predefine the functions $\phi(y)$ and $\eta(y)$, the epigenetic state $\beta_{n+1}$, given $\beta_n$, can be obtained by finding a random number of beta distributions with shape parameters $a(\beta_n)$ and $b(\beta_n)$. As an example, we took (Fig. 4.19a)

$$\phi(y) = \phi_0 + \phi_1 \frac{y^m}{y^m + y_0^m}, \quad (0 \le y \le 1), \tag{4.79}$$

and $\eta(y) \equiv \eta_0$, a trajectory of $\beta_n$ is shown in Fig. 4.19b.

The above numerical scheme provides a procedure for simulating the long-term gene expression process over multiple cell cycles. Figure 4.20 shows the cell-to-cell variance of gene expression with different sets of model parameters. The distribution profiles reveal the experimental observations of cell-to-cell variance in mammalian cells [17, 66].

Finally, based on the above arguments, we re-formulate the reaction rate equation (4.2) for the expression of a single gene over multiple cell cycles as (here, we let the gene copy number be 1)

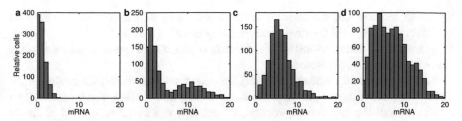

**Fig. 4.20** Distribution of mRNA numbers based on numerical simulations with different sets of parameters. The parameters are as follows: $\beta_m = 20$, $\gamma_m = 0.03$, $\phi_0 = 0.05$, $\eta = 60$, and **a** $h_0 = 0.1$, $h_1 = 0.2$, $\mu_0 = 1.0$, $\mu_1 = 2.0$, $\phi_1 = 1.1$, $y_0 = 0.56$, $m = 1.7$; **b** $h_0 = 0.4$, $h_1 = 0.2$, $\mu_0 = 2.0$, $\mu_1 = 1.0$, $\phi_1 = 1.0$, $y_0 = 0.4953$, $m = 2.7$; **c** $h_0 = 0.4$, $h_1 = 0.2$, $\mu_0 = 1.0$, $\mu_1 = 2.0$, $\phi_1 = 1.1$, $y_0 = 0.56$, $m = 1.8$; **d** $h_0 = 0.4$, $h_1 = 0.2$, $\mu_0 = 2.0$, $\mu_1 = 2.0$, $\phi_1 = 1.1$, $y_0 = 0.56$, $m = 1.8$. The results were obtained from independent runs of 1000 cells, each simulation were performed for 50 cycles, and the mRNA levels of each cell at cycle 50 were collected to determined the density of the transcription level. Here, the translation processes were omitted. Replotted from [66]

$$\begin{cases} \dfrac{dx_1}{dt} = \lambda_1^+(\beta_n)(1 - x_1) - \lambda_1^-(\beta_n)x_1 \\[2mm] \dfrac{dx_2}{dt} = \lambda_2 x_1 - \delta_2 x_2 \qquad\qquad (nT < t \le (n+1)T) \\[2mm] \dfrac{dx_3}{dt} = \lambda_3 x_2 - \delta_3 x_3 \end{cases} \tag{4.80}$$

and, given the functions $0 \le \phi(y) \le 1$ and $\eta(y) > 0$,

$$P(\beta_{n+1} = x|\beta_n = y) = \frac{x^{a(y)-1}(1 - x)^{b(y)-1}}{B(a(y), b(y))}, \tag{4.81}$$

$$a(y) = \eta(y)\phi(y), b(y) = \eta(y)(1 - \phi(y)), \quad (0 \le y \le 1).$$

Here, $T$ represents the timing of one cell cycle, and the values $x_i$ are reset at $t = nT (\forall n \in \mathbb{Z}^+)$ due to the particle distribution at mitosis. This equation gives a mathematical model framework to describe gene expression over multiple cell cycles. The model integrates the ordinary differential equations within a cell cycle with a random process between cycles. Moreover, we can extend equation (4.80) to stochastic differential equations to include random perturbations to the kinetic parameters [67].

## 4.8  Summary

This chapter introduces stochastic model frameworks for the kinetics of gene expression. We start from a simple model of transcription and translation in prokaryotes and introduce a chemical Langevin equation for the stochasticity of gene expression with both intrinsic and extrinsic noise perturbations to the parameters. The method to obtain the analytical distribution from the generating function is also introduced, from which the analytical mRNA distribution of the simple model is derived and is given by a gamma distribution.

Next, we introduce a model of gene expression in eukaryotes with random transitions between different promoter states. The analytical mRNA distribution can be obtained from the generating function, which is given by solving a set of linear differential equations. The promoter state transition is associated with both histone modifications and DNA methylation. We further present numerical schemes to model histone modifications and DNA methylation and show how epigenetic modifications can be incorporated into the transcription process through the regulation of chromatin structures. In particular, we found that the beta distribution can be used to describe the transition of epigenetic states across cell cycles and applied it to model the random process of epigenetic state transition over cell cycles. By considering the reassignment of epigenetic modifications during DNA replication, we are able to establish computational models for gene expression over many cell cycles. A model of DNA methylation with long-distance correlations between CpG sites is also presented. Finally, we introduce a unified model of gene expression that incorporates epigenetic modification with transcription initiation.

The modeling techniques introduced in this chapter are general ideas and frameworks to model gene expression. The real biological processes are much more complicated than what we have considered here. For example, details on how epigenetic regulation affects the transcription process remain unknown, the interactions between the cell cycle and gene expression are complicated, and different scale interactions are involved. Hence, the frameworks introduced here should be modified and extended when we study different scale problems.

# References

1. Kærn, M., Elston, T.C., Blake, W.J., Collins, J.J.: Stochasticity in gene expression: from theories to phenotypes. Nat. Rev. Genet. **6**, 451–464 (2005)
2. Paulsson, J.: Summing up the noise in gene networks. Nature **427**, 415–418 (2004)
3. Paulsson, J.: Models of stochastic gene expression. Phys. Life Rev. **2**, 157–175 (2005)
4. Ren, X., Kang, B., Zhang, Z.: Understanding tumor ecosystems by single-cell sequencing: promises and limitations. Genome Biol. **19**, 211 (2018)
5. Stuart, T., Satija, R.: Integrative single-cell analysis. Nat. Rev. Genet. **20**, 257–272 (2019)
6. Vera, M., Biswas, J., Senecal, A., Singer, R.H., Park, H.Y.: Single-cell and single-molecule analysis of gene expression regulation. Annu. Rev. Genet. **50**, 267–291 (2016)
7. Altschuler, S.J., Wu, L.F.: Cellular heterogeneity: do differences make a difference? Cell **141**, 559–563 (2010)
8. Lei, J.: Stochasticity in single gene expression with both intrinsic noise and fluctuation in kinetic parameters. J. Theor. Biol. **256**, 485–492 (2009)
9. Swain, P.S.: Intrinsic and extrinsic contributions to stochasticity in gene expression. Proc. Natl. Acad. Sci. USA **99**, 12795–12800 (2002)
10. Elowitz, M.B., Elowitz, M.B., Levine, A.J., Siggia, E.D., Swain, P.S.: Stochastic gene expression in a single cell. Science **297**, 1183–1186 (2002)

11. Ozbudak, E.M., Thattai, M., Kurtser, I., Grossman, A.D., van Oudenaarden, A.: Regulation of noise in the expression of a single gene. Nat. Genet. **31**, 69–73 (2002)
12. Shahrezaei, V., Swain, P.S.: Analytical distributions for stochastic gene expression. Proc. Natl. Acad. Sci. USA **105**, 17256–17261 (2008a)
13. Delgado, M.: Classroom note: the Lagrange-Charpit method. Siam Rev. **39**, 298–304 (1997)
14. Friedman, N., Cai, L., Xie, X.S.: Linking stochastic dynamics to population distribution: an analytical framework of gene expression. Phys. Rev. Lett. **97**, 168302 (2006)
15. Yu, J., Xiao, J., Ren, X., Lao, K., Xie, X.S.: Probing gene expression in live cells, one protein molecule at a time. Science **311**, 1600–1603 (2006)
16. Cai, L., Friedman, N., Xie, X.S.: Stochastic protein expression in individual cells at the single molecule level. Nature **440**, 358–362 (2006)
17. Battich, N., Stoeger, T., Pelkmans, L.: Control of transcript variability in single mammalian cells. Cell **163**, 1596–1610 (2015)
18. Kaufmann, B.B., van Oudenaarden, A.: Stochastic gene expression: from single molecules to the proteome. Curr. Opin. Genet. Dev. **17**, 107–112 (2007)
19. Rosenfeld, N., Young, J.W., Alon, U., Swain, P.S., Elowitz, M.B.: Gene regulation at the single-cell level. Science **307**, 1962–1965 (2005)
20. Sigal, A., Milo, R., Cohen, A., Geva-Zatorsky, N., Klein, Y., Liron, Y., Rosenfeld, N., Danon, T., Perzov, N., Alon, U.: Variability and memory of protein levels in human cells. Nature **444**, 643–646 (2006)
21. Shahrezaei, V., Ollivier, J.F., Swain, P.S.: Colored extrinsic fluctuations and stochastic gene expression. Mol. Syst. Biol. **4**, 1–9 (2008)
22. Orphanides, G., Reinberg, D.: A unified theory of gene expression. Cell **108**, 439–451 (2002)
23. Tupler, R., Perini, G., Green, M.R.: Expressing the human genome. Nature **409**, 832–833 (2001)
24. Proudfoot, N.J., Furger, A., Dye, M.J.: Integrating mRNA processing with transcription. Cell **108**, 501–512 (2002)
25. Mishra, S.K., Thakran, P.: Intron specificity in pre-mRNA splicing. Curr. Genet. **42**, 4640–8 (2018)
26. Reed, R.: Coupling transcription, splicing and mRNA export. Curr. Opin. Cell Biol. **15**, 326–331 (2003)
27. Reed, R., Hurt, E.: A Conserved mRNA Export Machinery Coupled to pre-mRNA Splicing. Cell **108**, 523–531 (2002)
28. Ramakrishnan, V.: Ribosome structure and the mechanism of translation. Cell **108**, 557–572 (2002)
29. Dever, T.E.: Gene-specific regulation by general translation factors. Cell **108**, 545–556 (2002)
30. Fersht, A.R., Daggett, V.: Protein folding and unfolding at atomic resolution. Cell **108**, 573–582 (2002)
31. Liu, J., Qian, C., Cao, X.: Post-translational modification control of innate immunity. Immunity **45**, 15–30 (2016)
32. Bode, A.M., Dong, Z.: Post-translational modification of p53 in tumorigenesis. Nat. Rev. Cancer **4**, 793–805 (2004)
33. Blake, W.J., Kærn, M., Cantor, C.R., Collins, J.J.: Noise in eukaryotic gene expression. Nature **422**, 633–637 (2003)
34. Zhang, J., Chen, L., Zhou, T.: Analytical distribution and Tunability of noise in a model of promoter progress. Biophys. J. **102**, 1247–1257 (2012)
35. Zhou, T., Zhang, J.: Analytical results for a multistate gene model. SIAM J. Appl. Math. **72**, 789–818 (2012)
36. Bintu, L., Yong, J., Antebi, Y.E., McCue, K., Kazuki, Y., Uno, N., Oshimura, M., Elowitz, M.B.: Dynamics of epigenetic regulation at the single-cell level. Science **351**, 720–724 (2016)
37. Kouzarides, T.: Chromatin modifications and their function. Cell **128**, 693–705 (2007)
38. Ben-Shahar, Y.: Epigenetic switch turns on genetic behavioral variations. Proc. Natl. Acad. Sci. USA **114**, 12365–12367 (2017)

39. Gaffney, D.J., McVicker, G., Pai, A.A., Fondufe-Mittendorf, Y.N., Lewellen, N., Michelini, K., Widom, J., Gilad, Y., Pritchard, J.K.: Controls of nucleosome positioning in the human genome. PLoS Genet. **8**, e1003036 (2012)
40. Lachner, M., O'Sullivan, R.J., Jenuwein, T.: An epigenetic road map for histone lysine methylation. J. Cell Sci. **116**, 2117–2124 (2003)
41. Zentner, G.E., Henikoff, S.: Regulation of nucleosome dynamics by histone modifications. Nat. Struct. Mol. Biol. **20**, 259–266 (2013)
42. Law, J.A., Jacobsen, S.E.: Establishing, maintaining and modifying DNA methylation patterns in plants and animals. Nat. Rev. Genet. **11**, 204–220 (2010)
43. Seisenberger, S., Peat, J.R., Hore, T.A., Santos, F., Dean, W., Reik, W.: Reprogramming DNA methylation in the mammalian life cycle: building and breaking epigenetic barriers. Philos. Trans. R Soc. Lond. B Biol. Sci. **368**, 20110330–20110330 (2013a)
44. Bird, A.: DNA methylation patterns and epigenetic memory. Genes. Dev. **16**, 6–21 (2002)
45. Wu, H., Zhang, Y.: Reversing DNA methylation: mechanisms, genomics, and biological functions. Cell **156**, 45–68 (2014)
46. Smith, Z.D., Meissner, A.: DNA methylation: roles in mammalian development. Nat. Rev. Genet. **14**, 204–220 (2013)
47. Klutstein, M., Nejman, D., Greenfield, R., Cedar, H.: DNA methylation in cancer and aging. Cancer Res. **76**, 3446–3450 (2016)
48. Witte, T., Plass, C., Gerhauser, C.: Pan-cancer patterns of DNA methylation. Genome Medic. 6:8 *6*, 66 (2014)
49. Huang, R., Lei, J.: Dynamics of gene expression with positive feedback to histone modifications at bivalent domains. Int. J. Mod. Phys. B **4**, 1850075 (2017)
50. Ku, W.L., Girvan, M., Yuan, G.-C., Sorrentino, F., Ott, E.: Modeling the dynamics of bivalent histone modifications. PLoS ONE **8**, e77944 (2013)
51. Kaelin, W.G., McKnight, S.L.: Influence of metabolism on epigenetics and disease. Cell **153**, 56–69 (2013)
52. Probst, A.V., Dunleavy, E., Almouzni, G.: Epigenetic inheritance during the cell cycle. Nat. Rev. Mol. Cell Biol. **10**, 192–206 (2009)
53. Ruthenburg, A.J., Allis, C.D., Wysocka, J.: Methylation of Lysine 4 on Histone H3: intricacy of writing and reading a single epigenetic mark. Mol. Cell **25**, 15–30 (2006)
54. Elledge, S.J.: Cell cycle checkpoints: preventing an identity crisis. Science **274**, 1664–1672 (1996)
55. Tyson, J.J., Novak, B., Odell, G.M., Chen, K., Thron, C.D.: Chemical kinetic theory: understanding cell-cycle regulation. Trends Biochem. Sci. **21**, 89–96 (1996)
56. Tyson, J.J., Csikasz-Nagy, A., Novak, B.: The dynamics of cell cycle regulation. Bioessays **24**, 1095–1109 (2002)
57. Ferrell, J.E.J.: Bistability, bifurcations, and Waddington's epigenetic landscape. Curr. Biol. **22**, R458–R466 (2012)
58. Song, Y., Ren, H., Lei, J.: Collaborations between CpG sites in DNA methylation. Int. J. Mod. Phys. B **31**, 1750243 (2017)
59. Zhang, L., Xie, W.J., Liu, S., Meng, L., Gu, C., Gao, Y.Q.: DNA methylation landscape reflects the spatial organization of chromatin in different cells. Biophys. J. **113**, 1395–1404 (2017)
60. Haerter, J.O., Lövkvist, C., Dodd, I.B., Sneppen, K.: Collaboration between CPG sites is needed for stable somatic inheritance of DNA methylation states. Nucleic. Acids Res. **42**, 2235–2244 (2014)
61. Chen, X., Skutt-Kakaria, K., Davison, J., Ou, Y.-L., Choi, E., Malik, P., Loeb, K., Wood, B., Georges, G., Torok-Storb, B., Paddison, P.J.: G9a/GLP-dependent histone H3K9me2 patterning during human hematopoietic stem cell lineage commitment. Genes Dev. **26**, 2499–2511 (2012)
62. Lehnertz, B., Ueda, Y., Derijck, A.A.H.A., Braunschweig, U., Perez-Burgos, L., Kubicek, S., Chen, T., Li, E., Jenuwein, T., Peters, A.H.F.M.: Suv39h-mediated histone H3 Lysine 9 methylation directs DNA methylation to major satellite repeats at Pericentric Heterochromatin. Curr. Biol. **13**, 1192–1200 (2003)

63. Nakamura, T., Liu, Y.-J., Nakashima, H., Umehara, H., Inoue, K., Matoba, S., Tachibana, M., Ogura, A., Shinkai, Y., Nakano, T.: PGC7 binds histone H3K9me2 to protect against conversion of 5mC to 5hmC in early embryos. Nature **486**, 415–419 (2012)
64. Habibi, E., Brinkman, A.B., Arand, J., Kroeze, L.I., Kerstens, H.H.D., Matarese, F., Lepikhov, K., Gut, M., Brun-Heath, I., Hubner, N.C., Benedetti, R., Altucci, L., Jansen, J.H., Walter, J., Gut, I.G., Marks, H., Stunnenberg, H.G.: Whole-genome bisulfite sequencing of two distinct interconvertible DNA methylomes of mouse embryonic stem cells. Cell Stem Cell **13**, 360–369 (2013)
65. Hathaway, N.A., Bell, O., Hodges, C., Miller, E.L., Neel, D.S., Crabtree, G.R.: Dynamics and memory of heterochromatin in living cells. Cell **149**, 1447–1460 (2012)
66. Jiao, X., Lei, J.: Dynamics of gene expression based on epigenetic modifications. Commun. Inform. Syst. **18**, 125–148 (2018)
67. Huang, R., Lei, J., Zhou, P.-Y.: Center for Applied Mathematics, MOE Key Laboratory of Bioinformatics, Tsinghua University, Beijing China, 100084: Cell-type switches induced by stochastic histone modification inheritance. Discrete & Continuous Dynamical Systems B **22**, 1–19 (2019)

# Chapter 5
# Mathematical Models for Gene Regulatory Network Dynamics

*Nature doesn't look at the world in terms of physics, chemistry or biology. Why should we?*

—Kerson Huang

## 5.1 Introduction

In Chap. 4, we introduced mathematical models for understanding the kinetics of single gene expression. In biological systems, the expression of a specific gene is modulated by many regulators, including various proteins, noncoding RNAs, small molecules, etc., so that the gene can be turned on or off when needed. For example, during early embryo development, proper gene regulation is important to ensure that appropriate genes are expressed at the proper time, which in turn is central to evolutionary developmental biology. Gene regulation can also help an organism response properly to the environmental stimuli. Moreover, gene expression products can further regulate the transcription of other genes to form a *gene regulatory network* (GRN).

The expression of genes in a gene regulatory network is dynamically regulated. In general, the system of a gene regulatory network can be considered biochemical reactions, and the gene expression dynamics can be modeled with the methods of either stochastic simulations or ordinary differential equations introduced in Chaps. 2–3. In this chapter, we introduce mathematical models used to study typical gene regulatory networks, including bistability, toggle switches, circadian oscillations, etc. Despite the simplicity, the methods introduced here can easily be extended to consider more complex regulatory networks.

© The Author(s), under exclusive license to Springer Nature Switzerland AG 2021    145
J. Lei, *Systems Biology*, Lecture Notes on Mathematical Modelling in the Life Sciences,
https://doi.org/10.1007/978-3-030-73033-8_5

## 5.2   Gene Regulation

The first identified system of gene regulation was the lac operon in *E. coli*, discovered by François Jacob and Jacques Monod in 1961 [1], in which enzymes involved in lactose metabolism are expressed only in the presence of lactose and the absence of glucose. Later, and increasing number of gene regulation systems were found in both prokaryotes and eukaryotes. In prokaryotic cells, gene expression is regulated by controlling the amount of transcription. The complexity of controlling the process of gene expression increased as eukaryotic cells evolved. The regulation of gene expression can occur at all stages within the process from transcription to translation. Regulation may occur at the epigenetic level when the DNA is uncoiled and loosened from nucleosomes to bind to transcription factors, the transcriptional level when RNA is transcribed, the posttranscriptional level when the RNA is processed and exported to the cytoplasm after it is transcribed, the translational level when RNA is translated into protein, or the posttranslational level after the protein has been made.

Transcription regulation is the most extensive strategy in the modulation of gene expression in both prokaryotes and eukaryotes. Transcription regulation mainly controls when the transcription process is turned on and off and modulates the amount of transcription. In prokaryotes, structural proteins with related functions are usually encoded together within the genome in a block called an operon and are transcribed together under the control of a single promoter. This forms a polycistronic transcript (Fig. 5.1). The promoter simultaneously controls the transcription of these structural genes since they are either all needed at the same time or none are needed. If more proteins are required, more transcription occurs. There are three types of regulatory

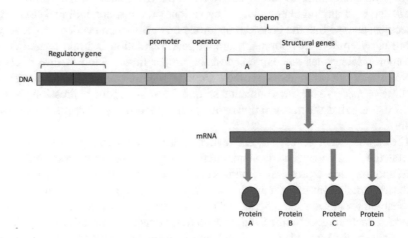

**Fig. 5.1** Illustration of the structure of an operon. In prokaryotes, structural genes of related functions are often organized together in the genome and transcribed together under the control of a single promoter. The regulatory region of an operon includes both the promoter and the operator. The structural genes are silent when a repressor binds to the operator, and transcription is enhanced when activators bind to the regulatory region. Replotted from https://courses.lumenlearning.com/

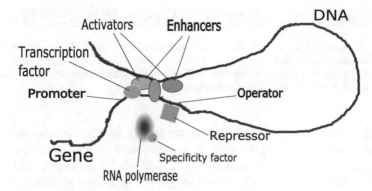

**Fig. 5.2** Illustration of transcription factors in eukaryotic cells. Activator proteins bind to pieces of DNA called enhancers. Their binding causes the DNA to bend and bring them near a gene promoter, even though they may be thousands of base pairs away. Other transcription factor proteins join activator proteins, forming a protein complex that binds to the gene promoter. Specificity factors make it easier for RNA polymerase to attach to the promoter and start transcribing a gene

molecules that can affect the expression of operons: repressors, activators, and inducers. Repressors are proteins that suppress the transcription of a gene in response to an external stimulus, whereas activators are proteins that increase the transcription of a gene in response to an external stimulus. Inducers are small molecules that either activate or repress transcription depending on the needs of the cell and the availability of substrate.

In eukaryotes, the transcription of a gene by RNA polymerase (RNAP) can be regulated by several mechanisms (Fig. 5.2). Various types of proteins are involved in transcription regulation. Specificity factors  mediate target recognition in RNAP to alter the RNAP for a given promoter or a set of promoters, making it more or less likely to bind to them. Repressors bind to the operator, the coding sequences on the DNA strand that are close to or overlapping the promoter region; they inhibit the binding of RNAP to the promoter and impede the expression of the gene. General transcription factors position RNAP at the transcription start site (TSS) of the protein-coding sequence and then release the polymerase to process mRNA transcription. Activators enhance the interaction between RNAP and a particular promoter, encouraging the expression of the gene. Activators do this by increasing the attraction of RNAP to the promoter by altering the DNA structure. In addition to promoters and operators, some other segments in the DNA sequence are also involved in transcription regulation. Enhancers are sites on the DNA helix that are bound by activators to loop the DNA and bring a specific promoter to the inhibition complex. Enhancers are much more common in eukaryotes than in prokaryotes, where only a few examples exist. Silencers are regions on the DNA sequences that can silence gene expression when they are bound by particular transcription factors.

Transcription factors (TFs) (or sequence-specific DNA-binding factors) are proteins that control the activity of a gene by determining whether the gene's DNA is transcribed into RNA. Transcription factors control when, where, and how efficiently

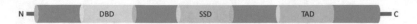

**Fig. 5.3** Schematic diagram of the amino acid sequence of a prototypical transcription factor. A prototypical transcription factor contains (1) a DNA-binding domain (DBD), (2) a signal-sensing domain (SSD), and (3) a transactivation domain (TAD). The order of placement and the number of domains may differ in various types of transcription factors

RNAP functions. In organisms, transcription factors are vital for normal development, as well as for routine cellular functions and responses to diseases. Transcription factors are modular in structure and contain the following domains (Fig. 5.3): a DNA-binding domain (DBD), which attaches to specific sequences of DNA adjacent to the regulated genes; a transactivation domain (TAD), which contains binding sites for other proteins such as transcription co-regulators; and an optional signal-sensing domain (SSD) (e.g., a ligand-binding domain), which senses and responds to external signals and transmits these signals to the rest of the transcription complex, resulting in the upregulation or downregulation of gene expression. Transcription factors can be grouped into classes based on their DBDs, such as basic helix-loop-helix [2], helix-turn-helix [3], and zinc fingers [4]. There are approximately 2600 proteins in the human genome that contain DNA-binding domains, and most of these are presumed to function as transcription factors [5]. Transcription factors work alone or with other proteins in a complex by promoting (as an activator) or blocking (as a repressor) the recruitment of RNA polymerase to specific genes. They may bind directly to special "promoter" regions of DNA or directly to the RNA polymerase molecule.

## 5.3  Positive Feedback and Bistability

Bistability is a common phenomenon in cellular behavior, and many cells show the coexistence of different phenotypes or responses under the same environment or external stimuli. This ability is very important for cell survival and the heterogeneity of multicellular tissues. For example, the potential to adapt to the changing environment through different phenotypes is crucial for bacteria to survive under harsh natural conditions. The potential of multiple phenotypes is often associated with the bistability (or multistability) of gene regulatory networks, which often involve positive feedback regulation. Here, we discuss the bistability of the lactose utilization network in *Escherichia coli* (*E. coli*) and introduce methods to analyze the conditions needed for bistability. Please refer to [6] for most of the content in this section.

### 5.3.1 Lac Operon

Many protein-coding genes in bacteria are clustered together in operons, which serve as transcriptional units that are coordinately regulated. The *lac* operon comprises three structural genes required for the uptake and metabolism of lactose and related sugars: *lacZ*, *lacY* and *lacA* (Fig. 5.4). The *lac* promoter P$_{lac}$, located at the 5' end of lacZ, directs the transcription of all three genes as a single mRNA (called a polycistronic message because it includes more than one gene); this mRNA is translated to give the three protein products. The *lacZ* gene encodes β-galactosidase, an enzyme that cleaves the sugar lactose into galactose and glucose, which are energy sources for the cell. The *lacY* gene encodes lactose permease (LacY), which facilitates the uptake of lactose and similar molecules, including thio-methylgalactoside (TMG), a nonmetabolizable lactose analogue. The *lacA* gene encodes thiogalactoside transacetylase, which rids the cell of toxic thiogalactosides that are also transported by LacY. These genes are expressed at high levels only when lactose is available and glucose is not.

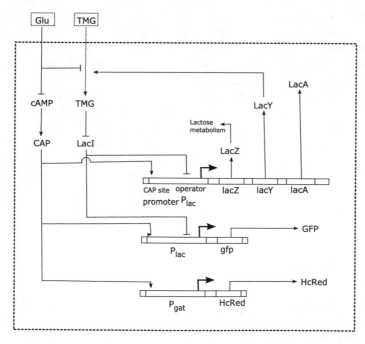

**Fig. 5.4** Gene regulation network of lactose. The uptake of TMG inhibits the lac repressor and induces the synthesis of LacY, which in turn promotes further TMG uptake, resulting in a positive feedback loop. The extracellular concentrations of TMG and glucose can be used to independently regulate the activities of LacI and CRP, the two regulatory inputs of the *lac* operon. The network's bistable response can be probed by a single copy of the green fluorescent protein (*GFP*) gene under the control of the *lac* promoter. Replotted from [6]

Two proteins are involved in the regulation of *lac* operon transcription: one is an activator protein called catabolite activator protein (CAP), or cAMP receptor protein (CRP),  and the other is a repressor called lac repressor. The lac repressor is encoded by the *lacI* gene, which is located near the other *lac* genes but is transcribed by its own promoter. The gene encoding CAP is located elsewhere on the bacterial chromosome and is not linked to the *lac* genes. Both CAP and the lac repressor are DNA-binding proteins, and each binds to a specific site on DNA at or near the *lac* promoter (Fig. 5.4). TMG can bind and inhibit the repressor LacI, whereas cAMP binds to CRP and triggers the activation of the *lac* genes. The concentration of cAMP drops in response to the uptake of various carbon sources, including glucose and lactose; glucose uptake also interferes with LacY activity, leading to the exclusion of TMG [7]. Thus, CAP mediates the effect of glucose, whereas the lac repressor mediates the lactose signal. The lac repressor can bind DNA and repress transcription only in the absence of lactose. CAP can bind DNA and activate the *lac* genes only in the absence of glucose. The combined effect of these two regulators ensures that the genes are expressed at significant levels when lactose is present and glucose is absent.

Figure 5.4 shows the lactose utilization network. The uptake of TMG inhibits the lac repressor and induces the synthesis of LacY, which in turn promotes further TMG uptake, resulting in a positive feedback loop. In experiments, one can measure the response of single cells, initially in a given state of *lac* expression, to exposure to various combinations of glucose and TMG levels. It is crucial to use cells with well-defined initial states, either uninduced or fully induced, because the bistable response is expected to depend on its history.

Experiments found that in the absence of glucose, the *lac* operon is not induced at low TMG concentration ($<3$ $\mu$M) and fully induced at high TMG concentrations ($>30$ $\mu$M) regardless of the cell's history. Between these switching thresholds, however, the cell response is hysteretic: TMG levels must exceed 30 $\mu$M to turn on initially uninduced cells but must drop below 3 $\mu$M to turn off initially induced cells [6]. When the TMG level approaches the boundaries of this bistable region, stochastic mechanisms cause growing numbers of cells to switch from their initial states, resulting in a bimodal distribution of green fluorescence levels.

## 5.3.2   Mathematical Formulation

In the lactose utilization network (Fig. 5.4), the *lac* gene regulatory network includes three genes (*lacZ*, *lacY*, and *lacA*), and three molecules (TMG, LacI, and LacY). LacY catalyses the uptake of TMG, which induces further expression of LacY through LacI. To simplify the model and focus on the core regulation, we considered the concentrations of TMG and LacY in a cell.

Let $x$ be the intracellular TMG concentration and $y$ be the concentration of LacY proteins. The dynamics of the *lac* operon are formulated as

$$
\begin{cases}
\dfrac{R}{R_T} = \dfrac{1}{1 + (x/x_0)^n} \\[2mm]
\tau_x \dfrac{dx}{dt} = \beta y - x \\[2mm]
\tau_y \dfrac{dy}{dt} = \alpha \dfrac{1}{1 + R/R_0} - y.
\end{cases}
\tag{5.1}
$$

Here, $R$ represents the concentration of active LacI proteins, and $R_T$ represents the total LacI concentration. The active LacI level depends on the intracellular TMG concentration through a Hill-type function with a Hill coefficient $n$. The LacY production rate is given by a Michaelis-Menten function of $R$, with the maximum production rate $\alpha$. When $R = R_T$, the production rate reaches the minimum level $\alpha/\rho$, where $\rho = 1 + R_T/R_0$ represents the ability of LacI to bind to and repress *lac* gene transcription. Here, $R_0$ is a constant for the LacI activity level when LacY production reaches $\alpha/2$, $\tau_y$ gives the average lifetime of LacY, i.e., the time constant of LacY degradation, and similarly, $\tau_x$ is the time constant of TMG degradation. The TMG uptake rate is proportional to the LacY concentration with the coefficient $\beta$. We can take the concentrations as relative levels (only the relative levels are important) and hence set $x_0 = 1$ for simplicity.

### 5.3.3   Steady State Analysis

To study the cell response to variance in the extracellular TMG concentration, we solved the steady state of (5.1) to find the relation between *lac* gene expression (LacY) and the TMG level. Letting the derivatives with respect to $t$ equal zero, we obtain the steady state given by the following algebraic equation:

$$
y = \alpha \frac{1 + (\beta y)^n}{\rho + (\beta y)^n}.
\tag{5.2}
$$

Here, the parameters $\rho$, $\alpha$ and $\beta$ (all positive real numbers) are dependent on the external signals of glucose and TMG concentrations. Specifically, the parameter $\beta$ depends on the TMG concentration $T$ through $\beta = 0.123 \times T^{0.6}$ [6].

We denote the right-hand side of (5.2) as $v(y)$. To understand the meaning of $v(y)$, let $dx/dt = 0$, and we have $x = \beta y$ from (5.1). Substituting $x = \beta y$ into the equation for $y$, the function $v(y)$ gives the production rate of LacY protein at the steady state.

If $\rho < 1$, $v(y)$ is a decreasing function, and (5.2) has only one positive root for any parameter $\beta > 0$; hence, there is no bistability.

In the case of the *lac* operon, we have $\rho = 1 + R_T/R_0 > 1$, and the production rate $v(y)$ is an increase function of the LacY concentration, which means a positive feedback loop and possible bistability.

In Eq. (5.2), if $n = 1$, there is only one root for all parameters $\rho$, $\alpha$, and $\beta$ and hence, there is no bistability. Thus, to obtain bistability, as seen in experiments, we should have $n > 1$, i.e., there is cooperation when TMG binds to and inhibits LacI activity. Here, we take $n = 2$.

If $n = 2$, Eq. (5.2) may have one, two, or three positive roots depending on the parameter values. In particular, bifurcation occurs in the critical situation when the equation has two roots. Thus, to achieve bifurcation, we find the condition in which (5.2) has two roots. From (5.2), it is easy to obtain the cubic equation

$$y^3 - \alpha y^2 + (\rho/\beta^2)y - (\alpha/\beta^2) = 0. \tag{5.3}$$

Assuming that the above equation has a multiple root $a$ and another root $\theta a$ ($\theta > 0$), we can rewrite the equation as

$$(y - a)(y - a)(y - \theta a) = y^3 - (2 + \theta)ay^2 + (1 + 2\theta)a^2 y - \theta a^3 = 0. \tag{5.4}$$

Hence, comparing the coefficients of (5.3) and (5.4), we obtain

$$\begin{cases} \alpha = (2 + \theta)a \\ \rho/\beta^2 = (1 + 2\theta)a^2 \\ \alpha/\beta^2 = \theta a^3, \end{cases}$$

which gives

$$\begin{cases} \rho = (1 + 2\theta)(1 + 2/\theta) \\ \alpha\beta = (2 + \theta)^{3/2}/\theta^{1/2} \end{cases} \tag{5.5}$$

after $a$ is eliminated. When $\theta$ varies from 0 to $+\infty$, Eq. (5.5) describes a curve in the $(\rho, \alpha\beta)$ plane, which gives the bifurcation curve corresponding to the boundary for bistability (Fig. 5.5a).

When we fix the parameters $\alpha$ and $\rho$ and vary the parameter $\beta$ (through the external TMG concentration $T$), the steady-state LacY concentration can be obtained by solving Eq. (5.2). Figure 5.5b shows the relation between LacY and TMG concentration at the steady state. There are two critical concentrations $T_1$ and $T_2$ ($T_1 < T_2$): when $T < T_1$, there is only one low-level LacY state (uninduced), and when $T > T_2$, the system has only a high-level LacY state (fully induced). When $T_1 < T < T_2$, the system has three steady states. However, not all states can be observed experimentally, and only the stable states can be observed. Hence, we need to analyze the stability of the steady states.

To analyze the stability, we denote the steady state as $(x^*, y^*)$, and let

$$x = \tilde{x} + x^*, \quad y = \tilde{y} + y^*.$$

Substituting them into (5.1) and expanding the equation into the first-order approximation, we obtain a linear equation for $\tilde{x}$ and $\tilde{y}$ as

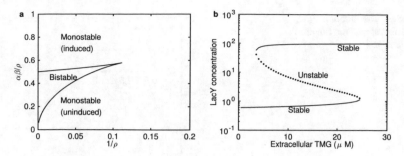

**Fig. 5.5** Bistability of the *lac* operon. **a** The parameter regions to yield bistability in the *lac* operon. **b** Dependence of LacY concentration (corresponding to the green fluorescence level) on the extracellular TMG concentration. Here $\alpha = 100$, $\rho = 167$, and $\beta$ is given by the TMG concentration $T$ through $\beta = 0.123 \times T^{0.6}$. Here, the relations between the parameters $\alpha$, $\beta$, and $\rho$ and the external glucose ($G$) and TMG ($T$) concentrations can be determined through the following method: adjust the glucose and TMG concentrations so that the system is in the critical state; hence, the parameters satisfy (5.5). Next, from the steady-state equation (5.2) and $\beta = y/x$, we obtain three equations for the three parameters $\alpha$, $\beta$, and $\rho$. Finally, we can determine the parameter values (depending on $G$ and $T$) through the observed concentrations of $x$ and $y$. Replotted from [6]

$$\begin{cases} \tau_x \dfrac{d\tilde{x}}{dt} = -\tilde{x} + \beta\tilde{y}, \\[2mm] \tau_y \dfrac{d\tilde{y}}{dt} = a\tilde{x} - \tilde{y}, \end{cases} \tag{5.6}$$

where

$$a = \frac{\partial}{\partial x}\left(\frac{\alpha}{1 + R/R_0}\right)\Bigg|_{x=x^*} = \frac{\partial}{\partial x}\left(\frac{\alpha(1 + x^n)}{\rho + x^n}\right)\Bigg|_{x=x^*}.$$

From the stability theory of ordinary differential equations (Sect. 2.2.2), the steady state is stable if and only if all eigenvalues of the coefficient matrix of (5.6) have negative real parts. The eigenvalues are given by the solutions $\lambda$ of the characteristic equation

$$\begin{vmatrix} \lambda + 1 & -\beta \\ -a & \lambda + 1 \end{vmatrix} = 0,$$

which gives

$$\lambda_{1,2} = -1 \pm \sqrt{a\beta}.$$

Thus, when $a\beta < 1$, the steady state is stable, and when $a\beta > 1$, the steady state is unstable. From (5.2), the steady state $x^*$ satisfies

$$\frac{x}{\beta} - \frac{a(1 + x^n)}{\rho + x^n} = 0,$$

**Fig. 5.6** The function $f(x)$. The inset shows the function at $0 < x < 1$

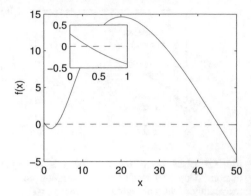

which yields

$$f(x) := \frac{a\beta(1 + x^n)}{\rho + x^n} - x = 0.$$

Thus, the stable condition $a\beta < 1$ is equivalent to the condition $f'(x^*) < 0$. When $T_1 < T < T_2$, the equation $f(x) = 0$ has 3 positive roots (Fig. 5.6). From the function plotted in Fig. 5.6, both steady states with the smallest $x^*$ and the largest $x^*$ are stable, while the state with the intermediate value $x^*$ is unstable. Hence, when $T_1 < T < T_2$, the system has two stable states, i.e., bistability.

Here, we discussed a simple genetic regulatory network with bistability: the regulation of the *lac* operon. From this example, two conditions are required to have bistability, a positive feedback loop and the cooperative effect of LacI regulation. Moreover, the parameters need to take values from a proper region to obtain bistability.

## 5.4   Noise Perturbation and Cell State Switches

We have seen an example of bistability in which a cell can be at one of two potential states under the same conditions. Moreover, the experiments in [6] showed that a cell can switch from one state to the other state when the TMG concentration is near the critical values $T_1$ or $T_2$. This phenomenon of cell state switches is important for a cell to adapt to changing environmental conditions. In the above section, we showed that positive feedback with cooperative regulation is necessary to achieve bistability. Nevertheless, to induce state switches, we need to introduce perturbations to destabilize a state so that the system can jump from one state to the other. In the case of the gene regulation network, fluctuations in gene expression that are induced by either intrinsic noise or external perturbations are possible origins to induce switches. Here, we show the methods to describe the switching processes.

### 5.4.1 Cell State Switches in the Lac Operon

In the above example of the *lac* operon, when the parameters $\alpha$, $\beta$ and $\rho$ are constants, the system approaches a deterministic stable steady state from any given initial state. To describe the cell state switches under external noise perturbation, we introduce a random perturbation to the system parameters.

According to the discussions in Sect. 3.4, when we consider a random perturbation of a parameter, such as the parameter $\alpha$, we perform the replacement

$$\alpha \to \alpha \eta(t), \tag{5.7}$$

where $\eta(t) = e^{\sigma \xi(t)}/\langle e^{\sigma \xi(t)} \rangle$. Here, $\xi(t)$ is a standard white noise, $\sigma$ is a positive constant for the noise strength, and the perturbation factor $\eta(t)$ is a log-normally distributed random number with a mean 1.

Now, we consider the effect of external noise on the *lac* operon and replace $\alpha$ in (5.1) with $\alpha \eta(t)$ to obtain the equation

$$\begin{cases} \tau_x \dfrac{dx}{dt} = \beta y - x, \\ \tau_y \dfrac{dy}{dt} = \alpha \eta(t) \dfrac{1 + (x/x_0)^n}{\rho + (x/x_0)^n} - y. \end{cases} \tag{5.8}$$

To investigate how external noise at work induces state switches, we solve the above equation numerically. First, to simulate the dynamics of a single cell, we take a set of parameters and an initial state (e.g., we take $y = 100$, $x = 100\beta$ for the induced state, or $y = 1$, $x = \beta$ for the uninduced state), solve Eq. (5.8) to $t = 10$, and examine the level of LacY proteins at $t = 10$. The simulation procedure can be repeated with different parameter sets to mimic the situation of a group of cells as in experiments. The distribution of the states of a group of cells shows the statistical behavior of the system under different extracellular TMG conditions. The simulation results are shown in Fig. 5.7. Here, we simulate 100 independent cells for a given parameter (extracellular TMG concentration). From Fig. 5.7, under certain conditions, the cells show state switches under external noise perturbations. We also note the noise strength $\sigma = 2$ in simulations, which implies strong noise with a large variance in $\eta(t)$:

$$\langle \eta(t)^2 \rangle \approx 28.$$

Hence, to observe the phenomena of state switches under noise perturbations, we often need to have strong noise perturbations, and the parameters take values near the bifurcation point.

From the above numerical simulation, it is not easy to see the bidirectional switches under the same parameter values unless the noise perturbation is strong enough. Hence, when we see bidirectional switches in biological systems, there should be some other mechanisms that regulation switch behaviors [8–10].

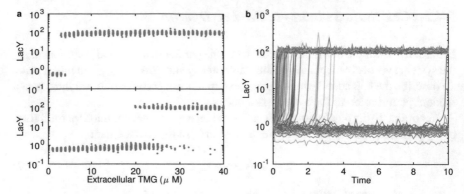

**Fig. 5.7** External noise-induced state switches in the *lac* operon. In simulations, for different TMG concentrations, we took 100 cells, with each cell starting from either the uninduced state or the induced state, and numerically solved the Eq. (5.8) to $t = 10$. **a** The states of each cell at $t = 10$. **b** The trajectories of 10 samples when $T = 25\ \mu$M. Here, we take $\sigma = 2$, $\tau_x = \tau_y = 50$, and other parameters, as shown in Fig. 5.5

## 5.4.2  Lysis-Lysogeny Decision of Bacteriophage λ

### 5.4.2.1  Promoter of the λ Repressor

In the above example of the *lac* operon, positive feedback is involved to ensure bistability, and noise perturbation can induce switches between the two stable states. Here, we introduce an example of lysis-lysogeny decisions in bacteriophage λ viruses through the regulation of the λ repressor [11]. Here, both positive feedback and negative feedback are involved in the regulation network, and bistability in the steady-state protein concentration arises naturally under certain conditions.

Bacteriophage λ (phage lambda) is a virus that infects *E. coli*. Upon infection, the phage can propagate in two ways: lytically or lysogenically. Lytic growth requires the replication of the phage DNA and the synthesis of new coat proteins. These components combine to form new phage particles that are released by lysis of the host cell. Lysogeny involves the integration of phage DNA into the bacteria, where it is passively replicated at each cell division, although it is a legitimate part of the bacterial genome. Lysogeny is extremely stable under normal circumstances, but dormant phages can efficiently switch to lytic growth if the cell is exposed to agents that damage DNA (and thus threaten the host cell's continued existence). This switch from lysogenic to lytic growth is called lysogenic induction. The choice of the development pathway depends on which of the two alternative programs of gene expression is adopted in the cell.

Bacteriophage λ has a 50-kb genome and approximately 50 genes. Most of these genes encode coat proteins and proteins involved in DNA replication, recombination, and lysis. Here, we merely concentrate on a few of them and consider a small area of the genome. This region contains two genes (cI and *cro*) and three promoters

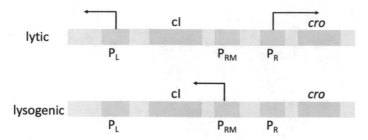

**Fig. 5.8** Transcription in the λ control regions during lytic and lysogenic growth. Arrows indicate the active promoters during lytic and lysogenic growth and the direction of transcription from each promoter

($P_R$, $P_L$ and $P_{RM}$). All the other phage genes are outside of this region and are transcribed directly from $P_R$, $P_L$, or other promoters whose activities are controlled by the products of genes transcribed from $P_R$ or $P_L$. The promoter $P_{RM}$ (promoter for repressor maintenance) transcribes only the cI gene. The promoters $P_L$ and $P_M$ are strong and constitutive; they bind RNAP effectively and direct transcription without help from an activator. In contrast, the promoter $P_{RM}$ is weak and only directs efficient transcription when an activator is bound upstream.

The two arrangements of gene expression are shown in Fig. 5.8: one renders growth lytic, and the other renders growth lysogenic. Lytic growth proceeds when $P_L$ and $P_R$ remain switched on while $P_{RM}$ is kept off. In contrast, lysogenic growth is a consequence of $P_L$ and $P_R$ being switched off and $P_{RM}$ being switched on.

The cI gene encodes a λ repressor, a protein that consists of two domains, an amino-terminal domain and a carboxy-terminal domain, joined by a flexible linker region. The amino-terminal domain contains the DNA-binding region that binds DNA as a dimer; the main dimerization contacts are made between the carboxy-terminal domains. A single dimer recognizes a 17-bp DNA sequence, and each monomer recognizes one half-site. Despite its name, the λ repressor can both activate and repress transcription. When functioning as a repressor, it binds to sites that overlap the promoter, whereas when functioning as an activator, it recruits polymerases through the amino-terminal domain.

Cro, which stands for *c*ontrol of *r*epressor and *o*ther things, only represses transcription by overlapping the promoter region. It is a single-domain protein and binds as a dimer to 17-bp DNA sequences.

There are six operator sites in the λ control region, and the λ repressor and Cro can each bind to any one of the operators. Three of these sites are found in the left-hand control region and three are found in the right-hand control region. We focus on the right-hand site region shown in Fig. 5.9a. The three binding sites in the right operator are called $O_{R1}$, $O_{OR2}$, and $O_{R3}$; these sites can bind either a dimer of repressor or a dimer of Cro with different affinities. The repressor binds $O_{R1}$ tenfold better than it binds $O_{R2}$, and $O_{R3}$ binds the repressor with approximately the same affinity as does $O_{R2}$. Cro, on the other hand, binds $O_{R3}$ with the highest affinity and only binds $O_{R2}$

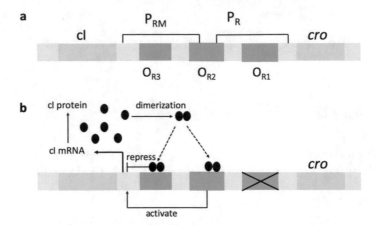

**Fig. 5.9** Binding sites in the promoter region. **a** Relative position and operator sites in $O_R$. **b** Regulatory pathway for the expression of the λ repressor. Here, we consider a mutant system whereby the operator site $O_{R1}$ is absent from the region. The cI gene expresses a repressor (CI), which in turn dimerizes and binds to DNA as a transcription factor. This binding can take place at one of the two binding sites, $O_{R2}$ or $O_{R3}$. Binding at $O_{R2}$ enhances transcription, which takes place downstream of $O_{R3}$, whereas binding at $O_{R3}$ represses transcription, effectively turning off production

and $O_{R1}$ when presented at tenfold higher concentrations [12]. The λ repressor binds DNA cooperatively. In addition to providing the dimerization contacts, the carboxy-terminal domain of the λ repressor mediates interactions between dimers. In this way, two dimers of the repressor can bind cooperatively to adjacent sites on DNA. For example, the repressor at $O_{R1}$ helps the repressor bind to the lower-affinity site $O_{R2}$ by cooperative binding. Consequently, $O_{R3}$ is not bound: the repressor bound cooperatively at $O_{R1}$ and $O_{R2}$ cannot simultaneously make contact with a third dimer at the adjacent site.

### 5.4.2.2  A Model for Repressor Expression

Here, we consider a mutant system whereby the operator site $O_{R1}$ is absent from the region (Fig. 5.9b) and introduce a mathematical model for lysis-lysogeny pathway dynamics [11]. In this pathway, the cI gene expresses a repressor (CI), which in turn dimerizes and binds to DNA to regulate its own transcription. The bindings can take place at either $O_{R2}$ or $O_{R3}$. Binding at $O_{R2}$ enhances transcription, whereas binding at $O_{R3}$ represses transcription, effectively turning off production. These reactions can be described by the following differential equations.

To establish the dynamical model, we divided the reactions into two categories: fast and slow. Fast reactions include the binding and dissociation of molecules at a time scale of a few seconds and are assumed to be in equilibrium with respect to slow reactions, which are on the order of minutes. Let $X$, $X_2$, and $D$ denote the repressor, repressor dimer, and DNA promoter site, respectively; then, we can write the equilibrium reactions

$$2X \underset{k_{-1}}{\overset{k_1}{\rightleftharpoons}} X_2,$$

$$D + X_2 \underset{k_{-2}}{\overset{k_2}{\rightleftharpoons}} DX_2,$$

$$D + X_2 \underset{k_{-3}}{\overset{k_3}{\rightleftharpoons}} DX_2^*,$$

$$DX_2 + X_2 \underset{k_{-4}}{\overset{k_4}{\rightleftharpoons}} DX_2 X_2,$$

where the $DX_2$ and $DX_2^*$ complexes denote binding to the $O_{R2}$ or $O_{R3}$ sites, respectively, $DX_2X_2$ denotes binding to both sites, and $k_i$ and $k_{-i}$ are reaction rates. In the discussions below, we denote $K_i = k_i/k_{-i}$ ($i = 1, 2, 3, 4$) for the forward equilibrium constants, and let $K_3 = \sigma_1 K_2$ and $K_4 = \sigma_2 K_2$, so $\sigma_1$ and $\sigma_2$ represent binding strengths relative to the dimer-$O_{R2}$ strength. Usually, it is difficult to measure the reaction rates directly; however, the equilibrium constants can be obtained by measuring the concentrations of complexes and reactants at the equilibrium state.

The slow reactions include transcription (mRNA production) and translation (protein production). Only when the repressor dimer binds to $O_{R2}$ is the transcription process active. Here, we assume that on average, $n$ protein molecules are produced per mRNA transcript. Hence, the production and degradation of proteins are represented by reaction channels (here, mRNA production and degradation are omitted)

$$DX_2 \overset{k_t}{\longrightarrow} DX_2 + nX,$$

$$X \overset{k_d}{\longrightarrow} \emptyset.$$

These reactions are one directional. Here, we omit the effect of RNAP by assuming that there are enough RNAPs.

Letting $X = [X]$, $Y = [X_2]$, $D = [D]$, $U = [DX_2]$, $V = [DX_2^*]$, and $Z = [DX_2X_2]$ be the concentrations of the complexes. We can write down a rate equation describing the evolution of the concentration of the repressor:

$$\frac{dX}{dT} = -2k_1 X^2 + 2k_{-1} Y + nk_t U - k_d X + r. \tag{5.9}$$

The parameter $r$ is the basal rate of the production of CI, i.e., the expression rate of the cI gene in the absence of a transcription factor.

Equation (5.9) is not closed, and we need to write the variables $Y$ and $U$ as a function of $X$ to obtain a closed-form equation. To this end, we apply the quasi-steady-state assumption to the fast reactions and write the algebraic expressions

$$
\begin{aligned}
Y &= K_1 X^2, \\
U &= K_2 D Y = K_1 K_2 D X^2, \\
V &= \sigma_1 K_2 D Y = \sigma_1 K_1 K_2 D X^2, \\
Z &= \sigma_2 K_2 U Y = \sigma_2 (K_1 K_2)^2 D X^4.
\end{aligned}
\tag{5.10}
$$

Moreover, the total number of promoters is a constant, denoted by $D_T$ for the concentration, and hence

$$
D_T = D + U + V + Z = D(1 + (1 + \sigma_1) K_1 K_2 X^2 + \sigma_2 K_1^2 K_2^2 X^4).
\tag{5.11}
$$

From (5.10) and (5.11), we can express $Y$ and $U$ through $X$ as

$$
\begin{aligned}
Y &= K_1 X^2, \\
U &= \frac{K_1 K_2 X^2}{1 + (1 + \sigma_1) K_1 K_2 X^2 + \sigma_2 K_1^2 K_2^2 X^4}.
\end{aligned}
$$

Substituting $Y$ and $U$ into (5.9), we obtain the following equation for the concentration of cI proteins:

$$
\frac{dX}{dT} = \frac{n k_t K_1 K_2 D_T X^2}{1 + (1 + \sigma_1) K_1 K_2 X^2 + \sigma_2 K_1^2 K_2^2 X^4} - k_d X + r.
\tag{5.12}
$$

Equation (5.12) includes 9 parameters. Before further analysis, we need to perform nondimensionalization. Nondimensionalization is the partial or full removal of units from an equation involving physical quantities by a suitable substitution of variables. This technique can simplify and parameterize problems where measured units are involved. It is closely related to dimensional analysis. In engineering and science, dimensional analysis is an important step that analyzes the relationship between different physical quantities (such as length, mass, time, and electric change) and units of measure (such as miles vs. kilometers or pounds vs. kilograms) and tracks these dimensions as calculations or comparisons. The conversion of units from one dimensional unit to another is often easier within the metric or SI system than in others due to the regular 10-base in all units.

To perform nondimensionalization, we first need to perform dimensional analysis to list all variables and parameters and their dimensions. Next, we need to select a reference for each independent dimension (such as concentration, length, time, and mass). Finally, we express the variables and parameters through the references and obtain the nondimensional variables and parameters. The most important step in nondimensionalization is to select proper references, and there is no standard method to do this. One may have to determine the reference variables according to the system under study and the problem concerned.

Now, considering Eq. (5.12), let $M$ and $T$ denote the units of concentration and time, respectively. All concentrations with dimensions $M$, $\sigma_1$ and $\sigma_2$ are nondimensional parameters. The dimensions of other parameters are (hereafter we write $[\cdot]$ as the dimension)

$$[K_1] = [K_2] = M^{-1}, [k_t] = T^{-1}, [k_d] = T^{-1}, [r] = MT^{-1}. \tag{5.13}$$

Hence, $(\sqrt{K_1 K_2})^{-1}$ has a dimension of concentration $M$, and $k_d^{-1}$ has a dimension of time $T$. Thus, we take $(\sqrt{K_1 K_2})^{-1}$ and $k_d^{-1}$ as the references for concentration and time, respectively. Therewith, the nondimensional variables are

$$x = X/(\sqrt{K_1 K_2})^{-1}, \quad t = T/k_d^{-1}.$$

Now, after substitutions, we can write the nondimensional equation

$$\frac{dx}{dt} = \frac{\alpha x^2}{1 + (1 + \sigma_1)x^2 + \sigma_2 x^4} - x + \gamma. \tag{5.14}$$

Here, $a$ and $\gamma$ are nondimensional parameters

$$\alpha = nk_t D_T \sqrt{K_1 K_2}/k_d, \quad \gamma = r\sqrt{K_1 K_2}/k_d. \tag{5.15}$$

In Eq. (5.14), the parameter $\alpha$ represents the ability of $\lambda$ repressor autoregulation by binding to the promoter binding sites, which gives the maximum protein production rate with respect to the basal level, and the parameter $\gamma$ represents the basal production level. The parameters $\sigma_i$ are the relative affinities of different complexes. Experimental findings suggest that $\sigma_1 \sim 1, \sigma_2 \sim 5$ [11]. Hence, in Eq. (5.14), we take $\sigma_1 = 1$ and $\sigma_2 = 5$, and $\alpha$ and $\gamma$ are main the control parameters that regulate the production or $\lambda$ repressor CI. Next, we analyze the effects of varying these two parameters.

### 5.4.2.3 Steady State Analysis

To study the steady state of (5.14), let

$$f(x) = \frac{\alpha x^2}{1 + (1 + \sigma_1)x^2 + \sigma_2 x^4} + \gamma;$$

the steady state is given by the equation

$$f(x) = x$$

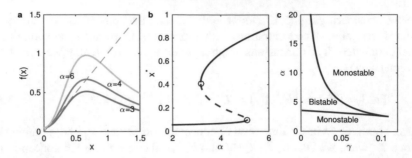

**Fig. 5.10** Bifurcation diagram of $\lambda$ repressor expression. **a** Plots of the functions $f(x)$ (solid lines) and $x$ (dashed line). Here, $\gamma = 0.05$, and $\alpha$ takes values (from top to bottom) of 6.0, 4.0, and 3.0. **b** Dependence of the steady state $x^*$ on the model parameter $\alpha$. The solid line represents the stable steady state, and the dashed line represents unstable steady state. The circles 'o' represent the two saddle node bifurcation points. **c** Bifurcation curve at the $\alpha$-$\gamma$ plane. Here, $\sigma_1 = 1, \sigma_2 = 5$

or the algebraic equation

$$\alpha x^2 + (\gamma - x)(1 + (1 + \sigma_1)x^2 + \sigma_2 x^4) = 0. \tag{5.16}$$

The equation cannot be solved exactly. The numerical solution shows that for a given $\gamma$, the number of roots depends on the parameter $\alpha$ (Fig. 5.10a). There are two critical values $\alpha_1 < \alpha_2$, so that there is only one steady state when either $\alpha < \alpha_1$ or $\alpha > \alpha_2$ is stable; when $\alpha_1 < \alpha < \alpha_2$, there are three steady states, two of which are stable, and the other (intermediate $x$ value) is unstable (Fig. 5.10b). Figure 5.10c shows the dependence of the critical values $\alpha_1$ and $\alpha_2$ on $\gamma$, which divide the $\alpha$-$\gamma$ plane into two regions, corresponding to only one stable state (either high or low CI protein concentration) or two stable states. For example, when $\gamma = 0.05$ and $3.2 < \alpha < 5.1$, the system has two stable steady states, either high or low CI protein concentration, corresponding to the two states of either lysogenic or lytic growth, respectively.

When the parameters take values from the region of bistability, external noise perturbation can induce switches between the two states. Here we consider the effects of external perturbation.

### 5.4.2.4   Random Perturbation and Lysis-Lysogeny Switches

From the above analysis, we take $\sigma_1 = 1, \sigma_2 = 5$ and $3.2 < \alpha < 5.1$ and consider the switches between two states under external noise perturbation.

We consider two types of perturbations. When there is a perturbation to the basal protein rate $r$, we replace the coefficient rate $r$ in (5.9) with $r\eta(t)$, where $\eta(t)$, as previously described, is a log-normal random process ($\eta(t) = e^{\sigma \xi(t)}/\langle e^{\sigma \xi(t)} \rangle$). Accordingly, the nondimensional parameter $\gamma$ becomes $\gamma \eta(t)$. Hence, we obtain the following random differential equation:

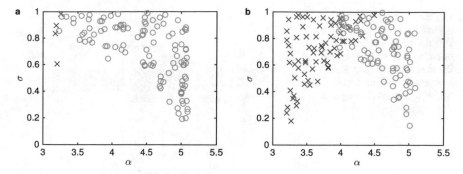

**Fig. 5.11** Switches of the λ repressor under external noise perturbation. Dots show the parameter values $(\alpha, \sigma)$ when switches from high to low ('x') or from low to high ('o') occur. **a** Noise perturbation to the basal expression level. **b** Noise perturbation to the protein production rate. The other parameters are the same as those in Fig. 5.10

$$\frac{dx}{dt} = \frac{\alpha x^2}{1 + (1 + \sigma_1)x^2 + \sigma_2 x^4} - x + \gamma \eta(t). \tag{5.17}$$

If we consider an external perturbation to the expression process, we need to replace the protein production rate $k_t$ with $k_t \eta(t)$. Accordingly, the nondimensional parameter $\alpha$ becomes $\alpha \eta(t)$ in the nondimensional equation, which gives

$$\frac{dx}{dt} = \frac{\alpha \eta(t) x^2}{1 + (1 + \sigma_1)x^2 + \sigma_2 x^4} - x + \gamma. \tag{5.18}$$

Equations (5.17) and (5.18) give formulations when there are noise perturbations to the system parameters. Let $3.2 < \alpha < 5.1$ and $0 < \sigma < 1$; we solve the equations numerically to investigate the probability of cell fate switches. Figure 5.11 shows the parameter value $(\alpha, \sigma)$ so that state switches occur when the system starts from either low or high cI gene expression. Model simulations show that switches can occur when the external noise is strong enough. When the noise perturbation is added to the basal level, it is easier to see the switches from low to high levels of expression than from high to low levels of expression; when the noise is added to the production rate, both direction switches can happen under the same noise strength, and bidirectional switches may occur when the noise strength is large enough ($\sigma > 0.7$ and $\alpha \approx 4.1$).

### 5.4.3   Genetic Toggle Switch Induced by Intrinsic Noise

We have shown examples of switches induced by external noise. In these two examples, cooperative positive feedback is required to produce bistability, and noise perturbation can destabilize one stable state and induce the system to switch to the other

**Fig. 5.12** Illustration of a mutual repression circuit. Protein $A$ binds to the operator of gene $B$ to repress the expression of gene $B$, and protein $B$ binds to the operator of gene $A$ to repress the expression of gene $A$

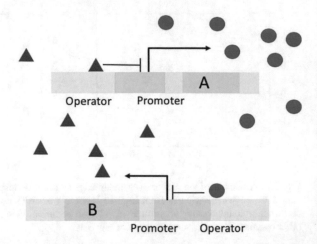

state. Here, we introduce an example of a genetic toggle switch without cooperative binding that can be induced by intrinsic noise [13].

Consider a mutual repression circuit of two genes ($A$ and $B$) (Fig. 5.12). Protein A binds to the operator of gene $B$ to repress the expression of gene $B$, and protein B binds to the operator of gene $A$ to repress the expression of gene $A$. The gene circuit is described by the following biochemical reactions:

$$A + D_B \underset{\alpha_1}{\overset{\alpha_0}{\rightleftharpoons}} D_B^*,$$
$$B + D_A \underset{\alpha_1}{\overset{\alpha_0}{\rightleftharpoons}} D_A^*,$$
$$D_A \overset{g_A}{\longrightarrow} A,$$
$$D_A \overset{g_B}{\longrightarrow} B,$$

(5.19)

where $D_A$ and $D_B$ are the free promoters of gene $A$ and gene $B$, respectively, and $D_A^*$ and $D_B^*$ are repressed promoters when the proteins are bound to the operators. Here, we assume that the reaction rates between proteins and the DNA sequence are the same for the two proteins. The genetic circuit forms positive feedback through mutual repression, and there is no cooperation between the proteins; hence, there is no bistability.

Let $[A]$ and $[B]$ represent the concentrations of proteins A and B, $[D_A^*]$ represent the fraction of operator A that is binding with protein B, and $[D_B^*]$ represent the fraction of operator B that is binding with protein A. Here, we assume that all genes have a single copy, and hence, $0 \leq [D_A^*], [D_B^*] \leq 1$, and $[D_A] = 1 - [D_A^*]$, $[D_B] = 1 - [D_B^*]$ are the fractions of free operators. The chemical rate equations for the above biochemical reactions are given by

$$\begin{cases} \dfrac{d[A]}{dt} = g_A(1 - [D_A^*]) - d_A[A] - \alpha_0[A](1 - [D_B^*]) + \alpha_1[D_B^*], \\[2mm] \dfrac{d[B]}{dt} = g_B(1 - [D_B^*]) - d_B[B] - \alpha_0[B](1 - [D_A^*]) + \alpha_1[D_A^*], \\[2mm] \dfrac{d[D_A^*]}{dt} = \alpha_0[B](1 - [D_A^*]) - \alpha_1[D_A^*], \\[2mm] \dfrac{d[D_B^*]}{dt} = \alpha_0[A](1 - [D_B^*]) - \alpha_1[D_B^*], \end{cases} \qquad (5.20)$$

where $g_X(X = A, B)$ are the maximum protein production rates, $d_X$ are degradation rates, $\alpha_0$ represents the binding rate between the proteins and the DNA, and $\alpha_1$ represents the dissociation rate.

From Eq. (5.20), there exists only one steady state, and the state is stable. Specifically, let $g_A = g_B = g$ and $d_A = d_B = d$, and define $k = \alpha_0/\alpha_1$ as the equilibrium binding constant between proteins and DNA. The steady state of (5.20) is given by

$$[A] = [B] = \frac{\sqrt{1 + 4kg/d} - 1}{2k}. \qquad (5.21)$$

Hence, while we describe the mutual repression circuit through the deterministic differential equation, there is only one stable state, which would not show bistability or switches.

Now, to consider the effect of intrinsic noise, we apply stochastic simulation based on the biochemical reactions in (5.19), which gives the random dynamics of the concentrations of proteins A and B (Fig. 5.13). From stochastic simulation, the system shows different dynamics with different parameters. When the equilibrium constant $k = \alpha_0/\alpha_1$ is very small, the protein numbers fluctuate around the mean level; however, when the constant $k$ is large, the protein numbers switch between low and high levels and show the typical dynamics of toggle switches. These results show that intrinsic noise can induce bimodal distribution and switches in genetic circuits without cooperative binding.

To analyze the role of fluctuations, we calculate the probability distribution

$$P(N_A, N_B) = \sum_{D_A^*, D_B^* = 0}^{1} P(N_A, N_B, D_A^*, D_B^*). \qquad (5.22)$$

We use the symmetric parameters $g = 0.05$ s$^{-1}$ and $d = 0.005$ s$^{-1}$ and vary the parameters $\alpha_0$ and $\alpha_1$ to represent different repression conditions. The population of free proteins depends only on the ratio $k = \alpha_0/\alpha_1$. From numerical simulations, the equilibrium constant $k$ is essential for the probability density function $P(N_A, N_B)$ (Fig. 5.14). Under conditions in which $k$ is small, the binding rates between protein and DNA are small; hence, the repression is weak, and the promoter sites are empty

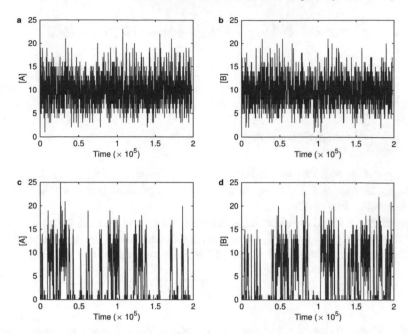

**Fig. 5.13** Stochastic simulation of the protein number dynamics. Here, we present the trajectories under weak repression (**a–b**) and strong repression (**c–d**). The parameters are $g_A = g_B = 0.05\,\text{s}^{-1}$, $d_A = d_B = 0.005\,\text{s}^{-1}$; weak repression ($k = 0.005$): $\alpha_0 = 0.005, \alpha_1 = 1.0$; strong repression ($k = 50$): $\alpha_0 = 1.0, \alpha_1 = 0.02$

most of the time. Thus, the steady steady exhibits the coexistence of $A$ and $B$ proteins in the cells. In this case, $P(N_A, N_B)$ exhibits a single symmetric peak, which corresponds to the stable steady state given by (5.21) from the deterministic equation (Fig. 5.14a). When the equilibrium constant $k$ is large, the affinities between proteins and DNA are strong, which yields strong repression. The distribution $P(N_A, N_B)$ exhibits three peaks: the $A$ population is suppressed in one peak, and the $B$ population is suppressed in another peak, as expected for a bistable system (Fig. 5.14b). The third peak appears near the origin, in which both populations of free proteins diminish. This peak represents a dead-lock situation caused by the fact that both $A$ and $B$ repressors are bound simultaneously.

From the above discussions, we can describe a genetic circuit through mathematical models of either stochastic simulation or deterministic ordinary differential equations, and the two models may not always give the same conclusions. In Chap. 3, we have seen that the chemical ensemble average equation is not equivalent to the reaction rate equation when there are high-order reactions. Hence, when there are high-order chemical reactions, the ordinary differential equation model can sometimes give results that differ from those of stochastic simulations. For example, Fig. 5.15 shows that the reaction rate equation model can give a correct ensemble

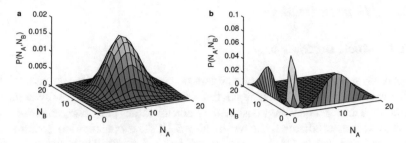

**Fig. 5.14** Probability density function $P(N_A, N_B)$ under different conditions. **a** Weak repression ($\alpha_0 = 0.005, \alpha_1 = 1.0, k = 0.005$). **b** Strong repression ($\alpha_0 = 1.0, \alpha_1 = 0.02, k = 50$). Replotted from [13]

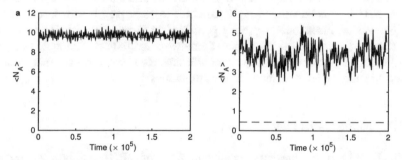

**Fig. 5.15** Dynamics of the average protein $A$ numbers under different conditions. **a** Weak repression ($k = 0.005$). **b** Strong repression ($k = 50$). The other parameters are the same as those in Fig. 5.13. The dashed line shows the steady state given by (5.21)

average only when the repression is weak. When the repression is strong, the steady-state solution obtained from the ordinary differential equation is much lower than that obtained from the stochastic simulation.

## 5.5 Negative Genetic Circuit and Oscillatory Dynamics

In the previous sections, we introduced examples of positive feedback genetic regulation networks that are associated with bistability; noise perturbations that can induce random switches between stable states. In biological systems, regular oscillations are important for many cellular behaviors, such as circadian rhythm, the cell cycle, and various biological clocks. In this section, we introduce examples of genetic networks that produce oscillatory dynamics.

### 5.5.1   Atkinson Oscillator

#### 5.5.1.1   Model Description

The Atkinson oscillator is an artificial genetic network that includes both positive feedback and negative feedback regulations and exhibits either toggle switching or oscillatory behavior under well-designed parameter values [14]. The genetic network of the Atkinson oscillator is shown in Fig. 5.16. The construction contains two genes, *glnG* and *lacI*, and their promoters, *glnAp2* and *glnKp*. The enhancer-binding protein NRI is encoded by the *glnG* gene; the phosphorylated form of the protein (NRI-P) binds to the promoters *glnAp2* and *glnKp* and promotes their transcription. The *glnAp2* promoter is repressed by LacI, which binds to two *lac* operator sites. Hence, NRI-P promotes the *glnAp2* promoter to form positive feedback, and the *glnKp* promoter fuses to the repressor gene *lacI* to form a negative feedback loop. Experiments found that the above genetic network exhibits either toggle switching or oscillatory behavior under different conditions. Here, we introduce mathematical models to analyze the regulation network dynamics.

#### 5.5.1.2   Mathematical Formulation

Let [lacI] and [LacI] represent the concentrations of mRNA and protein produced by the *lacI* gene and let [nri] and [NRI] represent the concentrations of mRNA and protein produced by the *glnG* gene. The concentration of the phosphorylated form

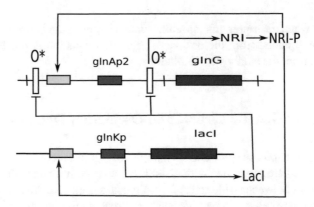

**Fig. 5.16** Illustration of the Atkinson oscillator. The genetic network includes an activator (NRI) and a repressor (LacI). The top construct contains the *glnAp2* promoter fused to the *glnG* gene. Transcription from *glnAp2* requires the phosphorylated form of the enhancer-binding protein NRI (*glnG* product). This promoter is repressed by LacI, which binds to two *lac* operator sites $O^*$. The bottom construct contains the *glnKp* promoter. The *glnKp* promoter also requires NRI-P for activation; however, the enhancer-binding sites are less potent than those at *glnAp2*. Replotted from [14]

of NRI is represented by [NRI-P]. The dynamics of the above genetic network are described by the ordinary differential equation model

$$
\frac{d[\text{lacI}]}{dt} = f_1([\text{NRI-P}]) - \delta_1[\text{lacI}],
$$

$$
\frac{d[\text{LacI}]}{dt} = \lambda_2[\text{lacI}] - \delta_2[\text{LacI}],
$$

$$
\frac{d[\text{nri}]}{dt} = f_2([\text{NRI-P}]) f_3([\text{LacI}]) - \delta_3[\text{nri}], \tag{5.23}
$$

$$
\frac{d[\text{NRI}]}{dt} = \lambda_4[\text{nri}] - \delta_4[\text{NRI}] - k_1[\text{NRI}] + k_{-1}[\text{NRI-P}],
$$

$$
\frac{d[\text{NRI-P}]}{dt} = k_1[\text{NRI}] - k_{-1}[\text{NRI-P}] - \delta_5[\text{NRI-P}].
$$

Here, $\lambda_i$ represents the translation rates, $\delta_i$ represents the degradation/dilution rates, and $k_1$ and $k_{-1}$ represent the reaction rates associated with NRI phosphorylation. The functions $f_i$ describe the regulation of promoter activity through the proteins and are given by Hill-type functions as

$$
f_1([\text{NRI-P}]) = \alpha_{1,0} + \alpha_{1,1} \frac{([\text{NRI-P}]/K_1)^{n_1}}{1 + ([\text{NRI-P}]/K_1)^{n_1}},
$$

$$
f_2([\text{NRI-P}]) = \alpha_{2,0} + \alpha_{2,1} \frac{([\text{NRI-P}]/K_2)^{n_2}}{1 + ([\text{NRI-P}]/K_2)^{n_2}},
$$

$$
f_3([\text{LacI}]) = \alpha_{3,1} \frac{1}{1 + ([\text{LacI}]/K_3)^{n_3}}.
$$

Here, we apply multiple functions for the regulation of NRI-P and LacI binding to the *glnAp2* promoter because the binding sites for the NRI-P and LacI proteins are independent of each other.

The phosphorylation process is fast compared with gene expression, and hence, we can apply the quasi-steady-state assumption to the phosphorylation process so that

$$
\frac{d[\text{NRI-P}]}{dt} = k_1[\text{NRI}] - k_{-1}[\text{NRI-P}] - \delta_5[\text{NRI-P}] \approx 0,
$$

which gives

$$
[\text{NRI-P}] = k_{\text{eq}}[\text{NRI}], \quad k_{\text{eq}} = \frac{k_1}{k_{-1} + \delta_5}. \tag{5.24}
$$

Substituting the above relation into (5.23), we obtain the dynamical equation for the *lacI* and *glnG* products.

To further simplify the model equation, we need to perform nondimensional analysis. First, from the function $f_3([\text{LacI}])$, the EC50 concentration of [LacI] is $K_3$. Hence, we take $K_3$ as the reference for [LacI], i.e.,

$$[\text{LacI}] \sim K_3. \tag{5.25}$$

Here, $\sim$ means 'about the same order as'. Similarly, from the function $f_1([\text{NRI-P}])$, we have

$$[\text{NRI-P}] \sim K_1. \tag{5.26}$$

From (5.24), we have

$$[\text{NRI}] \sim K_1/k_{\text{eq}}. \tag{5.27}$$

Hence, from (5.25)-(5.27), and when we let $d[\text{LacI}]/dt = d[\text{NRI}]/dt = 0$ in (5.23), we obtain

$$[\text{lacI}] \sim \delta_2 K_3/\lambda_2, \quad [\text{nri}] \sim (\delta_4 + \delta_5 k_{\text{eq}})(K_1/k_{\text{eq}})/\lambda_4 \tag{5.28}$$

at the equilibrium state.

From the above discussions, we introduce the following nondimensional variables:

$$x_1 = \frac{[\text{lacI}]}{\delta_2 K_3/\lambda_2}, \quad x_2 = \frac{[\text{LacI}]}{K_3}, \tag{5.29}$$

$$x_3 = \frac{[\text{nri}]}{(\delta_4 + \delta_5 k_{\text{eq}})(K_1/k_{\text{eq}})/\lambda_4}, \quad x_4 = \frac{[\text{NRI}]}{K_1/k_{\text{eq}}}, \quad \tilde{t} = \delta_2 t. \tag{5.30}$$

Here, we used the lifetime of the LacI protein as the reference for the time. We note that there are different methods to perform nondimensional analysis for a problem; one can try other ways to seek proper references for the quantities under consideration.

Based on the above nondimensional variables, we can write the equation with the following nondimensional parameters:

$$\beta_1 = \frac{\delta_1}{\delta_2}, \quad \beta_3 = \frac{\delta_3}{\delta_2}, \quad \beta_4 = \frac{(\delta_4 + \delta_5 k_{\text{eq}})}{\delta_2},$$

$$\alpha_1 = \frac{\alpha_{1,1}}{\alpha_{1,0}}, \quad \alpha_2 = \frac{\alpha_{2,1}}{\alpha_{2,0}}, \quad a = \frac{K_2}{K_1},$$

$$\lambda_1 = \frac{\alpha_{1,0}\lambda_2}{\delta_1 \delta_2 K_3}, \quad \lambda_3 = \frac{\lambda_4 k_{\text{eq}}\alpha_{2,0}\alpha_{3,1}}{\delta_3(\delta_4 + \delta_5 k_{\text{eq}})K_1}.$$

Now, the above model equations can be rewritten as the following form of nondimensional equations (here, we write $t$ for the nondimensional time variable):

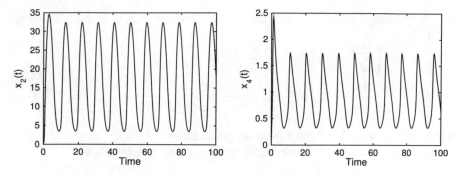

**Fig. 5.17** Simulation results of the Atkinson oscillator. Here, the parameters are $\beta_1 = \beta_3 = 30.0$, $\beta_4 = 1.0$, $\lambda_1 = \lambda_3 = 2.0$, $\alpha_1 = \alpha_2 = 20.0$, $\alpha_3 = 1.0$, $a = 1.0$, $n_1 = 4$, $n_2 = 5$, $n_3 = 1$, and the initial condition is $x_i(0) = 0.0$. Replotted from [14]

$$\frac{dx_1}{dt} = \beta_1 \left[ \lambda_1 \left( 1 + \alpha_1 \frac{x_4^{n_1}}{1 + x_4^{n_1}} \right) - x_1 \right],$$

$$\frac{dx_2}{dt} = x_1 - x_2,$$

$$\frac{dx_3}{dt} = \beta_3 \left[ \lambda_3 \left( 1 + \alpha_2 \frac{(x_4/a)^{n_2}}{1 + (x_4/a)^{n_2}} \right) \frac{1}{1 + x_2^{n_3}} - x_3 \right], \qquad (5.31)$$

$$\frac{dx_4}{dt} = \beta_4 (x_3 - x_4).$$

Based on Eq. (5.31) and the proper parameter values, numerical simulations show the oscillatory dynamics of mRNA and protein numbers (Fig. 5.17).

### 5.5.1.3 Existence of Periodic Solutions

To analyze the existence of periodic solutions, we further simplify the above equations to a planar system. In the above equations, we apply the quasi-steady-state assumption to the transcription process to obtain

$$dx_1/dt = dx_3/dt = 0,$$

which yields

$$x_1 = \lambda_1 \left( 1 + \alpha_1 \frac{x_4^{n_1}}{1 + x_4^{n_1}} \right),$$

$$x_3 = \lambda_3 \left( 1 + \alpha_2 \frac{(x_4/a)^{n_2}}{1 + (x_4/a)^{n_2}} \right) \frac{1}{1 + x_2^{n_3}}.$$

Hence, we simplify Eq. (5.31) into a second-order differential equation (planar system)

$$\begin{cases} \dfrac{dx_2}{dt} = \lambda_1 \left(1 + \alpha_1 \dfrac{x_4^{n_1}}{1 + x_4^{n_1}}\right) - x_2 \\[3mm] \dfrac{dx_4}{dt} = \beta_4 \left(\lambda_3 \left(1 + \alpha_2 \dfrac{(x_4/a)^{n_2}}{1 + (x_4/a)^{n_2}}\right) \dfrac{1}{1 + x_2^{n_3}} - x_4\right) \end{cases} \tag{5.32}$$

Here, we take $n_3 = 1$ and analyze the above equation.

First, letting $dx_2/dt = 0$ and $dx_4/dt = 0$, we obtain two nullclines:

$$x_2\text{-nullcline:}\quad x_2 = \lambda_1 \left(1 + \alpha_1 \frac{x_4^{n_1}}{1 + x_4^{n_1}}\right),$$

and

$$x_4\text{-nullcline:}\quad x_2 = \frac{\lambda_3}{x_4} \left(1 + \alpha_2 \frac{(x_4/a)^{n_2}}{1 + (x_4/a)^{n_2}}\right) - 1.$$

The solutions of (5.32), $x_2(t)$ and $x_4(t)$, have a nullcline when the orbit $(x_2(t), x_4(t))$ intersects with the above two nullclines in the phase plane. The intersections of the two nullclines give the steady state of the system (5.32). From the above expressions for the nullclines, in the $(x_4, x_2)$ phase plane, the $x_2$-nullcline increases with respect to $x_4$, while $x_2$ in the $x_4$-nullcline approaches infinity when $x_4 \to 0$ and approaches $-1$ when $x_4 \to +\infty$. Hence, there is at least one positive steady state, and two or three positive steady states are possible for different values of $a, \alpha_1$ or $\alpha_2$ (Fig. 5.18).

To analyze the stability of a steady state $(x_2^*, x_4^*)$, we calculate the trace of the coefficient matrix of the linearized dynamics around the state (please refer to Sect. 2.2.2):

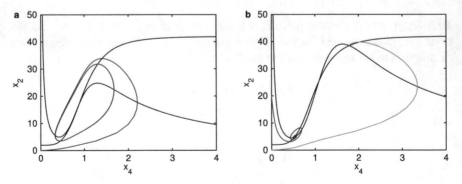

**Fig. 5.18** Phase plane portrait for Eq. (5.32). The black solid lines are nullclines, and the red or green solid lines are representative orbits of the system. **a** Oscillation, for $a = 1.0$, $\alpha_2 = 20$. **b** Bistability, for $a = 1.25$, $\alpha_2 = 40$. The other parameters are the same as those in Fig. 5.17. Replotted from [14]

$$\tau = \frac{\partial}{\partial x_2} \left[ \lambda_1 \left( 1 + \alpha_1 \frac{x_4^{n_1}}{1 + x_4^{n_1}} \right) - x_2 \right] \Bigg|_{(x_2^*, x_4^*)}$$

$$+ \frac{\partial}{\partial x_4} \left[ \beta_4 \left( \lambda_3 \left( 1 + \alpha_2 \frac{(x_4/a)^{n_2}}{1 + (x_4/a)^{n_2}} \right) \frac{1}{1 + x_2^{n_3}} - x_4 \right) \right] \Bigg|_{(x_2^*, x_4^*)}$$

$$= -1 - \beta_4 + n_2 \beta_4 \left( 1 - \frac{\lambda_3}{x_4^*(1 + x_2^{*n_3})} \right).$$

From previous discussions, when

$$\frac{(n_2 - 1)\beta_4 - 1}{n_2 \beta_4} > \frac{\lambda_3}{x_4^*(1 + x_2^{*n_3})},$$

the trace $\tau > 0$, and the steady state is unstable. Here, we note that the steady state $(x_2^*, x_4^*)$ depends on the parameters $a$ and $\alpha_2$. Hence, the condition $\tau > 0$ gives the parameter region in the $\alpha_2$-$a$ plane that exhibits an unstable steady state, as shown in Fig. 5.19b. In particular, when $(\alpha_2, a) = (20, 1.0)$, the system has only one steady state, and the state is unstable. Hence, there is a stable periodic solution (Fig. 5.18a).

In the bifurcation diagram, we are mainly interested in the parameters $a$ and $\alpha_2$. Both parameters are related to the regulation of the *gInAp2* and *gInKp* promoters by phosphorylated NRI. The parameter $a$ represents the ratio between the effective NRI-P concentrations for the two regulations, $\alpha_2$ represents the regulation strength of NRI to its own transcription, and a larger $\alpha_2$ means stronger regulation. Let $0 < \alpha_2 < 100$, $0 < a < 3$ and fix the other parameters. Figure 5.19 gives the dependence between the steady states and the parameters $(\alpha_2, a)$. When $\alpha_2$ is small ($\alpha_2 < 22$), i.e., the positive feedback is weak, there is only one steady state, and the steady state is unstable when $a \approx 1$. Hence, from the Poincaré-Bendixson theorem, there exists a stable periodic

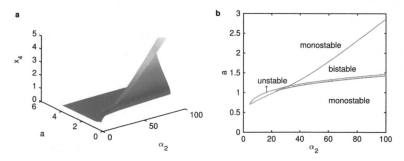

**Fig. 5.19** Bifurcation diagram of the Atkinson oscillator. **a** Dependence of $x_4$ at the steady state on the parameters $\alpha_2$ and $a$. **b** Bifurcation diagram in the $\alpha_2$-$a$ plane. The solid lines show the bifurcation curve between the monostable and bistable steady states, and the dashed curve shows the region where the steady state is unstable (if there are multiple steady states, we mean that the state with the lowest $x_4$ is unstable). The other parameters as the same as those in Fig. 5.17. Replotted from [14]

solution. Here, $a \approx 1$, which means $K_1 \approx K_2$, so the effective concentrations of NRI-P under the two regulations by the *gInAp2* and *gInKp* promoters are approximately the same level. This gives a biological condition to ensure the existence of a stable oscillatory state.

## 5.5.2    Noise Resistance in Genetic Oscillators

We have shown an example of oscillators introduced by the destabilization of the steady state. Here, we introduce an alternative mechanism of oscillators generated by random excitation. In these oscillators, the system is normally at a stable steady state that can be destabilized due to random perturbations. After destabilization, the system deviates from the original steady state, and due to system dissipation, it can eventually return to the original state and wait for the next excitation. Hence, the long-term behavior exhibits oscillatory-like dynamics.

### 5.5.2.1    Model Description and Formulation

Here, we study a minimal model of the circadian clock based on common positive and negative control elements found experimentally. The model is shown in Fig. 5.20; please refer to [15] for details. In this model, two genes, an activator A and a repressor R, are transcribed into mRNA and subsequently translated into protein. The activator A binds to both the $A$ and $R$ promoters to increase their transcription rates. The protein R binds to and sequesters the protein A and therefore acts as a repressor.

The biochemical reactions associated with the genetic circuit are listed in Fig. 5.20, along with the corresponding propensity functions and state-change vectors. Here, $c_i$ in Fig. 5.20 gives the rate constant of the reaction channel $R_i$. We note that in the reaction channel $R_{16}$, the complex breaks into R because of the degradation of A. Thus, $R_{16}$ is not the reversion process of $R_{15}$. Based on these biochemical reactions, the random dynamics of the system can be obtained from stochastic simulation following the Gillespie algorithm (Fig. 5.21).

Now, we establish the deterministic dynamics of the model. Let $D'_A$ and $D_A$ denote the number of activator genes with and without $A$ bound to its promoter, respectively; similarly, $D'_R$ and $D_R$ refer to the repressor promoter; $M_A$ and $M_R$ denote mRNAs of $A$ and $R$; $A$ and $R$ correspond to the activator and repressor proteins; and $C$ corresponds to the inactivated complex formed by $A$ and $R$. The above reactions give the following set of reaction rate equations:

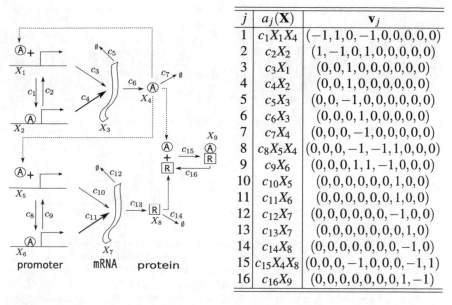

| $j$ | $a_j(\mathbf{X})$ | $\mathbf{v}_j$ |
|---|---|---|
| 1 | $c_1 X_1 X_4$ | $(-1,1,0,-1,0,0,0,0,0)$ |
| 2 | $c_2 X_2$ | $(1,-1,0,1,0,0,0,0,0)$ |
| 3 | $c_3 X_1$ | $(0,0,1,0,0,0,0,0,0)$ |
| 4 | $c_4 X_2$ | $(0,0,1,0,0,0,0,0,0)$ |
| 5 | $c_5 X_3$ | $(0,0,-1,0,0,0,0,0,0)$ |
| 6 | $c_6 X_3$ | $(0,0,0,1,0,0,0,0,0)$ |
| 7 | $c_7 X_4$ | $(0,0,0,-1,0,0,0,0,0)$ |
| 8 | $c_8 X_5 X_4$ | $(0,0,0,-1,-1,1,0,0,0)$ |
| 9 | $c_9 X_6$ | $(0,0,0,1,1,-1,0,0,0)$ |
| 10 | $c_{10} X_5$ | $(0,0,0,0,0,0,1,0,0)$ |
| 11 | $c_{11} X_6$ | $(0,0,0,0,0,0,1,0,0)$ |
| 12 | $c_{12} X_7$ | $(0,0,0,0,0,0,-1,0,0)$ |
| 13 | $c_{13} X_7$ | $(0,0,0,0,0,0,0,1,0)$ |
| 14 | $c_{14} X_8$ | $(0,0,0,0,0,0,0,-1,0)$ |
| 15 | $c_{15} X_4 X_8$ | $(0,0,0,-1,0,0,0,-1,1)$ |
| 16 | $c_{16} X_9$ | $(0,0,0,0,0,0,0,1,-1)$ |

**Fig. 5.20** The biochemical network of the circadian oscillator model. Left: illustration of the regulatory reactions. Right: propensities and state-change vectors of the reaction channels. Replotted from [15]

$$\frac{dD_A}{dt} = \theta_A D'_A - \gamma_A D_A A,$$

$$\frac{dD_R}{dt} = \theta_R D'_R - \gamma_R D_R A,$$

$$\frac{dD'_A}{dt} = \gamma_A D_A A - \theta_A D'_A,$$

$$\frac{dD'_R}{dt} = \gamma_R D_R A - \theta_R D'_R,$$

$$\frac{dM_A}{dt} = \alpha'_A D'_A + \alpha_A D_A - \delta_{M_A} M_A, \tag{5.33}$$

$$\frac{dA}{dt} = \beta_A M_A + \theta_A D'_A + \theta_R D'_R - A(\gamma_A D_A + \gamma_R D_R + \gamma_C R + \delta_A),$$

$$\frac{dM_R}{dt} = \alpha'_R D'_R + \alpha_R D_R - \delta_{M_R} M_R,$$

$$\frac{dR}{dt} = \beta_R M_R - \gamma_C A R + \delta_A C - \delta_R R,$$

$$\frac{dC}{dt} = \gamma_C A R - \delta_A C.$$

Here, the constants $\alpha_X$ and $\alpha'_X$ ($X = A, R$) denote the basal and activated rates of transcription, $\beta_X$ denotes the rates of translation, $\delta_X$ denotes the rates of spontaneous

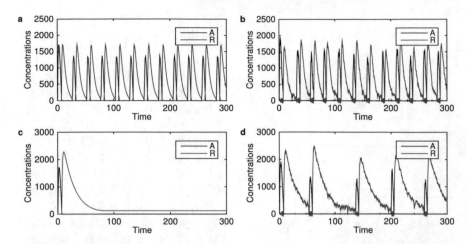

**Fig. 5.21** Oscillations in the repressor protein numbers obtained from numerical simulations of the deterministic (**a** and **c**) and stochastic (**b** and **d**) descriptions of the model. Parameters are $\alpha_A = 50 \text{ h}^{-1}$, $\alpha'_A = 500 \text{ h}^{-1}$, $\alpha_R = 0.01 \text{ h}^{-1}$, $\alpha'_R = 50 \text{ h}^{-1}$, $\beta_A = 40 \text{ h}^{-1}$, $\beta_R = 5 \text{ h}^{-1}$, $\delta_{M_A} = 10 \text{ h}^{-1}$, $\delta_{M_R} = 0.5 \text{ h}^{-1}$, $\delta_A = 1 \text{ h}^{-1}$, $\gamma_A = 1 \text{ mol}^{-1}\text{h}^{-1}$, $\gamma_R = 1 \text{ mol}^{-1}\text{h}^{-1}$, $\gamma_C = 2 \text{ mol}^{-1}\text{h}^{-1}$, $\theta_A = 50 \text{ h}^{-1}$, $\theta_R = 100 \text{ h}^{-1}$, where mol is the number of molecules. The initial conditions are $D_A = D_R = 1$ mol, $D'_A = D'_R = M_A = M_R = A = R = C = 0$, which required that the cell has a single copy of the activator and repressor genes: $D_A = D'_A = 1$ mol and $D_R + D'_R = 1$ mol. In **a** and **b**, we take $\delta_R = 0.05 \text{ h}^{-1}$, and in **c** and **d**, we take $\delta_R = 0.2 \text{ h}^{-1}$. The cellular volume is assumed to be in a state of unity so that the concentration and number of molecules are equivalent. Replotted from [15]

degradation, $\gamma_X$ denotes the rates of the binding of $A$ to other components, and $\theta_X$ denotes the rates of the unbinding of $A$ from those components. The cellular volume is assumed to be in a state of unity so that the concentration and number of molecules are equivalent. Of note, we assume that the complex breaks into $R$ because of the degradation of $A$; therefore, the parameter $\delta_A$ appears twice in the model equation.

The above equations can be solved numerically with the proper selected parameters. Numerical simulations show that the model exhibits periodic oscillation according to both deterministic and stochastic descriptions of the model (Fig. 5.21a, b). Nevertheless, we found that when the $R$ degradation rate $\delta_R$ is large, the deterministic model does not exhibit oscillations; however, the stochastic simulation gives oscillatory dynamics under the same parameters (Fig. 5.21c, d). These results show that deterministic and stochastic descriptions of the model may not give the same dynamics under certain parameters, and the random effect may benefit oscillatory dynamics. Here, we analyze the mechanism of oscillations due to random excitation.

### 5.5.2.2 Model Analysis

To obtain insight into the essential elements that are responsible for the oscillations, we simplify the deterministic rate equations as much as possible. Similar to previous examples, we make various quasi-equilibrium assumptions.

In the model Eq. (5.33), we assume that the processes of promoter state transition and transcription are fast processes, and hence

$$\frac{dD_A}{dt} = \frac{dD_R}{dt} = \frac{dD'_A}{dt} = \frac{dD'_R}{dt} = \frac{dM_A}{dt} = \frac{dM_R}{dt} = 0.$$

Therefore, we can write $D_A$, $D_R$, $D'_A$, $D'_R$, $M_A$ and $M_R$ through the concentration of protein A as

$$D_A = \frac{\theta_A}{\theta_A + \gamma_A A}, \quad D_R = \frac{\theta_R}{\theta_R + \gamma_R A},$$

$$D'_A = \frac{\gamma_A A}{\theta_A + \gamma_A A}, \quad D'_R = \frac{\gamma_R A}{\theta_R + \gamma_R A}, \tag{5.34}$$

$$M_A = \frac{1}{\delta_{M_A}} \frac{\alpha'_A \gamma_A A + \alpha_A \theta_A}{\theta_A + \gamma_A A}, \quad M_R = \frac{1}{\delta_{M_R}} \frac{\alpha'_A \gamma_R A + \alpha_R \theta_R}{\theta_R + \gamma_R A}.$$

Moreover, the process of binding and unbinding between the activator A and DNA is fast, and hence, we can assume that the protein A concentration is at quasi-equilibrium

$$\frac{dA}{dt} = \beta_A M_A + \theta_A D'_A + \theta_R D'_R - A(\gamma_A D_A + \gamma_R D_R + \gamma_C R + \delta_A) = 0,$$

which, together with (5.34), give

$$A = \frac{\beta_A}{\gamma_C R + \delta_A} \frac{1}{\delta_{M_A}} \frac{\alpha'_A \gamma A + \alpha_A \theta_A}{\theta_A + \gamma_A A}. \tag{5.35}$$

Solving Eq. (5.35), we obtain

$$A = \tilde{A}(R) := \frac{1}{2}(\alpha'_A \rho(R) - K_d) + \frac{1}{2}\sqrt{(\alpha'_A \rho(R) - K_d)^2 + 4\alpha_A \rho(R) K_d}, \tag{5.36}$$

where

$$\rho(R) = \frac{\beta_A}{\delta_{M_A}(\gamma_C R + \delta_A)}, \quad K_d = \theta_A/\gamma_A.$$

Finally, substituting (5.34)–(5.36) into (5.33), we obtain a system of second-order differential equations with the two slow variables $R$ and $C$

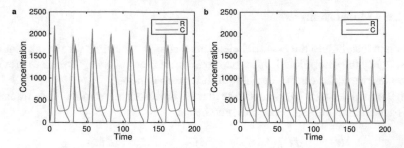

**Fig. 5.22** Oscillatory dynamics of the quantities $R$ and $S$. **a** Simulation results obtained from the complete system (5.33). **b** Simulation results obtained from the reduced system (5.37), which is simplified from (5.33) following the quasi-steady-state assumption. The parameters are the same as those in Fig. 5.21A. Replotted from [15]

$$\begin{cases} \dfrac{dR}{dt} = \dfrac{\beta_R}{\delta_{M_R}} \dfrac{\alpha_R\theta_R + \alpha'_R\gamma_R\tilde{A}(R)}{\theta_R + \gamma_R\tilde{A}(R)} - \gamma_C\tilde{A}(R)R + \delta_A C - \delta_R R, \\ \dfrac{dC}{dt} = \gamma_C\tilde{A}(R)R - \delta_A C. \end{cases} \tag{5.37}$$

To verify the simplified model, we solve (5.37) with the same parameters as those in Fig. 5.21 and compare the results with those obtained from the original Eq. (5.33). The repressor dynamics obtained from the two equations exhibit similar oscillatory dynamics, except slight differences in the amplitude and period (Fig. 5.22). Hence, we expect that the simplified model can provide insights into the mechanism of oscillations produced by random excitation.

Equation (5.37) defines second-order differential equations. We can apply the quantitative analysis methods introduced in Chap. 2. First, the steady states are obtained by solving

$$\frac{dR}{dt} = \frac{dC}{dt} = 0,$$

i.e., the equations

$$R = \frac{\beta_R}{\delta_R\delta_{M_R}} \frac{\alpha_R\theta_R + \alpha'_R\gamma_R\tilde{A}(R)}{\theta_R + \gamma_R\tilde{A}(R)}, \tag{5.38}$$

$$C = \frac{\gamma_C}{\delta_A}\tilde{A}(R)R. \tag{5.39}$$

From the function $\tilde{A}(R)$ given by (5.36), $\tilde{A}$ increases with $\rho(R)$, which in turn decreases with $R$. Hence, $\tilde{A}(R)$ is a decreasing function. Therefore, the right-hand side of (5.38) is a decreasing function, and Eq. (5.38) has a unique positive root, denoted by $R^*$. Hence, Eq. (5.37) has a unique steady state $(R^*, C^*)$, where $C^* = (\gamma_C/\delta_A)\tilde{A}(R^*)R^*$.

To analyze the stability of $(R^*, C^*)$, similar to the method described previously (Sect. 2.2.2), we calculate the trace of the coefficient matrix describing the linearized dynamics around the steady state, which is given by

$$
\begin{aligned}
\tau &= \left( \frac{\partial}{\partial R} \frac{dR}{dt} + \frac{\partial}{\partial C} \frac{dC}{dt} \right) \Bigg|_{(R,C)=(R^*,C^*)} \\
&= \left( \left[ \frac{\beta_R}{\delta_{M_R}} \frac{(\alpha'_R - \alpha_R)\theta_R \gamma_R}{(\theta_R + \gamma_R \tilde{A}(R))^2} - \gamma_C R \right] \frac{d\tilde{A}(R)}{dR} \right) \Bigg|_{R=R^*} - (\gamma_C \tilde{A}(R^*) + \delta_R + \delta_A).
\end{aligned}
$$

The steady state is unstable when $\tau > 0$. For the parameters in Fig. 5.21a, we have $(R^*, C^*) = (66.7491, 363.47)$ and $\tau = 0.81 > 0$; hence, the steady state is unstable, and the system has a stable periodic solution.

From the above analysis, we can determine the parameter region that may imply stable oscillatory dynamics based on the condition $\tau > 0$. Here, we are mainly interested in the degradation rate of the protein R ($\delta_R$). Since the repressor $R$ sequesters the activator A to repress system activation and induce oscillatory dynamics, the stability of R is important for the oscillations. We consider $\tau$ as a function of $\delta_R$ (here, we note that $R^*$ is dependent on $\delta_R$) and keep the other parameters unchanged. Numerical computation shows that $R^*$ decreases with $\delta_R$, and there is a critical value $\delta_R^* = 0.12$, so that when $\delta_R < \delta_R^*$, the steady state is unstable and the system has a stable periodic solution (Fig. 5.23a). When $\delta_R > \delta_R^*$, the steady state is stable. Hence, the system has a stable periodic solution when the degradation rate of R is small, i.e., the repression is strong. Similarly, we can also investigate the effects of the other parameters. Figure 5.23b shows the parameter region of the $\theta_A$–$\theta_R$ plane where there are periodic oscillations. Here, $\theta_A$ and $\theta_R$ represent the binding rates of the activator A and the repressor R with the DNA sequences, respectively.

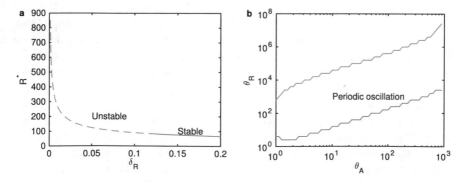

**Fig. 5.23** Bifurcation diagrams. **a** Dependence of the quantity $R^*$ at the steady state with the degradation $\delta_R$. The solid line shows the stable steady state, the dashed line shows the unstable steady state, and 'o' shows the critical point with $\delta_R = 0.12$. **b** The parameter region in the $\theta_A$-$\theta_R$ plane that shows periodic oscillation ($\tau > 0$). Replotted from [15]

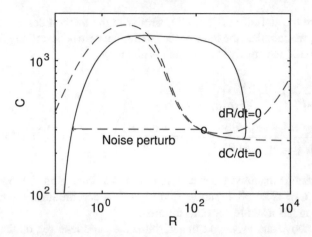

**Fig. 5.24** Phase portrait of the two-variable oscillator system (5.37) with parameter values given in Fig. 5.21c. The blue solid line illustrates the trajectory of the system, the red dashed lines show the $R$ and $C$ nullclines, and the blue dashed line shows the trajectory of moving away from the steady state due to random excitation. Replotted from [15]

Above analysis of the two variable oscillator system (5.37) has shown the condition to have periodic oscillations. Here, we further analyze the mechanism of producing oscillatory dynamics through random excitation when the parameter values are close to the critical point. To this end, we first analyze the vector field of (5.37) when the parameter is taken so that the steady state is stable ($\delta_R = 0.2$). Figure 5.24 shows the trajectory of the equation at the $C$-$R$ phase plane. Starting from low numbers of initial molecules near the origin of the phase plane, the trajectory rapidly shoots upwards along the $R$-nullcline. Here, the high levels of activator A rapidly induce the formation of the complex C. After reaching the maximum of the nullcline, the trajectory moves rapidly to the right and downwards, corresponding to a rapid increase in the repressor $R$ and a drop in $C$ due to the sequestering effect. The trajectory curves move past the $C$-nullcline and approach the fixed point along the $R$-nullcline (Fig. 5.24). In this case, there are no oscillations, as we can see in Fig. 5.21c. However, a perturbation of sufficient magnitude near the fixed point can push the system to the left of the fixed point so that the system state goes to the $R$-nullcline rapidly and initiates a new cycle. The trajectory in Fig. 5.24b comprises a fast phase corresponding to the rapid production of $C$ and $R$ and a slow phase corresponding to the slow degradation of $R$. These fast and slow dynamics are similar to the classic example of the Fitz Hugh-Nagumo model of action-potential transmission in neurons [16, 17]. The fast and slow legs correspond to the excitable and refractory phases of the system, respectively. For the circadian model discussed here, the intrinsic noise due to a small molecule number can perturb the fixed point of the fast and slow dynamics system and initiate a new cycle. This is a general mechanism for inducing a random periodic dynamics, as shown in Fig. 5.21d, due to random excitation.

### 5.5.3 Time-Delayed Negative Feedback

In the previous examples, negative feedback regulation was necessary for the existence of oscillatory dynamics. However, we note that the presence of negative feedback regulation may not always yield oscillation dynamics. For example, when the regulation network is absurdly simple with only one time-varying component, oscillatory dynamics are generally impossible (a first-order differential equation $dx/dt = f(x)$ has no periodic solution). From our discussions in Sect. 2.2.2, a two-component oscillator usually includes interlinked positive and negative feedback loops in order to exhibit robust biological oscillations [18]. A minimum oscillator without positive feedback loops usually includes three components, including the Goodwin oscillator and repressilator (Fig. 5.25). The intermediate steps in these genetic networks are important for the oscillations. Moreover, in previous genetic networks, we implicitly assume that there are no time delays in the processes of transcription, translation, or end product repression. However, there are surely many intermediate steps and delays in transcription and translation associated with mRNA and protein processing within the nucleus and cytoplasms, respectively [19–21]. There are also delays in the feedback loops because the end product must move into the nucleus, bind with transcription factors, and interact with the upstream regulatory sites to regulate the transcription of target genes. Mathematically, we can lump all these delays together and model the process with delay differential equations. This section mainly referred to Chap. 9 in [22].

#### 5.5.3.1 Formulation of the Delayed-Differential Equation Model

In general, when we lump all delays together, we can write a delay differential equation for negative feedback as

$$\frac{dX}{dt} = \frac{a}{1 + (Z/K)^p} - bX, \tag{5.40}$$

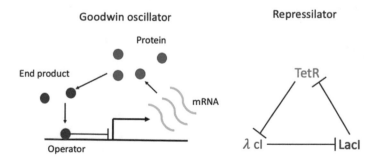

**Fig. 5.25** Typical genetic oscillators. Left: the Goodwin oscillator [23]; right: the repressilator [24]

where $Z(t)$ is the concentration of effective molecules at time $t$, which depends on the past of $X$. For example, for a discrete time lag

$$Z(t) = X(t - \tau),$$

$\tau$ is a constant for the time delay. Usually, the lag time is not a determined constant but a random number with a given probability density, and hence, $Z(t)$ can be expressed through a distributed time lag

$$Z(t) = \int_{-\infty}^{t} X(s)G_c^n(t - s)ds, \tag{5.41}$$

where $G_c^n(s)$ represents the probability of the lag time $s$, which is often given by a Poisson distribution with parameters $c$ and $n$

$$G_c^n(s) = \frac{c^{n+1}}{n!}s^n e^{-cs}. \tag{5.42}$$

The Poisson distribution density (5.42) has a maximum at $s = n/c$. As $n$ and $c$ increase, with $n/c$ fixed, the density approaches a delta function $\delta(s - n/c)$, and the distributed time lag approaches the discrete time lag with $\tau = n/c$ (Fig. 5.26); the distributed delay (5.41) tends to a discrete delay $Z(t - \tau)$.

To demonstrate the process of lumping all delays from the product $X$ to the effector $Z$, we consider the intermediate process of the state transitions of the product $X$. In the function $G_c^n(s)$, the parameter $n$ represents the number of intermediate states, and $c$ represents the reaction rate of state transitions. Then, all step processes form a linear negative feedback loop in Fig. 5.27. Let us introduce a family of $Z_j$'s

$$Z_j(t) = \int_{-\infty}^{t} X(s)G_c^j(t - s)ds, \quad (j = 0, 1, \cdots, n)$$

**Fig. 5.26** The shape of the probability density $G_c^n(s)$ of the distributed time lag in (5.42). Here, different curves (from top to bottom) represent the density function for $n = 10, 20, \cdots, 100$ and $n/c = 1$

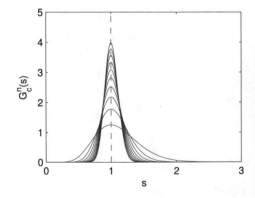

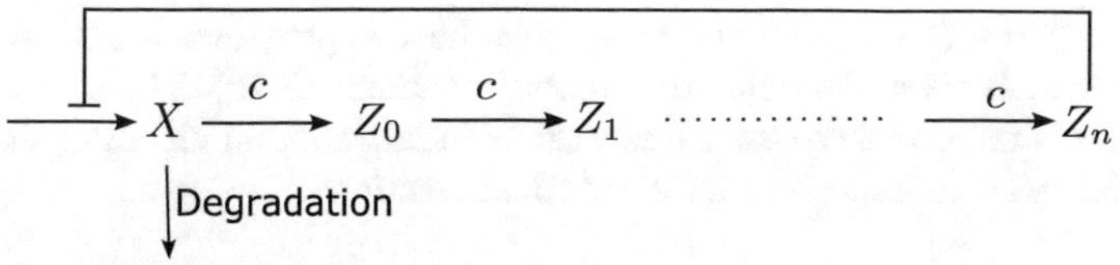

**Fig. 5.27** Multistep negative feedback loop

for the concentration of intermediate products. Here, $Z_n(t)$ gives the function $Z(t)$ in (5.40). When we differentiate these quantities, we have

$$\frac{dZ_j}{dt} = X(t)G_c^j(0) + \int_{-\infty}^t X(s)\frac{d}{dt}G_c^j(t-s)ds, \quad (j = 0, 1, \cdots, n).$$

Since $G_c^0(0) = c$ and $C_c^j(0) = 0$ $(j = 1, 2, \cdots, n)$, and the relation

$$\frac{d}{ds}G_c^j(s) = c(G_c^{j-1}(s) - G_c^j(s)),$$

we have

$$\frac{dZ_0}{dt} = cX(t) + \int_{-\infty}^t X(s)(-cG_c^0(t-s))ds,$$

$$\frac{dZ_j}{dt} = \int_{-\infty}^t X(s)(cG_c^{j-1}(t-s) - cG_c^j(t-s))ds, \quad (j = 1, \cdots, n).$$

Thus, together with (5.40), we write the above process as

$$\frac{dX}{dt} = \frac{a}{1 + (Z_n/K)^p} - bX, \tag{5.43}$$

$$\frac{dZ_0}{dt} = c(X - Z_0), \tag{5.44}$$

$$\frac{dZ_j}{dt} = c(Z_{j-1} - Z_j), \quad (j = 1, 2, \cdots, n). \tag{5.45}$$

These equations establish the connection between the delay differential equation and the ordinary differential equation models of the feedback loop in Fig. 5.27. The time scale of the state transition is approximately $1/c$ per step. We apply the quasi-steady-state equilibrium assumption to the intermediate states and have

$$Z_0(t) = X(t - 1/c), \quad Z_j(t) = Z_{j-1}(t - 1/c), \quad (j = 1, 2, \ldots, n).$$

Hence, $Z_n(t) = X(t - n/c)$. Let $\tau = n/c$; we have approximately $Z(t) = Z_n(t) = Z(t - \tau)$, which gives the delay differential equation (5.40). The above analyses show that it is reasonable to use a delay differential equation model to simplify the intermediate state in multiplestep biochemical reactions.

### 5.5.3.2   Stability Analysis of the Delay Differential Equation

Here, we analyze the stability of the steady state of Eq. (5.40) and obtain the conditions for the existence of periodic solutions. We only consider the situation with discrete delays. Let $x = X/K$ be the nondimensional concentration, $\delta = bK/a$ be the nondimensional degradation rate, and $at/K$ be the nondimensional time (also denoted by $t$ for simplicity). Equation (5.40) is rewritten as the following nondimensional form:

$$\frac{dx}{dt} = \frac{1}{1 + x_\tau^p} - \delta x, \quad x_\tau = x(t - \tau). \tag{5.46}$$

Here, $x_\tau = x(t - \tau)$. Equation (5.46) has a unique steady state $x(t) \equiv x^*$ given by an algebraic equation

$$\delta x^* ((x^*)^p + 1) - 1 = 0.$$

To analyze the stability, let $y(t) = x(t) - x^*$ and linearize the equation around the steady state $x^*$; thus, we have

$$\frac{dy}{dt} = -\phi y(t - \tau) - \delta y(t) + \text{higher order terms}, \quad \phi = \frac{px^{*p-1}}{(1 + x^{*p})^2}. \tag{5.47}$$

Here, $\phi > 0, \delta > 0$. We omit the higher-order terms and find the solution of (5.47) of form $y(t) = y_0 e^{\lambda t}$; the exponent $\lambda$ satisfies

$$\lambda + \delta = -\phi e^{-\lambda \tau}. \tag{5.48}$$

Thus, (5.48) gives the equation for the eigenvalues of (5.47). From the stability theory of delay differential equations (Sect. 2.3.2), the zero solution of (5.47) is stable only when all eigenvalues have negative real parts.

When $\tau = 0$, the zero solution is stable only when $\delta + \phi > 0$. When $\tau > 0$, Eq. (5.48) cannot be solved explicitly. Here, we assume that $\delta + \phi > 0$ and that there is a Hopf bifurcation delay $\tau_{\text{crit}}$ so that the zero solution becomes unstable when $\tau > \tau_{\text{crit}}$.

Given the parameters $(\delta, \phi)$, the solutions of Eq. (5.48) continuously depend on the delay $\tau$. When $\tau = 0$ (and $\delta + \phi > 0$), all eigenvalues have negative real parts, and when $\tau$ increases continuously ($\tau > 0$), the eigenvalues change continuously, and some eigenvalues cross the pure imaginal axis when $\tau = \tau_{\text{crit}}$, i.e., there are imaginal eigenvalues $\lambda = \pm \omega i$ when $\tau = \tau_{\text{crit}}$. Hence, we only need to solve the imaginal eigenvalues to find the critical delay.

Substituting $\lambda = \pm\omega i$ (here, $\omega$ is a positive real number) into (5.48) and separating the real parts and the imaginal parts, we have

$$\delta = -\phi\cos(\omega\tau), \quad \omega = \phi\sin(\omega\tau). \tag{5.49}$$

Hence,

$$\omega^2 = \phi^2 - \delta^2.$$

Thus, if $\phi^2 < \delta^2$, there is no positive real solution $\omega$, i.e., the critical delay $\tau_{\text{crit}} = +\infty$; if $\phi^2 > \delta^2$,

$$\omega = \sqrt{\phi^2 - \delta^2} = \delta\sqrt{(p/(1+(x^*)^{-p}))^2 - 1}, \tag{5.50}$$

and the critical delay is given by

$$\tau_{\text{crit}} = \frac{\cos^{-1}(-\delta/\phi)}{\omega} = \frac{\cos^{-1}(-(1+(x^*)^{-p})/p)}{\delta\sqrt{(p/(1+(x^*)^{-p}))^2 - 1}}. \tag{5.51}$$

When $\tau = \tau_{\text{crit}}$, the linearized Eq. (5.47) has periodic solutions $y(t) = y_0 e^{\pm\omega it}$, and the period is $2\pi/\omega$. Hence, the above equations also give the estimation of the periods when $\tau$ is near the critical delay (Fig. 5.28). We note that when $\tau$ is far away from the critical delay, the relationship between the period and parameters is unknown. From (5.50) and (5.51), Hopf bifurcation can only happen when $|\phi| > \delta$. In this case, $-1 < -(1+(x^*)^{-p})/p < 0$, and hence, $\cos^{-1}(-(1+(x^*)^{-p})/p)$ is meaningful and larger than $\pi/2$. Hence, to have a Hopf bifurcation and periodic solutions, we need to have $\tau > \tau_{\text{crit}} > \pi/(2\omega)$. This result indicates that in genetic negative feedback loops, multiple intermediate processes are required to have sustained oscillations.

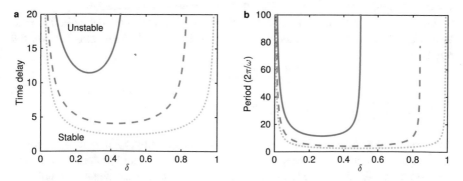

**Fig. 5.28** Hopf bifurcation of (5.46). Here, we plot $\tau_{\text{crit}}$ and period $(2\pi/\omega)$ as functions of $\delta$ for three values of $p$: 2 (solid line), 2.5 (dashed line), and 4 (dotted line). Replotted from Fig. 9.10 in [22]

## 5.5.4   Circadian Rhythms

We are all familiar with the importance of daily rhythms, such as the 24-hour sleep-wake cycle, body temperature, hormone secretion, and skin cell division. Such rhythms are driven by the circadian clock, and they are widely observed in all kinds of plants, animals, fungi, and cyanobacteria. Circadian rhythm can refer to any biological process that displays an endogenous, entrainable oscillation of approximately 24 h. Biologists have long been puzzled by the molecular basis of circadian rhythms until the discovery of clock genes, such as *period* (*per* for short) and *timeless* (*tim* for short) [25–27]. In this section, we introduce two simple mathematical models for the interactions between these clock genes to produce circadian rhythms. For further discussions on the predictive models of the circadian clock, please refer to [28].

### 5.5.4.1   A Simple Model of Circadian Rhythms

First, we introduce a simple mathematical model of circadian rhythms [29]. This model can be the basis of many biological oscillations.

The essential component of a circadian rhythm generator includes the transcription of a clock gene, a cascade of the production of the effective protein from its mRNA, and negative feedback from the effective protein to the production of its mRNA [29]. The protein production cascade involves translation and subsequent processing steps, such as phosphorylation, dimerization, transport, and nuclear entry. In a simplified model, we omit the processes involved in the production of the effective protein and characterize the full chain of reactions in terms of (1) the total duration and (2) the nonlinear relationship between the input and (delayed) output of the reaction chain (Fig. 5.29). Thus, the dynamics of the circadian rhythm generation can be represented by the concentrations of mRNA and the effective protein, and the protein production cascade and negative feedback are assumed to be nonlinear processes.

Let $M$ and $P$ be the relative concentrations of mRNA and the effective protein, respectively. The above model is described by the following delay differential equation:

$$\begin{cases} \dfrac{dM}{dt} = \dfrac{r_M}{1 + (P/K)^n} - q_M M, \\[3mm] \dfrac{dP}{dt} = r_P M_\tau^m - q_P P, \end{cases} \qquad M_\tau(t) = M(t - \tau). \qquad (5.52)$$

Here, $r_M$ denotes the scaled mRNA production rate constant, $r_P$ denotes the protein production rate constant, and $q_M$ and $q_P$ denote the mRNA and protein degradation rate constants. We apply the Hill-type function to represent the negative regulation of the effective protein on mRNA production, with $K$ being the scaled concentration constant and $M_\tau^m$ being the exponential function for the protein cascade from mRNA to the effective protein, including posttranslational modifications, phosphorylation and dimerization. The delay $\tau$ denotes the total duration of protein production from mRNA to the effective protein.

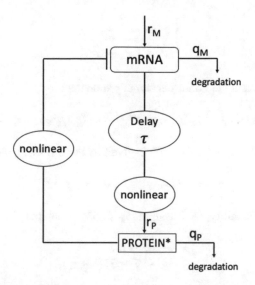

**Fig. 5.29** Simplified model of the biological elements in a circadian rhythm generator. The model includes the autoinhibition of protein production at the translational or transcriptional level and posttranslational processing, such as phosphorylation, dimerization, and transport. Here, we lump the posttranslational process through a delay $\tau$ and a nonlinear function in the protein production cascade and introduce nonlinear negative feedback from the effective protein to mRNA production. Replotted from [29]

Now, we analyze the model Eq. (5.52) to identify the condition needed to have oscillation solutions. It is easy to see that Eq. (5.52) has only one steady state. Similar to the situation of the second-order ordinary differential equation, the system exhibits oscillation solutions if the steady state is unstable. Hence, we need to analyze the stability of the steady state.

First, we perform nondimensional analysis. From (5.52), we take the EC50 concentration $K$ from the Hill-type function as the reference of the effective protein. At the equilibrium state, we have

$$M = \frac{r_M}{q_M} \frac{1}{1 + (P/M)^n} \sim \frac{r_M}{q_M},$$

where we can take $r_M/q_M$ as the reference of the mRNA concentration. Thus, we obtain the following nondimensional variables:

$$x = M/(r_M/q_M), \quad y = P/K, \quad \tilde{t} = q_M t.$$

Here, the lifetime of mRNA is taken as the reference of the time. Accordingly, we obtain the nondimensional equation

$$\begin{cases} x' = \dfrac{1}{1 + y^n} - x, \\ y' = rx_{\tau_c}^m - \delta y, \end{cases} \tag{5.53}$$

where $'$ means $\frac{d}{d\bar{t}}$, and the nondimensional parameters

$$r = (1/K)(r_P/q_M)(r_M/q_M)^m, \quad \delta = (q_P/q_M), \quad \tau_c = q_M\tau.$$

The unique equilibrium state $(x^*, y^*)$ is given by the following algebraic equation:

$$x^* = 1/(1 + y^{*n}), \quad y^* = (r/\delta)x^{*m}.$$

To analyze the stability, we linearize Eq. (5.53) to obtain

$$\begin{cases} \tilde{x}' = -\tilde{x} - a\tilde{y} \\ \tilde{y}' = -\delta\tilde{y} + b\tilde{x}(t - \tau_c), \end{cases} \tag{5.54}$$

where $\tilde{x} = x - x^*$, $\tilde{y} = y - y^*$, and

$$a = \frac{ny^{*(n-1)}}{(1 + y^{*n})^2}, \quad b = mrx^{*m-1}.$$

Now, we assume the exponential form solutions of (5.54)

$$\tilde{x} = c_1 e^{\lambda t}, \quad \tilde{y} = c_2 e^{\lambda t}, \quad (c_1, c_2 \neq 0).$$

Substituting this solution into (5.54), we obtain the equation for the coefficients of $c_1$ and $c_2$

$$\begin{bmatrix} 1+\lambda & a \\ -be^{\tau_c\lambda} & \delta+\lambda \end{bmatrix} \begin{bmatrix} c_1 \\ c_2 \end{bmatrix} = \begin{bmatrix} 0 \\ 0 \end{bmatrix}. \tag{5.55}$$

Equation (5.55) has a nonzero solution $(c_1, c_2)$ if and only if the determination of the coefficient matrix is zero. Hence, $\lambda$ satisfies the characteristic equation

$$f(\lambda) = (1 + \lambda)(\delta + \lambda) + abe^{-\tau_c\lambda} = 0. \tag{5.56}$$

The zero solution of (5.54) is stable if and only if all roots of the characteristic Eq. (5.56) have negative real parts.

Equation (5.56) cannot be solved exactly, and it is difficult to determine all roots of the equation to examine the stability of the zero solution. Nevertheless, similar to the discussion in Sect. 2.3.2, we can find the solutions of form $\lambda = \pm i\omega$ to determine the condition of Hopf bifurcation and identify the parameter regions for the stable zero solution of (5.54).

When $\tau_c = 0$ (without delay), the characteristic Eq. (5.56) has two roots

$$\lambda_{1,2} = \frac{-(\delta + 1) \pm \sqrt{(\delta + 1)^2 - 4(\delta + ab)}}{2},$$

both of which have negative real parts. Hence, in the case without delay (or in the case where the delay is small), the (unique) equilibrium state is stable, and there is no oscillation solution.

When $\tau_c > 0$, let $\lambda = i\omega$ ($\omega > 0$) in (5.55) and separate the real part and imaginal part of $f(i\omega)$ we obtain

$$\begin{cases} (ab) \cos \tau_c \omega = \omega^2 - \delta, \\ ab \sin \tau_c \omega = (1 + \delta)\omega. \end{cases}$$

Hence, $ab$ and $\tau_c$ satisfy

$$(ab)^2 = \omega^4 + (1 + \delta^2)\omega^2 + \delta^2 \tag{5.57}$$

and

$$(\omega^2 - \delta) \tan \tau_c \omega = (1 + \delta)\omega. \tag{5.58}$$

Given a parameter $\delta$, when $\omega$ varies over $(0, +\infty)$, (5.57)–(5.58) give a curve in the $(ab)$-$\tau_c$ plane. This curve separates the plane into two regions, corresponding to stable and unstable steady states (Fig. 5.30a). The region containing $\tau_c = 0$ corresponds to the stable steady state. We note that equations (5.57)-(5.58) can define an infinite number of curves; if $(ab, \tau_c)$ is a solution, $(ab, \tau_c + \pi/\omega)$ also solves

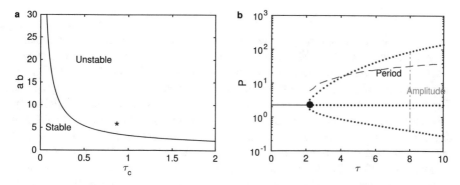

**Fig. 5.30** Bifurcation diagram of the circadian rhythm generator model (5.52). **a** The bifurcation curve in the $(ab)$-$\tau_c$ plane; '∗' shows the parameter used in the numerical simulation shown in Fig .5.31. **b** Dependencies of dynamical properties with the delay $\tau$. The black solid line shows the stable steady state (in protein concentration $P$), black dots show the unstable steady state, the black dot '●' shows the Hopf bifurcation, blue dots are the upper and lower bounds of the oscillation solutions, with periods given by the red dashed line, and the amplitude of a solution (for a given $\tau = 8$ h) is measured by the length of the green dashed-dot line. Replotted from [29]

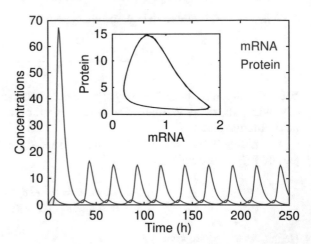

**Fig. 5.31** Simulation results of the circadian oscillation from model (5.52). The parameters are $r_M = 1.0\,\text{h}^{-1}, r_P = 1.0\,\text{h}^{-1}$, $q_M = 0.21\,\text{h}^{-1}$, $q_P = 0.21\,\text{h}^{-1}, n = 2.0$, $m = 3.0, \tau = 4.0\,\text{h}^{-1}$, and $K = 1$. The inset shows the oscillation orbit in the phase plane. Replotted from [29]

Eqs. (5.57)–(5.58). However, only the curve with the minimum $\tau_c$ gives the Hopf bifurcation, which corresponds to the first occurrence of pure imaginal eigenvalues.

From the bifurcation diagram, we can easily identify the parameters (from the region of unstable steady states) to show oscillation dynamics. Figure 5.30b shows the dependence on the period and amplitudes (in the protein concentrations $P$) of the oscillation solutions. We can see that both periods and amplitudes increase with the delay $\tau$. From these analyses, we can identify the parameters so that the system exhibits 24-hour circadian oscillations. The solution is shown in Fig. 5.31.

### 5.5.4.2 A Model Based on the Dimerization and Proteolysis of PER and TIM

The above circadian rhythm generator is a simplified model with delayed negative feedback to the clock gene. Here, we introduce a model for a specific regulatory network of clock genes: the dimerization and proteolysis of PER and TIM in *Drosophila melanogaster* [30]. Figure 5.32 shows the molecular mechanism of the two proteins. The PER and TIM proteins are encoded by the clock genes *per* and *tim*, respectively. PER monomers are rapidly phosphorylated by a casein-like kinase called DBT and degraded. DBT (encoded by the *double-time* gene) is presented at roughly constant levels during the rhythm. Experiments suggest that PER is stabilized when associated with TIM [31, 32]. Hence, when the transcription level is low, the PER proteins cannot form dimers effectively and are rapidly degraded through DBT; as the PER concentration increases, a greater proportion of proteins is dimerized and protected from DBT. Thus, the observed degradation rate of PER is a nonlinear function of its concentration. The phosphorylation of PER introduces positive feedback in PER accumulation. The PER dimers further form heterodimers with TIM in the cyto-

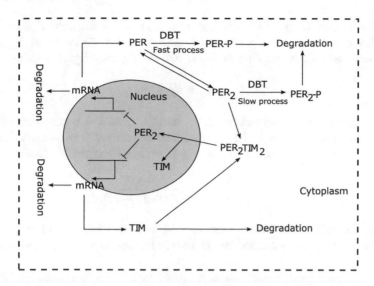

**Fig. 5.32** A simple molecular mechanism for the circadian clock in *Drosophila*. PER and TIM proteins are synthesized in the cytoplasm, where they may be destroyed by proteolysis or they may combine to form relatively stable heterodimers. Heteromeric complexes are transported into the nucleus, where they inhibit the transcription of *per* and *tim* mRNAs. Here, PER monomers are rapidly phosphorylated by DBT and then degraded, while PER dimers are poorer substrates for DBT. Replotted from [30]

plasm. The heteromeric complexes are transported into the nucleus and inhibit the transcription of *per* and *tim* to form a negative feedback loop.

Now, we establish an ordinary differential equation model for the mechanism shown in Fig. 5.32. First, we make some assumptions to reduce the number of variables included in the model. We note that PER and TIM mRNAs and proteins follow roughly similar time courses in vivo, and we lump them together into a single pool of clock proteins. In addition, we assume that the cytoplasmic and nuclear pools of dimeric proteins are in rapid equilibrium; hence, we do not separate the cytoplasm and the nucleus in the model. With these simplifying assumptions, the above mechanism can be described through three differential equations for mRNA concentration ($M$), monomer concentration ($P_1$), and dimer concentration ($P_2$).

From Fig. 5.32, the genetic network is described by the following differential equations:

$$\frac{dM}{dt} = \frac{v_m}{1 + (P_2/P_{\text{crit}})^2} - k_m M, \tag{5.59}$$

$$\frac{dP_1}{dt} = v_p M - 2k_a P_1^2 + 2k_d P_2 - \frac{k_{p1} P_1}{J_p + P_1 + r P_2} - k_{p3} P_1, \tag{5.60}$$

$$\frac{dP_2}{dt} = k_a P_1^2 - k_d P_2 - \frac{k_{p2} P_2}{J_p + P_1 + r P_2} - k_{p3} P_2. \tag{5.61}$$

Here, we assume cooperativity for the inhibition of mRNA transcription by the PER/TIM dimers and set the Hill coefficient as 2 and the EC50 as $P_{\text{crit}}$. Both monomers and dimers can bind to DBT, but $P_1$ is phosphorylated more rapidly, i.e.,

$$k_{p1} \gg k_{p2}. \tag{5.62}$$

The dimerization rate is represented by $k_a$, and the dissociation rate of the dimer is $k_d$. In addition to the degradation of monomers and dimers through DBT, they can also undergo first-order degradation at a rate of $k_{p3}$. Here, the Michaelis–Menten functions

$$\frac{k_{p1} P_1}{J_p + P_1 + r P_2}, \quad \frac{k_{p2} P_2}{J_p + P_1 + r P_2} \tag{5.63}$$

represent the degradation rates of monomers and dimers through DBT, respectively. Here, we show how the two functions are derived from quasi-steady-state equilibrium assumptions.

Assume that both monomer and dimer proteins are phosphorylated by DBT and destroyed by proteolysis. These processes are represented by the following reactions:

$$P_1 + \text{DBT} \underset{k_{-1}}{\overset{k_1}{\rightleftharpoons}} \text{DBT-}P_1 \overset{k'_1}{\longrightarrow} \text{DBT}, \quad P_2 + \text{DBT} \underset{k_{-2}}{\overset{k_2}{\rightleftharpoons}} \text{DBT-}P_2 \overset{k'_2}{\longrightarrow} \text{DBT}.$$

Let $D$ be the concentration of DBT, $D_1$ be the concentration of DBT-$P_1$, and $D_2$ be the concentration of DBT-$P_2$. We have the following chemical rate equations:

$$\frac{dP_1}{dt} = -k_1 P_1 D + k_{-1} D_1, \tag{5.64}$$

$$\frac{dP_2}{dt} = -k_2 P_2 D + k_{-2} D_2, \tag{5.65}$$

$$\frac{dD_1}{dt} = k_1 P_1 D - k_{-1} D_1 - k'_1 D_1, \tag{5.66}$$

$$\frac{dD_2}{dt} = k_2 P_2 D - k_{-2} D_2 - k'_2 D_2. \tag{5.67}$$

Here, the DBT protein exists in three difference states (the free state, binding with monomers, or binding with dimers), and the total concentration is unchanged. Hence,

$$D_T = D + D_1 + D_2$$

is a constant. Now, assuming the rapid process of degradation after binding to DBT, we apply the quasi-steady-state assumption to DBT-$P_1$ and DBT-$P_2$, which gives

$$\frac{dD_1}{dt} = \frac{dD_2}{dt} = 0.$$

The above assumptions give

$$D = \frac{D_T}{1 + k_{eq,1} P_1 + k_{eq,2} P_2}, \quad D_1 = k_{eq,1} P_1 D, \quad D_2 = k_{eq,2} P_2 D,$$

where

$$k_{eq,1} = \frac{k_1}{k_{-1} + k_1'}, \quad k_{eq,2} = \frac{k_2}{k_{-2} + k_2'}.$$

Therewith, the Eqs. (5.64)–(5.67) are rewritten as

$$\frac{dP_1}{dt} = -\frac{D_T k_1' P_1}{1/k_{eq,1} + P_1 + (k_{eq,2}/k_{eq,1}) P_2}, \tag{5.68}$$

$$\frac{dP_2}{dt} = -\frac{D_T k_2' (k_{eq,2}/k_{eq,1}) P_2}{1/k_{eq,1} + P_1 + (k_{eq,2}/k_{eq,1}) P_2}. \tag{5.69}$$

The right-hand sides of (5.68) and (5.69) give the proteolysis rates of monomers and dimers upon phosphorylation by DBT. Comparing the coefficients with the Michaelis-Menten functions in (5.63), we have

$$k_{p1} = D_T k_1', \quad k_{p2} = D_T k_2' (k_{eq,2}/k_{eq,1}), \quad J_p = 1/k_{eq}, \quad r = k_{eq,2}/k_{eq,1}.$$

From these relations, the condition (5.62) is equivalent to

$$k_1' k_{eq,1} \gg k_2' k_{eq,2}$$

or

$$\frac{k_1 k_1'}{k_{-1} + k_1'} \gg \frac{k_2 k_2'}{k_{-2} + k_2'}.$$

Thus, when $k_1 \gg k_2$ and $k_{-1} \ll k_1'$, $k_{-2} \ll k_2'$ (e.g., the monomer much more easily undergo phosphorylation by DBT than the dimer) and upon phosphorylation, the proteins are rapidly destroyed by proteolysis, and the condition (5.62) is satisfied.

Now, we can select the proper parameters to solve Eqs. (5.59)–(5.61). There are solutions that show 24-hour circadian oscillations (Fig. 5.33).

To analyze the mechanism of circadian oscillation, we first further simplify the model equations. The dimerization reactions are fast ($k_a$ and $k_d$ are large numbers); hence, the process $2P_1 \leftrightarrows P_2$ is always near equilibrium. Hence, we can apply the quasi-steady-state assumption to this dimerization reaction, which gives

$$k_a P_1^2 = k_d P_2$$

or

$$P_2 = K_{eq} P_1^2, \quad K_{eq} = k_a/k_d.$$

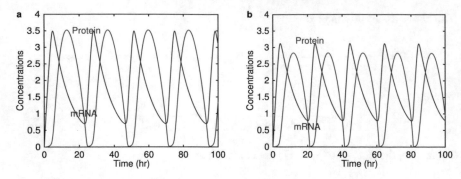

**Fig. 5.33** Oscillation solutions of model equations. **a** Solution obtained from the full version model (5.59)–(5.61). **b** Solution obtained from the simplified model (5.71)–(5.72). The parameters are $v_m = 1\,\mathrm{C_m h^{-1}}$, $k_m = 0.1\,\mathrm{h^{-1}}$, $v_p = 0.5\,\mathrm{C_p C_m h^{-1}}$, $k_{p1} = 10\,\mathrm{C_p h^{-1}}$, $k_{p2} = 0.03\,\mathrm{C_p h^{-1}}$, $k_{p3} = 0.1\,\mathrm{h^{-1}}$, $P_{\mathrm{crit}} = 0.1\,\mathrm{C_p}$, $J_p = 0.05\,\mathrm{C_p}$, $r = 1.2$, $k_a = 800\,\mathrm{C_p^{-1} h^{-1}}$, and $k_d = 4\,\mathrm{h^{-1}}$, where $\mathrm{C_m}$ and $\mathrm{C_p}$ represent the characteristic concentrations of mRNA and protein, respectively. Replotted from [30]

Let $P_T = P_1 + 2P_2$ be the total protein concentration. Here, the factor 2 means two proteins in each dimer. We can write $P_1$ and $P_2$ as functions of $P_T$, so that

$$P_1 = q P_T, \quad P_2 = \frac{1}{2}(1-p)P_T, \quad q = \frac{2}{1+\sqrt{1+8K_{\mathrm{eq}}P_T}}. \tag{5.70}$$

Now, we substitute (5.70) into (5.59)–(5.61) and have the following second-order differential equations:

$$\frac{dM}{dt} = \frac{v_m}{1+(P_T(1-q)/(2P_{\mathrm{crit}}))^2} - k_m M, \tag{5.71}$$

$$\frac{dP_T}{dt} = v_p M - \frac{k_{p2}(\alpha q P_T + P_T)}{J_p + q P_T + (r/2)(1-q)P_T} - k_{p3}P_T. \tag{5.72}$$

Here, $p$ depends on $P_T$ through (5.70), and $\alpha = (k_{p1} - k_{p2})/k_{p2}$. Figure 5.33 shows the numerical solutions based on both the full model (5.59)–(5.61) and the simplified model (5.71)–(5.72). The solutions obtained from both models are consistent with each other, and hence, the simplification is reasonable.

Let

$$f(P_T) = \frac{v_m}{1+(P_T(1-q)/(2P_{\mathrm{crit}}))^2},$$

$$g(P_T) = \frac{k_{p2}(\alpha q P_T + P_T)}{J_p + q P_T(r/2)(1-q)P_T} + k_{p3}P_T.$$

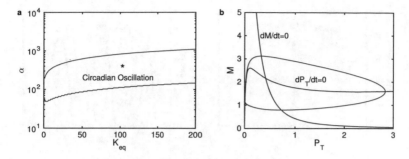

**Fig. 5.34** Bifurcation diagram. **a** Parameter region in the $\alpha$-$K_{eq}$ plane to have oscillation solutions. The star '*' shows the parameter $(K_{eq}, \alpha) = (100, 333)$. **b** $M$- and $P$-nullclines for the parameter values in Fig. 5.33. The close orbit is a stable limit cycle oscillation. Replotted from [30]

The steady states of the Eqs. (5.71)–(5.72) are given by the intersection point of the roots of the nullclines

$$M = f(P_T)/k_m, \text{ and } M = g(P_T)/v_p.$$

The nullclines, which are shown in Fig. 5.34, yield one positive equilibrium state. Let $(M^*, P^*)$ be the steady state. The coefficient matrix of the linearization around the steady state is

$$A = \begin{bmatrix} -k_m & -f'(P^*) \\ v_p & -g'(P^*) \end{bmatrix}.$$

When the trace $\mathrm{tr} = -(k_m + g'(P^*)) > 0$, or

$$g'(P^*) < -k_m, \tag{5.73}$$

the eigenvalues of $A$ have positive real parts, and the steady state is unstable. The condition (5.73) means that the slope of the $P_T$-nullcline $M = g(P_T)/v_p$ at the steady state is smaller than $(-k_m/v_p)$. In this case, the system has a stable period solution corresponding to the limit cycle in the phase plane (Fig. 5.34b).

The inequality (5.73) gives the condition needed to have an unstable steady state, and hence, the system exhibits stable oscillation solutions. Thus, based on (5.73), we can identify the parameters needed to yield periodic solutions. In the current study of clock gene regulation, the condition (5.62) that the degradation rate of monomers is larger than that of dimers is important for oscillatory dynamics. The condition (5.62) implies $\alpha \gg 1$ in the simplified model (5.72). To verify this point, we consider the dependence of $g'(P^*)$ on $\alpha$ and the equilibrium constant $K_{eq}$ and identify the parameter region in the $\alpha$-$K_{eq}$ plane to have oscillation solutions (Fig. 5.34a). From Fig. 5.34, the value of $\alpha$ is important for the existence of oscillation dynamics (for example, $100 < \alpha < 1000$), while $K_{eq}$ can take values from a

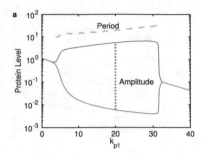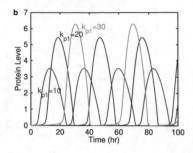

**Fig. 5.35** Dependence of the solution on the parameter $k_{p1}$. **a** Dependence of the period and amplitude of oscillation solutions on the parameter $k_{p1}$. **b** Numerical solutions with different values of $k_{p1}$ ($k_{p1} = 10, 20, 30$). Replotted from [30]

wide range. When $(K_{eq}, \alpha) = (100, 333)$, the corresponding solution is shown by the limit cycle in Fig. 5.34b.

In the study of circadian rhythms, we are often interested in the robustness of the period (24 h). Here, we can numerically solve the model equations with varying parameter values and investigate how the oscillation features may depend on the parameter values. Figure 5.35a shows the dependence of the period and amplitude on the monomer degradation rate $k_{p1}$. We can see that both the period and amplitude increase with $k_{p1}$. Figure 5.35b shows sample solutions with $k_{p1} = 10, 20, 30$. From the solutions, the duration of low levels of proteins increases with $k_{p1}$, which indicates that a longer time is required for proteins to accumulate enough dimers to repress the transcription process and produce oscillation dynamics.

## 5.6  Summary

Cell behaviors are controlled by complex genetic regulatory networks that include different levels of dynamic regulation at different stages of gene expression. The detailed components of the genetic circuits are often difficult to discover. A goal in systems biology is to understand how these components may govern cell behaviors or responses to external stimuli. Because of the complexity of genetic networks, such studies should eventually yield mathematical models that provide a reasonable understanding of biological behaviors. Currently, many works have attempted to model complex biological systems to quantitatively explore how cells make decisions or control their robust behaviors, such as apoptosis, differentiation, and circadian rhythm, or even a whole-cell computation model to predict phenotypes from genotypes [33–36]. In this chapter, we introduced basic methods to model and analyze simple genetic networks, e.g., positive feedbacks to yield bistability, toggle switching, and negative feedback loops to produce oscillation dynamics. The genetic motifs introduced in this chapter are building blocks of more complex networks, and the modeling techniques can be extended to studies of various genetic circuits.

# References

1. Jacob, F., Perrin, D., Sánchez, C., Monod, J.: Operon: a group of genes with the expression coordinated by an operator. C. R. Hebd. Seances Acad. Sci. **250**, 1727–1729 (1960)
2. Littlewood, T.D., Evan, G.I.: Transcription factors 2: helix-loop-helix. Protein Profile **2**, 621–702 (1995)
3. Wintjens, R., Rooman, M.: Structural classification of HTH DNA-binding domains and protein-DNA interaction modes. J. Mol. Biol. **262**, 294–313 (1996)
4. Laity, J.H., Lee, B.M., Wright, P.E.: Zinc finger proteins: new insights into structural and functional diversity. Curr. Opin. Struct. Biol. **11**, 39–46 (2001)
5. Babu, M.M., Luscombe, N.M., Aravind, L., Gerstein, M., Teichmann, S.A.: Structure and evolution of transcriptional regulatory networks. Curr. Opin. Struct. Biol. **14**, 283–291 (2004)
6. Ozbudak, E.M., Thattai, M., Lim, H.N., Shraiman, B.I., van Oudenaarden, A.: Multistability in the lactose utilization network of Escherichia coli. Nature **427**, 737–740 (2004)
7. Görke, B., Stülke, J.: Carbon catabolite repression in bacteria: many ways to make the most out of nutrients. Nat. Rev. Microbiol. **6**, 613–624 (2008)
8. Huang, R., Lei, J.: Zhou Pei-Yuan Center for Applied Mathematics, MOE Key Laboratory of Bioinformatics, Tsinghua University, Beijing China, 100084: Cell-type switches induced by stochastic histone modification inheritance. Discret. Contin. Dynam. Syst. B **22**, 1–19 (2019)
9. Lei, J., He, G., Liu, H., Nie, Q.: A delay model for noise-induced bi-directional switching. Nonlinearity **22**, 2845–2859 (2009)
10. Gupta, C., López, J.M., Ott, W., Josić, K., Bennett, M.R.: Transcriptional delay stabilizes bistable gene networks. Phys. Rev. Lett. **111**, 58104 (2013)
11. Hasty, J., Pradines, J., Dolnik, M., Collins, J.J.: Noise-based switches and amplifiers for gene expression. Proc. Natl. Acad. Sci. USA **97**, 2075–2080 (2000)
12. Watson, J.D., Baker, T.A., Bell, S.P., Gann, A., Levine, M., Losick, R.: Molecular Biology of the Gene. Sixth edn. Pearson Education (2008)
13. Lipshtat, A., Loinger, A., Balaban, N.Q., Biham, O.: Genetic toggle switch without cooperative binding. Phys. Rev. Lett. **96**, 4 (2006)
14. Atkinson, M.R., Savageau, M.A., Myers, J.T., Ninfa, A.J.: Development of genetic circuitry exhibiting toggle switch or oscillatory behavior in Escherichia coli. Cell **113**, 597–607 (2003)
15. Vilar, J.M.G., Kueh, H.Y., Barkai, N., Leibler, S.: Mechanisms of noise-resistance in genetic oscillators. Proc. Natl. Acad. Sci. USA **99**, 5988–5992 (2002)
16. Kaplan, D., Glass, L.: Understanding Noninear Dynamics. Springer, New York (1995)
17. Keener, J., Sneyd, J.: Mathematical Physiology. Springer, New York (1990)
18. Tsai, T.Y.-C., Choi, Y.S., Ma, W., Pomerening, J.R., Tang, C., Ferrell, J.E.: Robust, tunable biological oscillations from interlinked positive and negative feedback loops. Science **321**, 126–129 (2008)
19. Lemaire, V., Lee, C.F., Lei, J., Métivier, R., Glass, L.: Sequential recruitment and combinatorial assembling of multiprotein complexes in transcriptional activation. Phys. Rev. Lett. **96**, 4 (2006)
20. Zhang, X., Jin, H., Yang, Z., Lei, J.: Effects of elongation delay in transcription dynamics. Math. Biosci. Eng **11**, 1431–1448 (2014)
21. Xia, W., Lei, J.: Formulation of the protein synthesis rate with sequence information. MBE **15**, 507–522 (2018)
22. Fall, C., Marland, E., Wagner, J., Tyson, J. (eds.): Computational Cell Biology. Interdisciplinary Applied Mathematics. Springer, New York (2002)
23. Goodwin, B.C.: An entrainment model for timed enzyme syntheses in bacteria. Nature **209**, 479–481 (1966)
24. Elowitz, M., Leibler, S.: A synthetic oscillatory network of transcriptional regulators. Nature **403**, 335–338 (2000)
25. Konopka, R.J., Benzer, S.: Clock mutants of Drosophila melanogaster. Proc. Natl. Acad. Sci. USA **69**, 2112–2116 (1971)
26. Sehgal, A., Price, J.L., Man, B., Young, M.W.: Loss of circadian behavior rhythms and per RNA oscillations in Drosophila mutant timeless. Science **263**, 1603–1606 (1994)

27. Sehgal, A., Rothenfluh-Hilfiker, A., Hunter-Ensor, M., Chen, Y., Myers, M.P., Young, M.W.: Rhythmic expression of timeless: a basis for promoting circadian cycles in period gene autoregulation. Science **270**, 808–810 (1995)
28. Forger, D.B.: A detailed predictive model of the mammalian circadian clock. Proc. Natl. Acad. Sci. **100**, 14806–14811 (2003)
29. Scheper, T., Klinkenberg, D., Pennartz, C., van Pelt, J.: A mathematical model for the intracellular circadian rhythm generator. J. Neurosci. **19**, 40–47 (1999)
30. Tyson, J., Hong, C., Thron, C., Novak, B.: A simple model of circadian rhythms based on dimerization and proteolysis of PER and TIM. Biophys. J. **77**, 2411–2417 (1999)
31. Kloss, B., Price, J.L., Saez, L., Blau, J., Rothenfluh, A., Wesley, C.S., Young, M.W.: The Drosophila clock gene double-time encodes a protein closely related to human casein kinase Iepsilon. Cell **94**, 97–107 (1998)
32. Price, J.L., Blau, J., Rothenfluh, A., Abodeely, M., Kloss, B., Young, M.W.: double-time is a novel Drosophila clock gene that regulates PERIOD protein accumulation. Cell **94**, 83–95 (1998)
33. Karr, J.R., Sanghvi, J.C., Macklin, D.N., Gutschow, M.V., Jacobs, J.M., Jr., B. B., Glass, N.A.-G.J.I., Covert, M.W.: A whole-cell computational model predicts phenotype from genotype. Cell **150**, 389–401 (2012)
34. Novak, B., Tyson, J.J.: Design principles of biochemical oscillators. Nat. Rev. Mol. Cell Biol. **9**, 981–991 (2008)
35. Tyson, J.J., Novak, B.: A dynamical paradigm for molecular cell biology. Trends Cell Biol. **30**, 504–515 (2020)
36. Tomazou, M., Barahona, M., Polizzi, K.M., Stan, G.-B.: Computational re-design of synthetic genetic oscillators for independent amplitude and frequency modulation. Cell Syst. **6**, 508–520.e5 (2018)

# Dynamical Modeling of Stem Cell Regeneration

*Can algebraic formulae tell us more than reasoning about the behavior of complex biological systems?*

—Robert A. Weinberg

## 6.1 Introduction

Adult stem cells are present in most self-renewing tissues, including skin, intestinal epithelium, hematopoietic systems, etc.. The growth and regeneration of these adult stem cell pools are tightly controlled by feedback regulation at different levels and are crucial during normal development, tissue regeneration, wound healing, and healthy homeostasis in a lifespan. For example, the self-renewal and differentiation of hematopoietic stem cells (HSCs) are regulated through various feedback signaling molecules, including direct HSC-niche interactions and cytokines secreted from stromal cells [1–3]. Adult intestinal stem cells residing in a crypt niche are regulated by the paracrine secretion of growth factors and cytokines from the surrounding mesenchymal cells [4–6]. The mammalian olfactory epithelium, a self-renewing neural tissue, is regulated by negative feedback signals involving the diffusive molecules GDF11 and activin [7].

Stem cell biology is population biology. Many mathematical models of population dynamics have been widely studied to understand how stem cell regeneration is modulated in different contexts [7–17]. In most models, the dynamics of a homogeneous cell pool or the lineage of several homogeneous subpopulations are formulated through a set of differential equations. In recent years, the heterogeneity of stem cells has been highlighted due to novel experimental techniques at the single-cell level [18–23]. The heterogeneity mostly originates from random changes in epigenetic states at each cell cycle, including DNA methylation, histone modification, and the transcription of genes and noncoding RNAs. In the 1960s, Till et al. proposed a mathematical model of stem cell proliferation to consider the inherently random dynamics of individual cells based on a stochastic birth-death process [24]. This work provided

J. Lei, *Systems Biology*, Lecture Notes on Mathematical Modelling in the Life Sciences, https://doi.org/10.1007/978-3-030-73033-8_6

a new perspective on stem cell biology regarding the stochastic heterogenous dynamics of stem cell behavior [25]. Based on the stochastic epigenetic state transitions that occur during cell divisions, mathematical models for heterogeneous population dynamics have been developed with formulations of discrete dynamical systems or continuous differentiation-integral equations [26–29].

In this chapter, we first introduce a delay differential equation model derived from the age-structure model of $G_0$ phase stem cell regeneration. Analyses of the delay differential equation model are presented with applications to describe the abnormal cell growth that may be related to cancer development. We next introduce a general mathematical framework of heterogeneous stem cell regeneration, which is applied to model the population dynamics with cell heterogeneity and random transitions of epigenetic states. These types of models lead to single-cell-based simulation schemes that enable us to study the cell population dynamics from both scales of single-cell behavior and cell-to-cell variability. Finally, we introduce the application of stem cell regeneration modeling in studying a type of dynamic disease, dynamic hematological disease.

## 6.2    The $G_0$ Cell Cycle Model

The cell cycle or cell division cycle is the series of events that take place in a cell that leads to its division and duplication of its DNA (DNA replication) and to the production of two daughter cells (Fig. 6.1). In eukaryotes, the cell cycle is divided into three periods: interphase, the mitotic (M) phase, and cytokinesis. During interphase, the cell grows, accumulates nutrients needed for mitosis, prepares for cell division, and duplicates its DNA. During the mitotic phase, the chromosomes separate. During the final stage, cytokinesis, the chromosomes in the cytoplasm are separated into two new daughter cells. To ensure the proper division of a cell, there are various control mechanisms known as cell cycle checkpoints. We often divide the cell cycle process into four distinct phases: $G_1$ phase, S (synthesis) phase, $G_2$ phase, and $M$ (mitosis) phase. The $G_1$, S, and $G_2$ phases are collectively known as the interphase. At the postmitotic stages, cells temporarily leave the cycle, stop dividing and enter a state of quiescence called $G_0$ phase.

The concept of the $G_0$ phase was first introduced to explain how some cell populations, such as liver and bone marrow stem cells, alter their rate of cell production in response to a loss of cells [30, 31]. The mathematical model of the $G_0$ cell cycle was established by Burns and Tannock in 1970 [8]. The model assumed a resting phase ($G_0$) between cell cycles. During cell cycling, stem cells are classified into either resting or proliferating phases (Fig. 6.2). During each cell cycle, a cell in the proliferating phase either undergoes apoptosis or divides into two daughter cells; however, a cell in the resting phase either irreversibly differentiates into a terminally differentiated cell or returns to the proliferating phase. Resting phase cells can also be removed from the stem cell pool due to cell death or senescence.

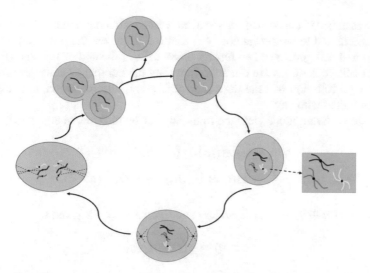

**Fig. 6.1 Life cycle of a cell**. The cell cycle is composed of interphase ($G_1$, S, and $G_2$ phases), followed by the mitotic phase (mitosis and cytokinesis) and the quiescent state of the $G_0$ phase

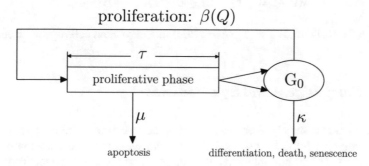

**Fig. 6.2 The $G_0$ model of stem cell regeneration**. During stem cell regeneration, cells in the resting phase either enter the proliferating phase at a rate of $\beta$ or are removed from the resting pool at a rate of $\kappa$ due to differentiation, senescence, or death. Proliferating cells undergo apoptosis at a rate of $\mu$

### 6.2.1 Age-Structured Model

We now present a typical partial differential equation (PDE) model of stem cell regeneration. We simplify the four-stages cell cycle (Fig. 6.1) as a two-phases process, as shown in Fig. 6.2. Stem cells are classified as resting phase ($G_0$) cells ($Q(t)$) or proliferative phase cells ($s(t, a)$). Resting phase cells can either re-enter the proliferative phase at a rate of $\beta(Q)$ that involves negative feedback or differentiate into downstream cell lines at a rate of $\kappa$. An age-structured quantity $s(t, a)$ is used to represent the population of proliferating stem cells, with age $a = 0$ for their time of entry into the proliferative state triggered by the $G_1$ checkpoint. Moreover, proliferating stem cells are assumed to undergo mitosis at a fixed time $\tau$ after entry into the proliferating compartment and to be lost randomly at a rate of $\mu$ during the proliferative phase. Each normal cell generates two resting phase cells at the end of mitosis. Here, the units for cell populations are often measured by the number of cells per unit body weight, e.g., cells/kg, and the rates of proliferation, differentiation, and apoptosis are often united with $day^{-1}$.

The above assumptions yield the following partial differential equations:

$$\nabla s(t, a) = -\mu s(t, a), \quad (t > 0, 0 \le a \le \tau) \tag{6.1}$$

$$\frac{dQ}{dt} = 2 s(t, \tau) - (\beta(Q) + \kappa)Q, \quad (t > 0). \tag{6.2}$$

Here, $\nabla = \partial/\partial t + \partial/\partial a$. The boundary condition at $a = 0$ is given by

$$s(t, 0) = \beta(Q(t))Q(t), \tag{6.3}$$

and the initial conditions are

$$s(0, a) = g(a), \quad (0 \le a \le \tau); \quad Q(0) = Q_0. \tag{6.4}$$

Equations (6.1)–(6.4) give a general PDE model of stem cell regeneration.

## 6.2.2 Delay Differential Equation Model

In Eq. (6.1), we are mostly interested in the resting phase cell number $Q(t)$. To derive a closed equation for $Q(t)$, we need to solve the age-structured model equation to obtain $s(t, \tau)$. This is possible when the apoptosis rate $\mu$ is a constant. To this end, we solve the first-order PDE

$$\nabla s(t, a) = -\mu s(t, a) \quad (t > 0, \quad 0 \le a \le \tau) \tag{6.5}$$

$$s(0, a) = g(a), \quad (0 \le a \le \tau) \tag{6.6}$$

$$s(t, 0) = \beta(Q(t))Q(t) \tag{6.7}$$

through the method of characteristic line.

Introduce a new variable $(x, y)$ so that

$$x = t - a, \quad y = a, \tag{6.8}$$

and denote $s(t, a) = q(x, y)$. The Eq. (6.5) is equivalent to

$$\frac{\partial q}{\partial y} = -\mu q(x, y), \tag{6.9}$$

which gives the solution

$$q(x, y) = q(x, 0)e^{-\mu y}.$$

In terms of the original variables $x$ and $y$, we have

$$s(t, a) = s(t - a, 0)e^{-\mu a}. \tag{6.10}$$

If $t > a$, we have $s(t - a, 0) = \beta(Q(t - a))Q(t - a)$, and hence

$$s(t, a) = \beta(Q(t - a))Q(t - a)e^{-\mu a}, \quad (t > a). \tag{6.11}$$

Now, we consider the situation $t \leq a$. Let $t = 0$ in (6.10), then

$$s(0, a) = s(-a, 0)e^{-\mu a},$$

which gives

$$s(-a, 0) = s(0, a)e^{\mu a} = g(a)e^{\mu a}$$

for any $0 < a < \tau$. Thus, $s(t - a, 0) = g(a - t)e^{\mu(a - t)}$ when $0 < t \leq a$, and hence

$$s(t, a) = s(t - a, 0)e^{-\mu a} = g(a - t)e^{-\mu t}, \quad (0 < t \leq a). \tag{6.12}$$

Combining (6.11) and (6.12), we have

$$s(t, a) = \begin{cases} g(a - t)e^{-\mu t}, & (0 < t \leq a) \\ \beta(Q(t - a))Q(t - a)e^{-\mu a}, & (t > a). \end{cases} \tag{6.13}$$

Let $a = \tau$, and substituting (6.13) into (6.2), we obtain a closed-form differential equation

$$\frac{dQ}{dt} = -(\beta(Q) + \kappa)Q + \begin{cases} 2 g(\tau - t)e^{-\mu t}, & 0 < t \leq \tau \\ 2e^{-\mu \tau}\beta(Q_\tau)Q_\tau, & t > \tau \end{cases} \tag{6.14}$$

Here, $Q_\tau = Q(t - \tau)$.

Finally, while we only consider the long-term behavior $(t > \tau)$ and shift the original time point to $\tau$, we have the delay differential equation (DDE) model

$$\frac{dQ}{dt} = -(\beta(Q) + \kappa)Q + 2e^{-\mu \tau}\beta(Q_\tau)Q_\tau. \tag{6.15}$$

This model describes the general population dynamics of stem cell regeneration.

From the above derivation, if the apoptosis rate $\mu$ is dependent on the age $a$, i.e., $\mu = \mu(a)$, the factor $e^{-\mu \tau}$ in (6.15) should be replaced by $e^{-\int_0^\tau \mu(a)da}$. Moreover, if the duration of the proliferation phase is varied so that $\tau$ is a random number with probability density $\rho(\tau)$, Eq. (6.15) should be revised as a delay differential equation with distributed delay:

$$\frac{dQ}{dt} = -(\beta(Q) + \kappa)Q + 2 \int_0^{+\infty} \rho(\tau)e^{-\int_0^\tau \mu(a)da}\beta(Q(t - \tau))Q(t - \tau)d\tau. \quad (6.16)$$

The above technique is important in the mathematical modeling of life science for connecting an age-structured model with a delay differential equation model.

### 6.2.3 Formulation of the Proliferation Rate

In the above equations, the overall effect of feedback regulation from the cell population to the proliferation rate is given by the function $\beta(Q)$. Biologically, the self-renewal ability of a stem cell is determined by both the microenvironment conditions and intracellular signaling pathways, e.g., growth factors and various types of cytokines. For example, fibroblast growth factors (FGFs) are a family of growth factors that play key roles in the proliferation and differentiation processes of a wide variety of cells and tissues [32]. Transforming growth factor-beta (TGF-$\beta$) is a multifunctional cytokine that belongs to the transforming growth factor superfamily and includes four different isoforms (TGF-$\beta$ 1 to 4). The transforming growth factor superfamily includes many other signaling proteins produced by all white blood cell lineages. TGF-$\beta$ is secreted by many cell types; activated TGF-$\beta$ can form a serine/threonine kinase complex with other factors and bind to TGF-$\beta$ receptors to trigger a signaling cascade. This leads to the activation of different downstream substrates and regulatory proteins, inducing the transcription of different target genes that function in the differentiation, chemotaxis, proliferation, and activation of many immune cells [33, 34]. TGF-$\beta$ plays a crucial role in the regulation of the cell cycle by blocking progression through the $G_1$ phase. It mediates $G_1$ cell cycle arrest by inducing or activating cyclin-dependent kinase (CDK) inhibitors and by inhibiting factors required for CDK activation [35]. TGF-$\beta$ signaling regulates tumorigenesis, and in human cancer, its signaling pathway is often modified during tumor progression and has dual roles as both a tumor suppressor and a tumor promoter. Early during tumorigenesis, TGF-$\beta$ acts to inhibit tumor cell growth and induce apoptosis; however, during the late stage of cancer, it acts as a major inducer of immune suppression and as a driver of epithelial-mesenchymal transition (EMT) in tumor cells [36–38].

Even though the exact activation pathways that regulate stem cell regeneration are poorly understood, one can assume that the cellular response is determined by both the positive and negative signals of proliferation. Assuming that positive growth factors are secreted by the niche and growth factor inhibitors are secreted by the cells, different types of cytokines bind to cell surface receptors to regulate cell behavior, and we can derive the dependence of the proliferation rate $\beta(Q)$ [11, 39].

Let [L] be the concentration of ligands for growth factor inhibitors, [R] be the density of free receptors, [R*] be the density of activated receptors, and $Q$ be the concentration of stem cells. The total number of receptors is

$$[R] + [R^*] = mQ, \tag{6.17}$$

where $m$ is the average number of receptors per cell. If $n$ ligands are required to activate one receptor, we assume that ligands bind to the receptor following the law of mass action

$$[R] + n[L] \leftrightarrows [R^*]. \tag{6.18}$$

At the equilibrium, we have

$$[R][L]^n = K[R^*], \tag{6.19}$$

where $K$ is the equilibrium constant. We assume that the activated receptor inhibits cell proliferation so that the proliferation rate $\beta$ is proportional to the fraction of free receptors on a cell,

$$\beta = \beta_0 \frac{[R]}{mQ}. \tag{6.20}$$

From Eqs. (6.17) and (6.19), we obtain

$$\frac{[R]}{mQ} = \frac{K}{K + [L]^n}. \tag{6.21}$$

When the ligands are secreted from stem cells, and are cleaned at a constant rate, the ligand concentration is proportional to the cell number, which gives $[L] = \sigma Q$. Thus, we obtain the final form of the proliferation rate

$$\beta(Q) = \beta_0 \frac{\theta^n}{\theta^n + Q^n}, \tag{6.22}$$

where $\theta = \sqrt[n]{K}/\sigma$ is the 50% effective concentration (EC50).

From (6.22), the proliferation rate is a decreasing function of the cell population; this is important to maintain the homeostasis of systems. Here, the coefficient $\beta_0$ measures the maximum proliferation in the absence of antigrowth signals, which is mostly determined by the mitogenic growth signals. The EC50 coefficient $\theta$ represents the combined effects from the secretion of antigrowth cytokines and cytokine receptors and the activation of the downstream pathways.

In normal tissues, both mitogenic growth signals and antiproliferative signals are required to maintain cellular quiescence and tissue homeostasis. Nevertheless, the capabilities of self-sufficiency in growth signals and insensitivity to antigrowth signals are two important hallmarks of cancers [36]. Many oncogenes in cancer cells can mimic normal growth signals, thereby reducing their dependence on stimulation from their normal tissue microenvironment. Moreover, the antiproliferative signal pathway can be disrupted in a variety of ways in different types of human tumors. Thus, to model abnormal growth in cancers, we can modify the proliferation rate (6.22) as

$$\beta(Q) = \beta_0 \frac{\theta^n}{\theta^n + Q^n} + \beta_1. \tag{6.23}$$

Here, the nonzero constant $\beta_1$ represents the proliferation rate that is independent of antigrowth signals. For a more realistic model, cell growth can be restricted by physical interactions, and hence, $\beta_1$ may decrease to zero when the cell number $Q$ becomes large enough.

## 6.2.4 Stability of the G₀ Cell Cycle Model

The existence and stability of the equilibrium state of the $G_0$ cell cycle model are essential for the healthy physiological state.

From the delay differential equation (6.15), the equilibrium state solution $Q(t) \equiv Q^*$ satisfies

$$-(\beta(Q^*) + \kappa)Q^* + 2e^{-\mu\tau}\beta(Q^*)Q^* = 0. \tag{6.24}$$

It is easy to see that there is always a zero solution $Q^* = 0$. When the proliferation rate $\beta(Q)$ is a decreasing function, let

$$\beta(0) = \beta_0 + \beta_1, \quad \lim_{Q \to +\infty} \beta(Q) = \beta_1; \tag{6.25}$$

if and only if

$$\beta_0 > \frac{\kappa}{2e^{-\mu\tau} - 1} - \beta_1 > 0, \tag{6.26}$$

there exists a unique positive solution $Q^* > 0$, which is given by

$$\beta(Q^*) = \frac{\kappa}{2e^{-\mu\tau} - 1}. \tag{6.27}$$

To study the stability of the equilibrium state $Q(t) \equiv Q^*$, let $x(t) = Q(t) - Q^*$ and linearize (6.15) at $x = 0$; we obtain

$$\frac{dx}{dt} = -ax + bx_\tau, \tag{6.28}$$

where

$$a = (\beta(Q^*) + \kappa + \beta'(Q^*)Q^*), \quad b = 2e^{-\mu\tau}(\beta(Q^*) + \beta'(Q^*)Q^*). \tag{6.29}$$

From the discussions in Sect. 2.3.2, for any $\tau \geq 0$, the parameter region in the $(a, b)$-plane such that the zero solution of (6.28) is stable is given by

$$S = \{(a, b) \in \mathbb{R}^2 | a \sec \omega\tau < b < a, \text{ where } \omega = -a \tan \omega\tau, a > -\frac{1}{\tau}, \omega \in (0, \frac{\pi}{\tau})\}.$$
(6.30)

Now, we apply (6.30) to discuss the stability of the equilibrium states.

### 6.2.4.1   The Equilibrium State $Q^* = 0$

Now, we consider the zero equilibrium state $Q^* = 0$, which gives

$$a = \beta_0 + \beta_1 + \kappa > 0, \quad b = 2e^{-\mu\tau}(\beta_0 + \beta_1) > 0.$$

The region $S$ given by (6.30) implies that the zero solution is stable if and only if

$$\beta_0 < \frac{\kappa}{2e^{-\mu\tau} - 1} - \beta_1.$$
(6.31)

Interestingly, from the condition (6.26), if a positive equilibrium state exists, the zero solution is unstable. Moreover, if $\beta_1 = 0$ and the zero solution is unstable, there will always exist a positive equilibrium state.

### 6.2.4.2   The Positive Equilibrium State $Q^* > 0$

Now, we consider the positive equilibrium state $Q(t) \equiv Q^* > 0$. In this case,

$$a = (\beta + \kappa - \hat{\beta}), \quad b = 2e^{-\mu\tau}(\beta - \hat{\beta}),$$
(6.32)

where $\beta = \beta(Q^*) > 0$ and $\hat{\beta} = -\beta'(Q^*)Q^* > 0$. At the equilibrium state, we have

$$2e^{-\mu\tau} > 1, \quad \beta = \frac{\kappa}{2e^{-\mu\tau} - 1}.$$
(6.33)

Hence, it is easy to verify $b < a$; thus, the Hopf bifurcation for the equilibrium state is given by

$$b = a \sec \omega\tau, \quad \text{where } \omega = -a \tan \omega\tau.$$
(6.34)

To obtain the bifurcation curve, Eqs. (6.33) and (6.32) yield

$$a = 2\beta e^{-\mu\tau} - \hat{\beta}.$$

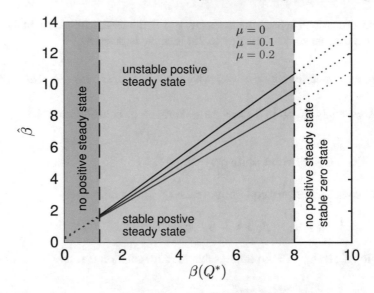

**Fig. 6.3 Bifurcation diagram of the** $G_0$ **cell cycle model** (6.15). The solid lines show the bifurcation curves $\hat{\beta}_{crit} = f(\beta)$. Here, $\tau = 2.8$ days, $\beta_0 = 8.0$ day$^{-1}$, $\beta_1 = 1.2$ day$^{-1}$ (please refer to Sect. 6.2.5)

Hence, the bifurcation curve (6.34) can be rewritten as

$$\begin{cases} \hat{\beta} = -\dfrac{2e^{-\mu\tau}(\sec\omega\tau - 1)}{2e^{-\mu\tau} - \sec\omega\tau}\beta \\[4mm] \omega = -\beta\dfrac{2e^{-\mu\tau}(2e^{-\mu\tau} - 1)}{2e^{-\mu\tau} - \sec\omega\tau}\tan\omega\tau. \end{cases} \tag{6.35}$$

The solution of (6.35) gives the bifurcation curve $\hat{\beta}_{crit} = f(\beta)$ (Fig. 6.3). The curve shows a Hopf bifurcation of the system; when $\hat{\beta} > \hat{\beta}_{crit} = f(\beta)$, the equilibrium state becomes unstable, and a periodic solution may emerge. Here, we note that $\hat{\beta}$ depends on the system parameters through the equilibrium state cell number $Q^*$.

In summary, Fig. 6.3 shows the bifurcation diagram given by the values of $\beta(Q^*)$ and $\hat{\beta} = -\beta(Q^*)Q^*$ at the equilibrium state. A positive equilibrium state exists only when the condition of (6.26) is satisfied (Fig. 6.3, yellow region). Under the condition of (6.31), there is only one equilibrium state, the trivial zero equilibrium state, and the state is stable. When there is a positive equilibrium state, the stability of this state is separated by the bifurcation curve that is defined by Eq. (6.35). The three solid lines in Fig. 6.3 show the bifurcation curves corresponding to different values of the apoptosis rate $\mu$.

As an interesting result from Fig. 6.3, when

$$\frac{\kappa}{2e^{-\mu\tau} - 1} < \beta_1, \tag{6.36}$$

there is no finite equilibrium state of the equation, and $Q(t) \to +\infty$ when $t \to +\infty$. Hence, there is a unique stable steady state at $Q(t) = +\infty$. Consequently, uncontrolled malignant growth is possible when $\beta_1 > 0$ (selfsufficiency growth) and the stem cell removal rate $\kappa$ decreases to a value below the threshold defined by (6.36). Hence, the simple $G_0$ cell cycle model implies three possible conditions of abnormal cell growth: dysregulation in the pathways of differentiation and/or senescence (decreasing $\kappa$), sustained proliferative signaling (increasing $\beta_1$), and evasion of apoptosis (decreasing $\mu$). These conditions are well-known hallmarks of cancer [36]. Moreover, major oncogenic signaling pathways obtained from an integrated analysis of genetic alterations in The Cancer Genome Atlas (TCGA) show direct connections to the coefficients $\beta(Q)$, $\mu$ and $\kappa$ [40, 41].

### 6.2.5 Hematopoietic Stem Cell Regeneration

Now, we apply the above $G_0$ cell cycle model to hemopoiesis. All mature circulating blood cells arise from morphologically undifferentiated cells in bone marrow, which form hematopoietic stem cells (HSCs). The existence of the $G_0$ phase was first proposed in 1970 by Burns and Tannock [8], and the $G_0$ cell cycle model was later applied to investigate the origin of aplastic anemia and periodic hematopoiesis [11]. The model equation for resting phase HSCs is given by

$$\frac{dQ}{dt} = -(\beta(Q) + \kappa)Q + 2e^{-\mu\tau}\beta(Q_\tau)Q_\tau, \tag{6.37}$$

where

$$\beta(Q) = \beta_0 \frac{\theta^n}{\theta^n + Q^n} + \beta_1. \tag{6.38}$$

In this case, the positive steady state $Q^*$ is given by

$$Q^* = \theta \sqrt[n]{\frac{\beta_0}{\eta - \beta_1} - 1}, \quad \eta = \frac{\kappa}{2e^{-\mu\tau} - 1}, \tag{6.39}$$

with the condition

$$2e^{-\mu\tau} > 1, \quad 0 < \eta - \beta_1 < \beta_0. \tag{6.40}$$

**Table 6.1** Default parameter values for HSCs

| Parameter name | Value used | Unit |
|---|---|---|
| $Q^*$ | 1.53 | $\times 10^6$ cells/kg |
| $\mu$ | 0.07 | day$^{-1}$ |
| $\kappa$ | 0.02 | day$^{-1}$ |
| $\tau$ | 2.8 | days |
| $\beta_0$ | 8.0 | day$^{-1}$ |
| $\beta_1$ | 0 | day$^{-1}$ |
| $n$ | 2 | – |
| $\theta$ | 0.096 | $\times 10^6$ cells/kg |

The model parameters can be estimated from different sources. The value of $Q^*$ can be derived from various data. There are 8 stem cells per $10^5$ nucleated bone marrow cells in cats [42], while in mice, there are approximately $1 - 50$ stem cells per $10^5$ nucleated bone marrow cells [43, 44]. There are approximately $1.4 \times 10^{10}$ nucleated bone marrow cells per kg of body mass in mice, which implies that there are between $1.4 \times 10^5$ and $7 \times 10^6$ HSCs per kg of body mass in mice or $1.1 \times 10^6$ stem cells per kg of body mass in cats.

The other parameters have been estimated based on the continuous labeling method [12]. In the mice, the apoptosis rate is approximately $\mu \in (0.69, 0.228)$ day$^{-1}$, the re-entry rate of resting phase cells to the proliferative phase at the steady state is $\beta(Q^*) \in (0.020, 0.053)$ day$^{-1}$, the duration of the proliferative phase is $\tau \in (1.41, 4.25)$ days, and the steady-state rate $\kappa$ of removing differentiation into a downstream cell line from $Q$ is between 0.010 and 0.024 days$^{-1}$ [12]. The duration of the proliferative phase of the HSC cell cycle has been estimated as $\tau = 2.8$ days in cats [42] and between 1.4 and 4.3 days in mice [12].

The parameter values may vary over quite a large region. Here, we assume a resting steady-state HSC number as $Q^* = 1.53 \times 10^6$ cells/kg, an apoptosis rate of $\mu = 0.07$ day$^{-1}$, a proliferation rate at steady-state $\beta(Q^*) = 0.06$ day$^{-1}$, a differentiation rate of $\kappa = 0.02$ day$^{-1}$, and a proliferation duration of $\tau = 2.8$ days [39, 45, 46]. To estimate the parameters in the Hill function (6.38), we assume a Hill coefficient $n = 2$, and $\beta_1 = 0$ for normal HSCs. Values of $2.0 \sim 2.5$ day$^{-1}$ were found for re-entry rate into the cell cycle, suggesting a value of $\beta_0 \geq 2.5$ day$^{-1}$. We take $\beta_0 = 8.0$ day$^{-1}$ [39]. These allow us to compute the value of $\theta = 0.096 \times 10^6$ cells/kg. The default values are listed in Table 6.1.

Based on the default values, one can discuss how changes in each parameter may affect the stability of the positive steady state.

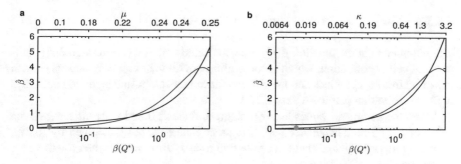

**Fig. 6.4 Bifurcation diagram of the delay differential equation model for HSC dynamics. a.**
Dependence of the bifurcation curve $\hat{\beta}_{\text{crit}} = f(\beta)$ (black line) and $\hat{\beta}$ (red line) on the apoptosis
rate $\mu$. **b.** Dependence of the bifurcation curve $\hat{\beta}_{\text{crit}} = f(\beta)$ (block line) and $\hat{\beta}$ (red line) on the
differentiation rate $\kappa$. Here, $\beta(Q^*)$ and $\hat{\beta} = -\beta'(Q^*)Q^*$ take values at the steady state, and green
circles mark the value corresponding to the default value. The other parameters are set as the default
values shown in Table 6.1

### 6.2.5.1   Changing the Apoptosis Rate $\mu$

The rate $\mu$ is mainly associated with apoptosis during the proliferative phase, which
is mostly regulated by DNA damage responses. When we vary $\mu$ ($\mu < \tau^{-1} \log 2$), the
steady state $Q^*$ varies according to (6.27), and hence, $\beta(Q^*)$ and $\hat{\beta} = -\beta'(Q^*)Q^*$ are
varied. Figure 6.4a shows the dependence of $\hat{\beta}$ on various $\mu$ values (red line) together
with the bifurcation curve shown in Fig. 6.3. From Fig. 6.4, decreasing the apoptosis
rate $\mu$ would not destabilize the positive steady state. Nevertheless, increasing the
apoptosis rate can destabilize the positive steady state by Hopf bifurcation and induce
oscillations in HSCs. This was considered as an origin of periodic hematopoiesis that
shows significant oscillations in blood cell numbers [11, 39].

### 6.2.5.2   Changing the Differentiation Rate $\kappa$

The rate $\kappa$ is mainly associated with the irreversible removal of stem cells from the
resting phase, including stem cell differentiation, death, and senescence. Increasing
$\kappa$ usually results a decrease in the steady state $Q^*$. Figure 6.4b shows the dependence
of $\hat{\beta}$ on $\kappa$ as well as the bifurcation curve shown in Fig. 6.3. Similar to the situation
of the changes in $\mu$, decreasing $\kappa$ would not destabilize the positive steady state,
while increasing $\kappa$ may destabilize the positive steady state by Hopf bifurcation.
Increasing the differentiation rate $\kappa$ has been proposed as a mechanism of periodic
hematopoiesis [11, 39].

### 6.2.5.3  Varying the Proliferation Rate Function $\beta(Q)$

The parameters in the function $\beta(Q)$ are associated with the various signaling pathways of cell proliferation, which involve microenvironment growth factors, growth factor inhibitors, cytokines and the corresponding receptors, and signaling activities in the proliferation pathway (Sect. 6.2.3).

When the parameter values in $\beta(Q)$ change, since $\kappa$, $\mu$ and $\tau$ are unchanged, the value of $\beta(Q^*)$ remains unchanged. The steady-state value $Q^*$ is given by (6.39), which is proportional to $\theta$ (here, we note that $\sigma = 1/\theta^n$ represents the growth factor inhibition ability). Moreover,

$$\hat{\beta} = -\beta'(Q^*)Q^* = n\left[1 - \frac{\eta - \beta_1}{\beta_0}\right](\eta - \beta_1), \quad \eta = \frac{\kappa}{2e^{-\mu\tau} - 1}. \tag{6.41}$$

Comparing the bifurcation curve $\hat{\beta}_{\text{crit}}$ in Fig. 6.4 and $\hat{\beta}$ given by (6.41), we conclude the following:

1. Varying the parameter $\theta$ leads to a linear change in $Q^*$; however, it does not affect the stability of $Q^*$. Biologically, when growth inhibition is ineffective so that $\theta \to +\infty$, the steady state $Q^*$ approaches infinity, which implies uncontrolled growth.
2. We usually have $\eta \ll \beta_0$. Hence, varying the parameter $\beta_0$ usually does not change the stability of $Q^*$. This result suggests that cell growth is robust with respect to growth factor secretion from the niche.
3. When $\beta_1$ increases from 0 to $\eta$, the value of $\hat{\beta}$ decreases from $n\eta(1 - \eta/\beta_0)$ to 0, and hence, the stability of $Q^*$ remains unchanged. However, the positive steady state does not exist when $\beta_1 > \eta$. Biologically, this result indicates that the capability of self-sufficiency in cancer stem cells can result in uncontrolled tumor cell growth in patients.

Based on the above analysis, we show sample solutions for various parameter values (Fig. 6.5). From these sample solutions, increasing either $\mu$ or $\kappa$ can induce oscillatory solutions with stem cell numbers below default values; increasing $\beta_1$ above a threshold results in the blow-up of the solution. We note that when the solution is oscillatory, the oscillation patterns are dependent on the sources of changes in the parameter values.

## 6.3  Heterogeneous Stem Cell Population Dynamics

In the above $G_0$ cell cycle model, all cells are assumed to be homogeneous, and the cell population dynamics are described by either a partial differential equation (age-structured model) or a delay differential equation. However, cell heterogeneity is essential for an organism, and cells differ by their epigenetic states, including transcriptomes, patterns of DNA methylation, and nucleosome histone modifications.

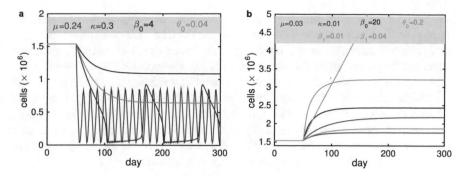

**Fig. 6.5  Sample solutions of Eq.** (6.37) **with various parameter values. a.** Parameters to reduce the stem cell number. **b.** Parameters to increase the stem cell number. The changed parameters are shown in the figure, and the other values are the same as the default values in Table 6.1. In all simulations, we first set the parameters as their default values, $Q(t) \equiv Q^*$ for $t < \tau$, solve the equation to $t = 50$, then vary a parameter value accordingly and continue the numerical solution to $t = 300$

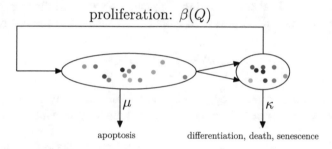

**Fig. 6.6  Illustration of heterogeneous stem cell regeneration.** During stem cell regeneration, cells in the resting phase either enter the proliferative phase at a rate of $\beta$ or are removed from the resting pool at a rate of $\kappa$. Proliferating cells undergo apoptosis at a rate of $\mu$. The heterogeneous cells are represented by dots of different colors, and the rates of proliferation, differentiation, and apoptosis for each cell are dependent on its epigenetic state

During the proliferative phase, each mother cell duplicates its genome and regenerates the epigenetic states therein so that the epigenetic states randomly change over cell divisions [47, 48] (Fig. 6.6). The epigenetic states are not perfectly inherited, which may lead to dynamic heterogeneity in daughter cells that have a state distinguished from the state of mother cells [22, 49, 50].

## 6.3.1  Mathematical Formulation

To establish the formulation of heterogeneous stem cell regeneration, we introduce $\Omega$ to denote a space for all possible epigenetic states of resting phase stem cells, and the epigenetic state of a cell is represented by $\mathbf{x} \in \Omega$. Here, the epigenetic state of a cell can be any heritable quantity of gene activity whose changes are not caused by changes in the DNA sequence; for example, the expression levels, pattern of DNA methylation, state of histone modifications, or expression levels of cell surface marker genes that may distinguish differentiated cells from stem cells. Experimentally, the quantities can be measured by single-cell sequencing techniques, such as single-cell RNA sequencing [51], single-cell ChIP sequencing [52], or single-cell genome-wide bisulfite sequencing [53]. In applications, we usually do not use $\mathbf{x}$ to refer to the whole genome data. Instead, we may consider $\mathbf{x}$ as a low-dimensional quantity associated with specific genes or pathway activities that may affect the signaling pathways controlling cell cycle progression, apoptosis, or cell growth.

### 6.3.1.1   Delay Differential-Integral Equation Model of Heterogeneous Stem Cell Regeneration

The population dynamics of the stem cell regeneration system are measured by the number of stem cells at time $t$ during the resting phase with different epigenetic states $\mathbf{x}$, denoted as $Q(t, \mathbf{x})$. The total number of stem cells in the resting phase is then given by

$$Q(t) = \int_\Omega Q(t, \mathbf{x})d\mathbf{x}. \tag{6.42}$$

We always assume continuous epigenetic states. Mathematically, it is easy to extend the discussions to discrete situations by replacing the integral with summation over all $\mathbf{x}$ in $\Omega$.

Each cell in the resting phase re-enters the proliferative phase at a rate of $\beta$. The proliferation rate $\beta$ is dependent on the total cell population through various types of cytokines [7, 14, 39] as well as the epigenetic state $\mathbf{x}$ of each individual cell. Thus, the proliferation rate has the form $\beta(c, \mathbf{x})$, where $c$ measures the effective concentration of growth inhibition cytokines

$$c(t) = \int_\Omega Q(t, \mathbf{x})\zeta(\mathbf{x})d\mathbf{x}, \tag{6.43}$$

where $\zeta(\mathbf{x})$ is the rate of cytokine secretion. Both the apoptosis rate $\mu$ and the differentiation rate $\kappa$ are dependent on the epigenetic state $\mathbf{x}$ and are denoted as $\mu(\mathbf{x})$ and $\kappa(\mathbf{x})$, respectively. The timing of cell cycling $\tau$ also depends on the state $\mathbf{x}$. Finally, the epigenetic state of a proliferating cell undergoes state transition during the proliferative process (from the $G_1$ phase to the $M$ phase) due to stochastic dynamics in DNA methylation and histone modification [47, 48]. Here, we do not model the detailed biochemical reactions during cell cycling; instead, we introduce a transition function $p(\mathbf{x}, \mathbf{y})$ to represent the probability that a mother cell of state $\mathbf{y}$ gives a daughter cell of state $\mathbf{x}$ (at the end of the mitosis phase). It is obvious to have the normalization condition

$$\int_{\Omega} p(\mathbf{x}, \mathbf{y})d\mathbf{x} = 1, \qquad \forall \mathbf{y} \in \Omega. \tag{6.44}$$

In a small time interval $[t, t + \Delta t]$, a cell in the resting phase with state $\mathbf{x}$ either re-enters the proliferative phase with a probability of $\beta(c, \mathbf{x})\Delta t$, is removed with a probability of $\kappa(\mathbf{x})\Delta t$, or remains at the resting phase. When a cell in state $\mathbf{x}$ enters the proliferative phase ($G_1$ phase) at $t - \tau(\mathbf{x})$, it either undergoes apoptosis with a probability of $1 - e^{-\mu(\mathbf{x})\tau(\mathbf{x})}$ during the proliferative phase or, if it survives (with a probability of $e^{-\mu(\mathbf{x})\tau(\mathbf{x})}$), it produces two daughter cells at time $t$. Each daughter cell gains the epigenetic state $\mathbf{x}$ with a probability of $p(\mathbf{x}, \mathbf{y})$ given the state $\mathbf{y}$ of the mother cell. These processes lead to the following iteration equation for the resting phase cell number (also refer to [26] and [27]):

$$Q(t + \Delta t, \mathbf{x}) = Q(t, \mathbf{x}) - Q(t, \mathbf{x})(\beta(c, \mathbf{x}) + \kappa(\mathbf{x}))\Delta t$$
$$+ 2\int_{\Omega} \beta(c_{\tau(\mathbf{y})}, \mathbf{y})Q(t - \tau(\mathbf{y}), \mathbf{y})e^{-\mu(\mathbf{y})\tau(\mathbf{y})}p(\mathbf{x}, \mathbf{y})d\mathbf{y}\Delta t.$$

This gives the following delay differential-integral equation for all $\mathbf{x} \in \Omega$:

$$\begin{cases} \dfrac{\partial Q(t, \mathbf{x})}{\partial t} = -Q(t, \mathbf{x})(\beta(c, \mathbf{x}) + \kappa(\mathbf{x})) \\ \qquad + 2\displaystyle\int_{\Omega} \beta(c_{\tau(\mathbf{y})}, \mathbf{y})Q(t - \tau(\mathbf{y}), \mathbf{y})e^{-\mu(\mathbf{y})\tau(\mathbf{y})}p(\mathbf{x}, \mathbf{y})d\mathbf{y}, \\ c(t) = \displaystyle\int_{\Omega} Q(t, \mathbf{x})\zeta(\mathbf{x})d\mathbf{x}. \end{cases} \tag{6.45}$$

Here, $c_{\tau} = c(t - \tau)$, and the coefficient 2 means that each mother cell generates two daughter cells in cell division. The differential-integral equation (6.45) gives the basic dynamical equation to model stem cell regeneration with epigenetic transition.

The above equation can also be derived from the extension of the age-structure model (6.1)–(6.2), which reads

$$\nabla s(t, a, \mathbf{x}) = -\mu(\mathbf{x})s(t, a, \mathbf{x}), \quad (t > 0, 0 < a < \tau(\mathbf{x})) \tag{6.46}$$

$$\frac{\partial Q(t, \mathbf{x})}{\partial t} = 2\int_{\Omega} s(t, \tau(\mathbf{y}), \mathbf{y})p(\mathbf{x}, \mathbf{y})d\mathbf{y}$$
$$- (\beta(c(t), \mathbf{x}) + \kappa(\mathbf{x}))Q(t, \mathbf{x}), \quad (t > 0) \tag{6.47}$$

and

$$s(t, 0, \mathbf{x}) = \beta(c(t), \mathbf{x})Q(t, \mathbf{x}).$$

Here, $\nabla = \partial/\partial t + \partial/\partial a$. Similar to the previous argument, (6.46) can be solved through the characteristic line method, which gives

$$s(t, \tau(\mathbf{x}), \mathbf{x}) = \beta(c_{\tau(\mathbf{x})}, \mathbf{x})Q(t - \tau(\mathbf{x}), \mathbf{x})e^{-\mu(\mathbf{x})\tau(\mathbf{x})}.$$

Thus, substituting $s(t, \tau(\mathbf{x}), \mathbf{x})$ into (6.47), we re-obtain the differential-integral equation (6.45).

In the Eq. (6.45), $\beta(c, \mathbf{x})$, $\kappa(\mathbf{x})$, and the $\mu(\mathbf{x})$ are rates of proliferation, differentiation, and apoptosis, respectively, and $p(\mathbf{x}, \mathbf{y})$ is the transition function. Hence, we always have the following:

$$\beta(c, \mathbf{x}), \kappa(\mathbf{x}), \mu(\mathbf{x}), \tau(\mathbf{x}), \zeta(\mathbf{x}) \geq 0, p(\mathbf{x}, \mathbf{y}) \geq 0. \tag{6.48}$$

Furthermore, the function $\beta(c, \mathbf{x})$ is continuous with $c \in \mathbb{R}^+$. Biologically, Eq. (6.45) connects gene expression at the single-cell level ($\mathbf{x}$), population dynamic properties ($\beta(c, \mathbf{x})$, $\kappa(\mathbf{x})$, and $\mu(\mathbf{x})$), the cell cycle ($\tau(\mathbf{x})$), and epigenetic modification inheritance ($p(\mathbf{x}, \mathbf{y})$). Thus, this equation gives a multiscale model for stem cell regeneration and can be applied to different problems related to cell regeneration, such as development, aging, and tumor development. For further discussions and open problems related to Eq. (6.45), please refer to [29, 41].

### 6.3.1.2   The Transition Function $p(\mathbf{x}, \mathbf{y})$

It is in general difficult to determine the transition function $p(\mathbf{x}, \mathbf{y})$, which may depend on the biochemical details within cell divisions. Here, we propose a phenomenological function to represent the stochastic inheritance of epigenetic states [28].

First, when $\mathbf{x} = (x_1, x_2, \cdots, x_n)$ represents the expression levels of $n$ independent marker genes, we assume

$$p(\mathbf{x}, \mathbf{y}) = \prod_{i=1}^{n} p_i(x_i, \mathbf{y}), \tag{6.49}$$

with $p_i(x_i, \mathbf{y})$ being the transition function for each gene.

The epigenetic state of a daughter cell is a random number with a distribution depending on that of the mother cell. In Sect. 4.6.1, we found that the nucleosome modification level of daughter cells, when normalized to the interval [0, 1], can be described by a beta-distributed random number dependent on the state of the mother cell. Therefore, we generalize these findings and define the inheritance function $p_i(x_i, \mathbf{y})$ through the beta distribution density function as follows:

$$p_i(x_i, \mathbf{y}) = \frac{x_i^{a_i(\mathbf{y})-1}(1 - x_i)^{b_i(\mathbf{y})-1}}{B(a_i(\mathbf{y}), b_i(\mathbf{y}))}, \quad B(a, b) = \frac{\Gamma(a)\Gamma(b)}{\Gamma(a + b)}, \tag{6.50}$$

where $a_i$ and $b_i$ are shape parameters that depend on the state of the mother cell.

The beta distribution is one of the few common distributions defined on a finite interval and is parameterized by two positive shape parameters that appear as positive exponents of the random variable. As the shape parameters vary, the beta distribution can take different shapes, including strictly decreasing ($a \leq 1, b > 1$), strictly increasing ($a > 1, b \leq 1$), U-shaped ($a < 1, b < 1$), or unimodal ($a > 1, b > 1$).

For a random variable $X$ that is beta-distributed with parameters $a$ and $b$, the corresponding mean and variance are

$$E[X] = \frac{a}{a+b}, \quad \mathrm{Var}[X] = \frac{ab}{(a+b)^2(a+b-1)}.$$

Thus, if we assume that there are functions $\phi_i(\mathbf{y})$ and $\eta_i(\mathbf{y})$ so that the mean and variance of $x_i$, given the state $\mathbf{y}$, are

$$E(x_i|\mathbf{y}) = \phi_i(\mathbf{y}), \quad \mathrm{Var}(x_i|\mathbf{y}) = \frac{1}{1+\eta_i(\mathbf{y})}\phi_i(\mathbf{y})(1-\phi_i(\mathbf{y})),$$

the shape parameters are given by

$$a_i(\mathbf{y}) = \eta_i(\mathbf{y})\phi_i(\mathbf{y}), \quad b_i(\mathbf{y}) = \eta_i(\mathbf{y})(1-\phi_i(\mathbf{y})). \tag{6.51}$$

Here, the functions $\phi_i(\mathbf{y})$ and $\eta_i(\mathbf{y})$ always satisfy

$$0 < \phi_i(\mathbf{y}) < 1, \quad \eta_i(\mathbf{y}) > 0. \tag{6.52}$$

Hence, the transition function can be determined through the predefined functions $\phi_i(\mathbf{y})$ and $\eta_i(\mathbf{y})$, often through data-driven modeling or assumptions that satisfy (6.52).

Given the functions $\phi_i(\mathbf{y})$ and $\eta_i(\mathbf{y})$ that satisfy (6.52) and the shape parameters $a_i(\mathbf{y})$ and $b_i(\mathbf{y})$ defined by (6.51), the transition function $p(\mathbf{x}, \mathbf{y})$ is given by

$$p(\mathbf{x}, \mathbf{y}) = \prod_{i=1}^{n} \frac{x_i^{a_i(\mathbf{y})-1}(1-x_i)^{b_i(\mathbf{y})-1}}{B(a_i(\mathbf{y}), b_i(\mathbf{y}))}. \tag{6.53}$$

Please note that the selection of the beta distribution is only one of the possible options used to define the transition function. For instance, the analytic distributions in Sect. 4.3.1 suggest that the gamma distribution may also be a proper choice to describe the transition of protein numbers.

### 6.3.1.3 Remarks

*Remark 1* When the delay $\tau$ is not considered explicitly, Eq. (6.45) should be rewritten as the following differential-integral equation:

$$\frac{\partial Q(t, \mathbf{x})}{\partial t} = -Q(t, \mathbf{x})(\beta(c, \mathbf{x}) + \kappa(\mathbf{x})) + 2\int_{\Omega} \beta(c, \mathbf{y})Q(t, \mathbf{y})e^{-\mu(\mathbf{y})}p(\mathbf{x}, \mathbf{y})d\mathbf{y}. \tag{6.54}$$

Here, we note that by omitting the delay, we do not simply set $\mu = 0$ in Eq. (6.45); we only omit the delays in $Q(t - \tau(\mathbf{y}), \mathbf{y})$ and $c_{\tau(\mathbf{y})}$ and replace $e^{-\mu(\mathbf{y})\tau(\mathbf{y})}$ with $e^{-\mu(\mathbf{y})}$, which represents the survival probability of a cell in state $\mathbf{y}$ at the end of the proliferative phase, and $\mu(\mathbf{y})$ represents the apoptosis rate of a cell during the proliferative phase.

*Remark 2* The assumption of a beta distribution (6.50) should not be considered definitive in the proposed model. The beta distribution is only an approximation under certain circumstances and should be improved and generalized from different aspects. For example, a gamma distribution may be more appropriate for describing the transcriptional levels [54, 55]. Gene expression patterns of paired daughter cells from individual stem cell divisions can also be studied through machine learning methods [56].

*Remark 3* Integrating equation (6.45) with respect to all $\mathbf{x} \in \Omega$, we obtain the equation for the total cell number $Q(t) = \int_\Omega Q(t, \mathbf{x})d\mathbf{x}$:

$$\frac{dQ}{dt} = -\int_\Omega Q(t, \mathbf{x})(\beta(c, \mathbf{x}) + \kappa(\mathbf{x}))d\mathbf{x} + 2\int_\Omega \beta(c_{\tau(\mathbf{x})}, \mathbf{x})Q(t - \tau(\mathbf{x}), \mathbf{x})e^{-\mu(\mathbf{x})\tau(\mathbf{x})}d\mathbf{x}.$$

(6.55)

Moreover, if we omit the heterogeneity, all rate functions are independent of $\mathbf{x}$ and $c(t) = Q(t)$ and have a delay differential equation

$$\frac{dQ}{dt} = -(\beta(Q) + \kappa)Q + 2e^{-\mu\tau}\beta(Q_\tau)Q_\tau.$$

(6.56)

Thus, we re-obtain the delay differential equation of stem cell regeneration in Eq. (6.15).

*Remark 4* When the relative cell number is defined

$$f(t, \mathbf{x}) = \frac{Q(t, \mathbf{x})}{Q(t)},$$

we can derive the equation for $f(t, \mathbf{x})$ as

$$\frac{\partial f(t, \mathbf{x})}{\partial t} = -f(t, \mathbf{x})\int_\Omega f(t, \mathbf{y})[(\beta(c, \mathbf{x}) + \kappa(\mathbf{x})) - (\beta(c, \mathbf{y}) + \kappa(\mathbf{y}))]d\mathbf{y} \quad (6.57)$$

$$+ \frac{2}{Q(t)}\int_\Omega \beta(c_{\tau(\mathbf{y})}, \mathbf{y})Q(t - \tau(\mathbf{y}), \mathbf{y})e^{-\mu(\mathbf{y})\tau(\mathbf{y})}(p(\mathbf{x}, \mathbf{y}) - f(t, \mathbf{x}))d\mathbf{y}$$

Equations (6.55) and (6.57) provide the evolution dynamics of the relative numbers of cells that can be obtained from experiments by single-cell sequencing or flow cytometry. These set equations can be a theoretical basis for inferring population dynamics from single-cell RNA-sequencing time series data [57].

At the state of homeostasis, the relative cell number

$$f(\mathbf{x}) = \lim_{t \to +\infty} f(t, \mathbf{x})$$

satisfies the integral equations

$$
\begin{cases}
-f(\mathbf{x}) \int_\Omega f(\mathbf{y}) \Big( (\beta(c, \mathbf{x}) + \kappa(\mathbf{x})) - (\beta(c, \mathbf{y}) + \kappa(\mathbf{y})) \Big) d\mathbf{y} \\
\qquad + 2 \int_\Omega \beta(c, \mathbf{y}) f(\mathbf{y}) e^{-\mu(\mathbf{y})\tau(\mathbf{y})} (p(\mathbf{x}, \mathbf{y}) - f(\mathbf{x})) d\mathbf{y} = 0, \\
\int_\Omega Q(\mathbf{x}) \Big( 2\beta(c, \mathbf{x}) e^{-\mu(\mathbf{x})\tau(\mathbf{x})} - (\beta(c, \mathbf{x}) + \kappa(\mathbf{x})) \Big) d\mathbf{x} = 0, \\
\qquad\qquad\qquad\qquad c - \int_\Omega Q(\mathbf{x}) \zeta(\mathbf{x}) d\mathbf{x} = 0.
\end{cases}
$$

The function $f(\mathbf{x})$ gives the density of cells with different epigenetic states at homeostasis, which can be directly associated with experimental data obtained from single-cell techniques, such as single-cell sequencing or flow cytometry.

*Remark 5* The model Eq. (6.45) is a delay differential-integral equation that contains nonlocal transitions between different epigenetic states. This type of equation is different from most physical problems because of the principle of locality. Mathematically, many basic problems remain open for the Eq. (6.45). Here, we list some of them that are important for understanding of the biological processes of stem cell regeneration (please refer to [29] for detailed discussions).

To formulate the simple form of mathematical problems, we omit the delay and consider the equation with only one epigenetic state dimension. This gives the following equation

$$
\frac{\partial Q(t, x)}{\partial t} = -Q(t, x)(\beta(c, x) + \kappa(x)) + 2 \int_\Omega \beta(c, y) Q(t, y) e^{-\mu(y)} p(x, y) dy,
$$
$$
c(t) = \int_\Omega Q(t, x) \zeta(x) dx.
$$

$$(6.58)$$

Here, $x \in \Omega \subset \mathbb{R}^+$, and we assume

$$\beta(x) \geq 0, \quad \kappa(x) \geq 0, \quad \mu(x) \geq 0, \quad \zeta(x) \geq 0, \quad p(x, y) \geq 0, \qquad (6.59)$$

and

$$\int_\Omega p(x, y) dx = 1, \quad \forall y \in \Omega. \qquad (6.60)$$

Based on Eq. (6.58), the problem of the existence and uniqueness of the steady-state solution leads to a nonlinear eigenvalue problem

$$L_c[Q(x)] = 2 \int_\Omega \beta(c, y) e^{-\mu(x)} p(x, y) Q(y) dy - (\beta(c, x) + \kappa(x)) Q(x) = 0. \quad (6.61)$$

We need to find a positive eigenvalue $c$ for the operator $L_c$ so that the corresponding eigenfunction $Q(x)$ is nonnegative for all $x \in \Omega$.

Biologically, the steady-state solution corresponds to the possible persistent state. The existence and stability of steady states is biologically important for understanding the persistence of different states.

We further define the fraction of cells with a given epigenetic state $x$ as

$$f(t, x) = \frac{Q(t, x)}{Q(t)}, \quad Q(t) = \int_\Omega Q(t, x) dx,$$

and the entropy of the multicellular system

$$E(t) = - \int_\Omega f(t, x) \log f(t, x) dx.$$

The evolution dynamics of the entropy is given by

$$\frac{dE}{dt} = \int_\Omega \frac{\partial f(t, x)}{\partial t} (1 + \log f(t, x)) dx, \quad (6.62)$$

and the equation for $f(t, x)$ reads

$$\frac{\partial f(t, x)}{\partial x} = -f(t, x) \int_\Omega f(t, y)((\beta(c, x) + \kappa(x)) - (\beta(c, y) + \kappa(y))) dy$$

$$+ 2 \int_\Omega f(t, y) \beta(c, y) e^{-\mu(y)} (p(x, y) - f(t, x)) dy, \quad (6.63)$$

$$\frac{dQ}{dt} = Q \int_\Omega f(t, x)(\beta(c, x)(2e^{-\mu(x)} - 1) - \kappa(x)) dx, \quad (6.64)$$

$$c = Q(t) \int_\Omega f(t, x) \zeta(x) dx. \quad (6.65)$$

These equations give the dynamics of entropy changes during the developmental process of tissue growth.

When the state transition is localized, i.e.,

$$p(x, y) = \varphi(y - x) \quad (6.66)$$

so that $\varphi(r) > 0$ only when $|r| < \varepsilon \ll 1$, we let

$$a = \int_{-\infty}^{+\infty} r \varphi(r) dr, \quad D = \int_{-\infty}^{+\infty} r^2 \varphi(r) dr.$$

Substituting (6.66) into (6.63) and expanding the function $\varphi$ to the second-order approximation, we obtain the following differential-integral equation:

$$\frac{\partial f(t, x)}{\partial x} = \frac{\partial}{\partial x} \left[ \left( D\frac{\partial}{\partial x} + 2a \right) (\beta(c, x)e^{-\mu(x)}f(t, x)) \right]$$

$$+ f(t, x) \left( \gamma(c, x) - \int_{\Omega} f(t, y)\gamma(c, y)dy \right), \tag{6.67}$$

where

$$\gamma(c, x) = \beta(c, x)(2e^{-\mu(x)} - 1) - \kappa(x).$$

This gives a nonlinear diffusion equation model for the dynamics of the amount of epigenetic state spreading of a cell population. When we consider the steady-state solution $(f(t, x) = f(x))$, which corresponds to the stationary epigenetic state distribution, we have a second-order differential equation

$$\frac{d}{dx} \left[ (D\frac{d}{dx} + 2a)(\beta(\lambda, x)e^{-\mu(x)}f(x)) \right] + \gamma(\lambda, x)f(x) = 0 \tag{6.68}$$

with the boundary condition (here, $\Omega \subset \mathbb{R}$)

$$f(x) \geq 0, \quad \int_{\Omega} f(x)dx = 1, \quad \lambda = Q^* \int_{\Omega} f(x)\zeta(x)dx.$$

This equation defines a nonlinear eigenvalue problem for the eigenvalue $\lambda$ that is associated with the stationary state cell number $Q^*$.

## 6.3.2 Mathematical Model of Cell Lineages

### 6.3.2.1 Discrete Cell Lineages

Heterogeneity in stem cell regeneration can be described by mathematical models of cell lineages. The cell lineage is important in tissue and organ development, maintenance, and regeneration. The highly terminally differentiated cells that characterize the mature functions of tissues are seen as the end products of orderly, tissue-specific sequences of cell divisions, during which progenitor cells pass through distinct stages that are marked by the expression of stage-specific genes [58–61]. The functions of lineages are often presented in terms of the progressive allocation of developmental potential. Stem cells are often found at the starting points of lineages, particularly those in self-renewing tissues such as blood, epidermis, and the intestinal lining. Stem cells are characterized by both multipotency and their ability to maintain their own numbers through self-replication. Lineage intermediates often display "transit-amplifying" behavior, i.e., they are capable of at least some degree of self-replication.

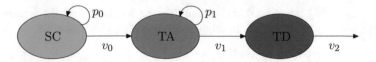

**Fig. 6.7 Model of cell lineage behavior.** Schematic representation of an unbranched lineage that begins with stem cells (SCs), progresses through a transit-amplifying (TA) stage, and ends with terminally differentiated (TD) cells. The parameters $p_i$ and $v_i$ are the rates of net proliferation and differentiation, respectively. The turnover of the terminally differentiated cells is represented with a cell death rate constant $v_2$

At the end points of lineages, there are often terminally differentiated cells that are no longer dividing [7, 62].

A model diagram of discrete cell lineages is shown in Fig. 6.7. The lineage begins with stem cells (SCs), progresses through a transit-amplifying (TA) stage, and ends with terminally differentiated (TD) cells. The net proliferation of stem cells is assumed to be smaller than that of TA cells ($p_0 < p_1$). The differentiation rates from stem cells to TA cells and from TA cells to TD cells are $v_0$ and $v_1$, respectively. The TD cells are no longer dividing and are lost with a cell death rate of $v_2$. Let $x_0$, $x_1$, and $x_2$ be the numbers of stem cells, TA cells, and TD cells, respectively; the lineage behavior is represented by a set of ordinary differential equations

$$\frac{dx_0}{dt} = p_0 x_0 - v_0 x_0, \tag{6.69}$$

$$\frac{dx_1}{dt} = v_0 x_0 + p_1 x_1 - v_1 x_1, \tag{6.70}$$

$$\frac{dx_2}{dt} = v_1 x_1 - v_2 x_2. \tag{6.71}$$

In the absence of feedback controls, the rates $p_i$ and $v_i$ are independent of cell populations. In this case, the model system (6.69)–(6.71) is severely fragile because $p_0$ must be *exactly* equal to $v_0$ for a nonzero steady state to exist. Moreover, the steady-state solution of the output, the TD cell number

$$x_2 = \frac{v_0 v_1}{v_2(p_1 - v_1)} x_0 \tag{6.72}$$

is linear or more than linear depending on the system parameters. Hence, a cell lineage model without feedback control is a *bad* biological system in terms of tissue growth and maintenance [7].

Many tissues and organs grow to precise sizes and when injured, regenerate accurately and rapidly [6, 63, 64]. Hence, the growth and regeneration of tissue stem cells must be tightly controlled. In various studies, feedback regulation from terminally differentiation cells to the proliferation or differentiation of progenitor cells was introduced; such feedback can be implemented when terminally differentiated cells secrete molecules that decrease the probability of progenitor cell replication or

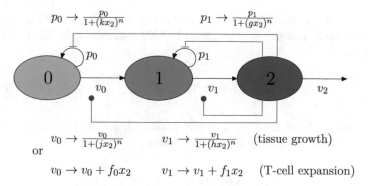

$$p_0 \rightarrow \frac{p_0}{1+(kx_2)^n} \qquad\qquad p_1 \rightarrow \frac{p_1}{1+(gx_2)^n}$$

$$\text{or} \quad v_0 \rightarrow \frac{v_0}{1+(jx_2)^n} \qquad v_1 \rightarrow \frac{v_1}{1+(hx_2)^n} \quad \text{(tissue growth)}$$

$$v_0 \rightarrow v_0 + f_0x_2 \qquad v_1 \rightarrow v_1 + f_1x_2 \quad \text{(T-cell expansion)}$$

**Fig. 6.8 Feedbacks in cell lineage behavior**. Negative feedbacks from terminally differentiated cells to progenitor cell proliferation and differentiation are modeled by decreasing functions of the cell number $x_2$. In the case of T-cell expansion, the positive regulation from terminally differentiated cells to progenitor cell differentiation can be described by a linear dependence of the differentiation rate on the terminal mature cell number $x_2$. Replotted from [7]

differentiation [7]. In mathematical models, the feedback regulations are described by the dependence of rates $p_i$, $v_i (i = 0, 1)$ on the terminally differentiated cells $x_2$ through decreasing functions (Fig. 6.8).

In the immune system, specific T-cell numbers rapidly increase in response to cognate antigens and then steeply decline, approaching relatively stable frequencies that are higher than those in the naive cell population [65, 66]. To this end, terminally differentiated cells may increase differentiation rate of progenitor cells to form positive feedback [62] (Fig. 6.8).

### 6.3.2.2 Continuous Cell Lineages

In the above discrete cell lineage model, the cell lineages include well-defined discrete cell types from stem cells to terminally differentiated cells, and the differentiation process is mostly irreversible. However, in recent years, an increasing number of single-cell sequencing data have suggested that cell state transitions are reversible and show a continuous spectrum in stem cell differentiation [67–71]. To model the reversible cell fate transition and continuous spectrum in stem cell differentiation, we can apply the continuous cell lineage model based on the previously introduced heterogeneous stem cell regeneration model.

To select a proper epigenetic state variable for the cell fate of HSCs, we note CD34 , a member of the single-pass transmembrane sialomucin proteins family that expressed in early hematopoietic and vascular-associated tissues. CD34 expression (CD34$^+$) is often a maker of HSCs; however, little is known about its exact function [72]. To model the cell lineage of hematopoietic stem cells, in Eq. (6.54), we can set $x$ as the expression level of CD34 (state $x$ may refer to alternative marker genes or simply the stemness index when we consider other types of stem cells). Moreover, the proliferation rate $\beta$ and differentiation rate $\kappa(x)$ are dependent on CD34 expression, and the apoptosis rate $\mu$ is independent of CD34 expression, so Eq. (6.54) can be rewritten as

$$\frac{\partial Q(t, x)}{\partial t} = -Q(t, x)(\beta(c, x) + \kappa(x)) + 2e^{-\mu} \int_0^{+\infty} \beta(c, y)Q(t, y)p(x, y)dy.$$

(6.73)

Here, we have omitted the delay $\tau$. Furthermore, while CD34 is mainly associated with the stemness, we assume $\beta(c, x)$ as a form of

$$\beta(c, x) = \beta(x)f(c), \quad f(c) = \frac{\theta^n}{\theta^n + c^n}.$$

(6.74)

Thus, Eq. (6.73) becomes

$$\begin{cases} \dfrac{\partial Q(t, x)}{\partial t} = -Q(t, x)(\beta(x)f(c) + \kappa(x)) \\ \qquad\qquad + 2e^{-\mu}f(c) \displaystyle\int_0^{+\infty} \beta(y)Q(t, y)p(x, y)dy \\ c(t) = \displaystyle\int_0^{+\infty} Q(t, x)\zeta(x)dx. \end{cases}$$

(6.75)

Equation (6.75) models the dynamics of self-renewal cells. To consider the number of TD cells, we introduce $L(t)$ for TD cells, and (6.75) is extended to

$$\begin{cases} \dfrac{\partial Q(t, x)}{\partial t} = -Q(t, x)(\beta(x)f(c) + \kappa(x)) \\ \qquad\qquad + 2e^{-\mu}f(c) \displaystyle\int_0^{+\infty} \beta(y)Q(t, y)p(x, y)dy \\ \dfrac{dL}{dt} = \displaystyle\int_0^{+\infty} \kappa(x)Q(t, x)dx - \delta L, \\ c(t) = \displaystyle\int_0^{+\infty} Q(t, x)\zeta(x)dx. \end{cases}$$

(6.76)

There are rare results based on the model Eq. (6.76). Here, we show a numerical simulation of the model to demonstrate how this model can be applied to describe the dynamics of cell lineages. Here, we assume $0 < x < 1$ and take the rate functions

$$\beta(x) = \beta_0 \frac{3.8x}{1 + (2.5x)^6}, \quad \kappa(x) = \frac{\kappa_0}{1 + (2.5x)^6},$$

(6.77)

$$\zeta(x) = \frac{(1.8x)^8}{1 + (1.8x)^8}, \quad f(c) = \frac{\theta^2}{\theta^2 + c^2}.$$

(6.78)

**Fig. 6.9 The functions**
$\beta(x)$, $\kappa(x)$ and $\zeta(x)$. The
dashed line separates stem
cells with a low proliferation
rate and progenitors with the
ability of rapid proliferation

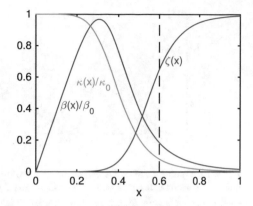

Figure 6.9 shows the plots of the functions $\beta(x)$, $\kappa(x)$ and $\zeta(x)$. Based on the function $\beta(x)$, the cells are classified into stem cells ($x > 0.6$, low proliferation rate) and progenitors ($x < 0.6$, high proliferation rate). Thus, the numbers of stem cells and progenitors are respectively given by

$$Q_{\text{stem cell}}(t) = \int_{0.6}^{1} Q(t, x)dx, \quad Q_{\text{progenitor}}(t) = \int_{0}^{0.6} Q(t, x)dx.$$

We take the transition function $p(x, y)$ following the beta distribution as

$$p(x, y) = \frac{x^{a(y)-1}(1 - x)^{b(y)-1}}{B(a(y), b(y))}, \quad B(a, b) = \frac{\Gamma(a)\Gamma(b)}{\Gamma(a + b)}, \tag{6.79}$$

where

$$a(y) = (m - 1)\phi(y), \quad b(y) = (m - 1)(1 - \phi(y)), \quad \phi(y) = 0.1 + 0.8y, \quad m = 60. \tag{6.80}$$

Here, $y_0$ gives the average of $x$ for a given $y$:

$$\phi(y) = \int_{0}^{1} xp(x, y)dx. \tag{6.81}$$

Figure 6.10 shows the results based on the above formulations. Simulations show that the continuous cell lineage model can reproduce the results of discrete cell lineage models well, and the system is robust with respect to random perturbations. The population size and distribution of CD34 expressions in self-renewal cells converge to a stable steady state from randomly distributed expression levels. Here, the steady state of marker gene expression follows a single-mode distribution with a wide range of epigenetic states, which is in agreement with the observations in many experiments [67, 73–75].

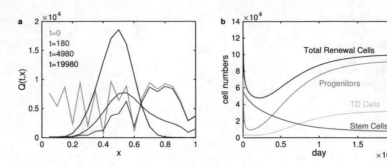

**Fig. 6.10 Dynamics of cell lineages. a.**Time evolution of the cell numbers $Q(t, x)$. **b.** Time evolution of various cell types: stem cells $Q_{\text{stem cell}}(t)$, progenitors $Q_{\text{progenitor}}(t)$, total renewal cells $Q_{\text{total}}(t)$, and TD cells $L(t)$. Here, the parameters are $\beta_0 = 20$ day$^{-1}$, $\theta = 0.069 \times 10^4$ cells/kg, $\mu = 0.08$ day$^{-1}$, $\kappa_0 = 0.02$ day$^{-1}$, and $\delta = 0.001$ day$^{-1}$

**Fig. 6.11 Time evolution of two-mode cell lineages.** Here, $\phi(y) = 0.1 +$ $0.8\dfrac{(2y)^2}{1 + (2y)^2}$

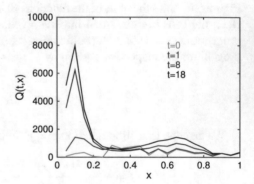

Here, the transition function $p(x, y)$ can be adjusted to mimic the dynamic phenomena of cell type transitions. For example, we can mimic the situation with two-mode cell types through a function (Fig. 6.11)

$$\phi(y) = \int_0^1 xp(x, y)dx = 0.1 + 0.8\frac{(2y)^2}{1 + (2y)^2}. \tag{6.82}$$

The readers are referred to refer to [41] for further discussions.

## 6.3.3 A One-Dimensional Variable to Represent Transcriptome Heterogeneity

In the above mathematical model of heterogeneous stem cell regeneration, we propose a high-dimensional variable **x** for the epigenetic state of a cell. However, in real applications and mathematical analysis, it would be convenient to represent the state of a cell with a low-dimensional variable. From the viewpoint of complex systems, such variables can be referred to as the macroscopic state of a cell that is described by a large number of molecular components. Advances in molecular biology have given rise to a multitude of single-cell sequencing technologies that provide opportunities to obtain the direct measurements of the molecular details within a cell [76, 77]. Single-cell RNA sequencing (scRNA-seq) is a widely used single-cell sequencing approach that allows us to investigate the transcriptomes at the level of individual cells. It is fundamental to define the state of a cell based on such microscopic information.

In [78], the concept of single-cell entropy (scEntropy) was proposed to measure the order of the cellular transcriptome profile from scRNA-seq data. scEntropy measures the order of cellular transcription with respect to a reference level; larger entropy means lower order in the transcriptions.

The scEntropy of a cell is defined as the information entropy of the difference in transcription between the cell and a predefined reference cell. Given an $N \times M$ gene expression matrix with $N$ cells and $M$ genes, as well as the gene expression vector $\mathbf{r}$ of the reference cell, let $\mathbf{x}_i$ $(i = 1, \cdots, N)$ be the gene expression vector of the $i^{\text{th}}$ cell. The calculation of the scEntropy of $\mathbf{x}_i$ with reference to $\mathbf{r}$, $S(\mathbf{x}_i|\mathbf{r})$, includes two steps:

(1) calculate the difference between $\mathbf{x}_i$ and $\mathbf{r}$

$$\mathbf{y}_i = \mathbf{x}_i - \mathbf{r} = (y_{i1}, y_{i2}, \cdots, y_{iM});$$

(2) the entropy $S(\mathbf{x}_i|\mathbf{r})$ is given by the information entropy of the signal sequence $\mathbf{y}_i$, i.e.,

$$S(\mathbf{x}_i|\mathbf{r}) = - \int p_i(y) \log_2 p_i(y) dy,$$

where $p_i(y)$ is the distribution density of the components $y_{ij}$ in $\mathbf{y}_i$.

From the above procedure, the definition of scEntropy is straightforward and parameter free, with the reference cell as the only degree of freedom. Here, we note that the reference cell $\mathbf{r}$ is a variable in the entropy $S(\mathbf{x}_i|\mathbf{r})$, which means that the baseline transcriptome has a minimum entropy of zero. The biological interpretation of $S(\mathbf{x}_i|\mathbf{r})$ is closely dependent on the selection of the reference cell, which is often defined preliminarily according to the specific biological problem. For example, we can take the reference cell as a zygote in early embryo development, the average transcription in normal cells compared with that in malignant cells, or pluripotent stem cells in the differentiation process. scEntropy can be used to quantify the process of early embryo development, to describe the dynamics of the reprogramming and differentiation of induced pluripotent stem cells and to distinguish malignant cells from nonmalignant cells in various types of tumors [78, 79].

Taking scEntropy as a one-dimensional variable to denote the macroscopic state of a cell, we can write the equation for heterogeneous stem cell regeneration as (here, $x$ represents the scEntropy)

$$\frac{\partial Q(t, x)}{\partial t} = -Q(t, x)(\beta(c, x) + \kappa(x))$$

$$+ 2 \int_{\Omega} \beta(c_{\tau(y)}, y)Q(t - \tau(y), y)e^{-\mu(y)\tau(y)}p(x, y)dy \qquad (6.83)$$

$$c(t) = \int_{\Omega} Q(t, x)\zeta(x)dx.$$

Here, $x \in \Omega \subset \mathbb{R}^+$, $c_{\tau} = c(t - \tau)$, and we always assume

$$\beta(x) \geq 0, \kappa(x) \geq 0, \mu(x) \geq 0, \zeta(x) \geq 0, \tau(x) \geq 0, p(x, y) \geq 0 \qquad (6.84)$$

and

$$\int_{\Omega} p(x, y)dx = 1, \quad \forall y \in \Omega. \qquad (6.85)$$

In this equation, different cells are characterized by their kinetic rates of proliferation, apoptosis, differentiation, senescence, etc.. These dynamic features define the *kinetotype* of a cell, which is analogous to the genotype, epigenotype, or phenotype [41].


## 6.4 Mathematical Models of Dynamic Hematological Disease

The persistent regeneration of stem cells provides a source to replenish dying cells and repair damaged tissue to maintain normal population dynamics and ensure regular physiological functions. Nevertheless, the dysregulation of in stem cell regeneration may result in severe diseases. In the case of the hematological system, all blood cells arise from a common origin in the bone marrow, hematopoietic stem cells (HSCs). HSCs are morphologically undifferentiated cells that can either proliferate or differentiate to produce all types of blood cells (erythrocytes, neutrophils and platelets). The proliferation of hematopoietic stem cells and progenitor cells is controlled by a negative feedback system mediated by hematopoietic cytokines, such as erythropoietin (EPO), granulocyte colony-stimulating factor (G-CSF), and thrombopoietin (TPO). EPO is the hormone that mediates red blood cell (RBC) production, G-CSF controls the regulation of neutrophils, and TPO (known as c-mpl ligand or megakaryocyte growth and development factor) is the primary regulator of thrombopoiesis. Hematopoiesis is a homeostatic system, and consequently, most disorders of its regulation lead to transient or chronic failures in the production of either all or only one blood cell type. Among many diseases that affect blood cells, some are characterized by predictable oscillations in one or more cellular elements of the blood. They are called periodic or dynamical diseases [80]. The investigation of their dynamic characteristic offers an opportunity to enrich our knowledge about the regulatory processes controlling blood cell production and may suggest better therapeutic strategies [81, 82].

## 6.4.1  Dynamic Hematological Disease

Dysregulation of hematopoiesis can induce various types of hematological disease. Classic examples of dynamic hematopological diseases include cyclical neutropenia (CN) [83–85], periodic chronic myelogenous leukemia (PCML) [46, 86], periodic autoimmune hemolytic anemia (AIHA) [87, 88], and cyclical thrombocytopenia (CT) [89–91]. These diseases show various features in terms of the oscillation patterns of blood cells. PCML and CN involve fluctuations in all major blood cell lines with the same period in a given subject, while CT and periodic AIHA patients show significant oscillations in only one type of cell count. Different mechanisms are responsible for the different oscillation patterns in these dynamic hematological diseases [92].

Neutrophils are critical to the immune response, and low absolute neutrophil counts in the blood can lead to infections. CN is characterized by oscillations in the number of neutrophils from normal to very low levels (less than $0.5 \times 10^9$ cells/L, also called severe neutropenia). Through family studies and linkage analysis, mutations in the gene encoding neutrophil elastase (ELA2) are recognized as the cause of CN [93]. The period of these oscillations is usually approximately 3 weeks for humans, although periods of up to 45 days were also observed [94]. One major characteristic of CN is that oscillations are present not only in neutrophils but also in platelets, monocytes and reticulocytes [84]. For CN patients, the period of severe neutropenia usually lasts for approximately 3–5 days within every 3-week period [85, 95]. CN also occurs in grey collies with the same characteristics as those in humans and with oscillation periods on the order of 11 to 16 days [84, 94, 96]. This animal model has provided extensive experimental data that would be difficult, if not impossible, to obtain in humans. It is believed that CN arises from the stem cell compartment in the bone marrow [11, 39].

Platelets are blood cells whose function is to take part in the clotting process, and the term thrombocytopenia denotes a situation of reduced platelet (thrombocyte) count. CT is a rare hematological disorder characterized by periodic cycling in platelet counts. In CT, platelet counts generally oscillate from very low levels ($1 \times 10^9$ cells/L) to normal ($150 - 450 \times 10^9$ platelets/L blood) or above normal levels ($2000 \times 10^9$ cells/L) [89]. These oscillations have been observed with periods varying from 13 to 65 days among different patients [89]. In most known CT cases, oscillations appear only in platelets and not in white or red blood cells. However, a case report of CT documented statistically significant oscillations in neutrophil counts with the same period (39 days) as the platelet oscillations [91]. There are two proposed origins of CT. One is of autoimmune origin and is most prevalent in females [97], and the other is of amegakaryocytic origin and is more common in males [98].

Examples of periodic anemia are relatively rare in the clinical literature, though there are a few well documented examples [99, 100]. Periodic AIHA is a rare form of hemolytic anemia in humans characterized by oscillatory erythrocyte numbers around a depressed level [99]. Although periodic fluctuations of erythroid precursors in the bone marrow are well documented in cyclical neutropenia and some cases of periodic leukemia (see below), the rarity of reports of actual periodic anemia is presumably due to the extremely long lifespan of circulating erythrocytes in humans. There are, however, well documented examples of cyclical anemia in mice following either the administration of a single dose of $^{89}$Sr [101–103] or after whole body irradiation [104, 105]. The origin of the disease is unclear. Periodic AIHA, with a period of 16 to 17 days in hemoglobin and reticulocyte counts, has been induced in rabbits by using red blood cell autoantibodies [106].

Leukemia is a cancer of the blood or bone marrow characterized by an abnormal proliferation of blood cells, usually leukocytes. Chronic myelogenous leukemia (CML) is distinguished from other leukemias by the presence of a genetic abnormality in blood cells, called the Philadelphia chromosome, which is a translocation between chromosomes 9 and 22 that leads to the formation of the BCR-ABL fusion protein [107]. This protein is thought to be responsible for the dysfunctional regulation of myelocyte growth and other features of CML [108]. PCML is a dynamic disease characterized by oscillations in circulating cell numbers that occur primarily in leukocytes, but may also occur in platelets and reticulocytes [86]. The leukocyte count varies periodically, typically between values of 30 and 200 × 10$^9$ cells/L, with a period ranging from 40 to 80 days. In addition, the oscillation of platelets and reticulocytes may occur with the same period as the leukocytes around normal or elevated numbers [86, 109]. As in cyclical neutropenia, the hypothesis that the disease originates from the stem cell compartment is supported by the presence of oscillations in more than one cell lineage.

## 6.4.2 Mathematical Models of Hematopoiesis

### 6.4.2.1 Model Illustration

The regulation of blood cell production is complicated, and its understanding is constantly evolving; however, the broad outlines are clear [84, 110]. Here, we introduce a general framework for the mathematical model of hematopoiesis. The model is formulated as a set of delay differential equations.

Figure 6.12 shows a schematic representation of hematopoiesis. There are four linages, including hematopoietic stem cells, and three differentiated cell lines, leukocytes, erythrocytes, and platelets.

Hematopoietic stem cells (HSCs) are the stem cells that give rise to other blood cells. This process is called hematopoiesis and occurs in the bone marrow in adults. The HSC regeneration process is similar to that of other stem cells and can be described through the $G_0$ cell cycle model in Sect. 6.2. Resting phase HSCs can develop into mixed myeloid progenitor cells, which further differentiate into precursors of any of the three cell lines, leukocytes (white blood cells, WBCs), erythrocytes (red blood cells, RBCs), or thrombocytes (platelets). The rates of differentiation into these three cell lines depend on the numbers of circulating cells of the relevant types that encapsulate the feedback between the circulating cell numbers and the production.

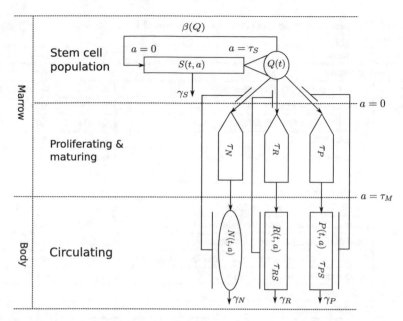

**Fig. 6.12 A schematic representation of an age-structured model of hematopoiesis.** See the text for details and notations. Replotted from [45]

There are two stages for each of the circulating cell lines after differentiation: amplification/maturation of precursor cells in the bone marrow and circulation of mature cells throughout the whole body. In the stage of amplification/maturation, precursor cells undergo many stages of cell division and randomly die so that the number of precursors increases rapidly in a period: one stem cell is capable of producing approximately $10^6$ mature blood cells after 20 cell divisions [111]. In the circulation stage, mature blood cells are randomly removed at a fixed rate. In addition, circulating erythrocytes and platelets are actively destroyed at a fixed time from entry into the circulating compartment [112, 113]. The proliferation and differentiation of hematopoietic cells and the function of mature blood cells are regulated by a variety of cytokines, including erythropoietin (EPO) [114], which mediates the regulation of erythrocyte production, granulocyte colony-stimulating factor (G-CSF) [115], which regulates the neutrophil number, and thrombopoietin (TPO) [116], which regulates the production of platelets and other cell lineages.

### 6.4.2.2 Mathematical Formulation

Many mathematical tools have been used in hematological modeling, including differential equations (partial, ordinary or delay), stochastic processes, Boolean networks, Bayesian theory, multivariate statistics, decision trees. See [117] for a review of the development of modeling blood cell dynamics. Here, an extension of the pre-

**Table 6.2** Variables used in the model equations and typical values for hematologically normal individuals [45, 46]

| Variable | Definition | Value | Unit |
|---|---|---|---|
| *Stem cell compartment* | | | |
| $Q(t)$ | Population of resting phase stem cells | 1.12 | $10^6$ cells/kg |
| $s(t, a)$ | Population of proliferative phase stem cells | – | cells/kg |
| $\beta$ | Rate of re-entering the proliferative phase | 0.0433 | $day^{-1}$ |
| $\tau_S$ | Duration of mitosis | 2.83 | days |
| $\gamma_S$ | Apoptosis rate of proliferating stem cells | 0.1013 | $day^{-1}$ |
| *Neutrophil compartment* | | | |
| $n(t, a)$ | Population of neutrophils | – | cells/kg |
| $N(t)$ | Population of circulating neutrophils | 5.59 | $10^8$ cells/kg |
| $\kappa_N$ | Differentiation rate from stem cells to neutrophils | 0.0077369 | $day^{-1}$ |
| $\eta_N$ | Amplification rate of neutrophil precursor cells | 2.2887 | $day^{-1}$ |
| $\tau_N$ | Duration of neutrophil precursor amplification/maturation | 12.6 | days |
| $\gamma_N$ | Apoptosis rate of circulating neutrophils | 2.4 | $day^{-1}$ |
| *Erythrocyte compartment* | | | |
| $r(t, a)$ | Population of erythrocytes | – | cells/kg |
| $R(t)$ | Population of circulating erythrocytes | 3.5 | $10^{11}$ cells/kg |
| $\kappa_R$ | Differentiation rate from stem cells to erythrocytes | 0.005271 | $day^{-1}$ |
| $\eta_R$ | Amplification rate of erythrocyte precursor cells | 1.8114 | $day^{-1}$ |
| $\tau_R$ | Duration of erythrocyte precursor amplification/maturation | 6 | days |
| $\gamma_R$ | Apoptosis rate of circulating erythrocytes | 0.001 | $day^{-1}$ |
| $\tau_{RS}$ | Life time of circulating erythrocytes | 120 | days |
| $\tau_{Rsum}$ | $\tau_R + \tau_{RS}$ | 126 | days |
| *Platelet compartment* | | | |
| $p(t, a)$ | Population of platelets | – | cells/kg |
| $P(t)$ | Population of circulating platelets | 1.3924 | $10^{10}$ cells/kg |
| $\kappa_P$ | Differentiation rate from stem cells to platelets | 0.0087074 | $day^{-1}$ |
| $\eta_P$ | Amplification rate of platelet precursor cells | 1.7928 | $day^{-1}$ |
| $\tau_P$ | Duration of platelet precursor amplification/maturation | 7 | days |
| $\gamma_P$ | Apoptosis rate of circulating platelets | 0.15 | $day^{-1}$ |
| $\tau_{PS}$ | Lifetime of circulating platelets | 9.5 | days |
| $\tau_{Psum}$ | $\tau_P + \tau_{PS}$ | 16.5 | days |

vious $G_0$ cell cycle model is introduced to include different cell lines, which provides a natural way to model hematopoietic dynamics.

We refer the model illustrated in Fig. 6.12. The variables used in the following equations and typical values for hematologically normal individuals are summarized in Table 6.2.

Let $Q(t)$ (cells/kg) be the population of resting phase stem cells and $s(t, a)$ (cells/kg) be the population of stem cells in the proliferative phase, with age $a = 0$ for their time of entry into the proliferative state. For the other three cell lines, let $n(t, a)$, $r(t, a)$, and $p(t, a)$ (cells/kg) represent the populations of leukocytes, erythrocytes, and platelets, respectively, with age $a = 0$ for the time point of differentiation from stem cells.

Resting phase HSCs can either re-enter the proliferative phase at a rate of $\beta(Q)$ (day$^{-1}$) that involves negative feedback or differentiate into any of the three cell lines, leukocytes, erythrocytes, or platelets, at rates of $\kappa_N$ (day$^{-1}$), $\kappa_R$ (day$^{-1}$), or $\kappa_P$ (day$^{-1}$), respectively. The proliferating stem cells are assumed to undergo mitosis at a fixed time $\tau_S$ after entry into the proliferating compartment and to be lost randomly at a rate of $\gamma_S$ (day$^{-1}$) during the proliferative phase. Each normal cell generates two resting phase cells at the end of mitosis. All the differentiation rates $\kappa_N(N)$, $\kappa_R(R)$, and $\kappa_P(P)$ involve negative feedback loops, with the rates depending on the population of the corresponding circulating cells $N(t)$, $R(t)$, and $P(t)$, respectively. After differentiation, the precursor cells undergo many stages of cell division and randomly die in a period of $\tau_{NM}$ (days) for leukocytes, $\tau_{RM}$ (days) for erythrocytes, and $\tau_{PM}$ (days) for platelets. The net proliferation rates of the precursor cells are represented by $\eta_N$ (day$^{-1}$) for leukocytes, $\eta_R$ (day$^{-1}$) for erythrocytes, and $\eta_P$ (day$^{-1}$) for platelets. Thus, increasing the apoptosis rate of precursor cells can result in a decrease in the proliferation rates. Circulating cells leukocytes, erythrocytes, and platelets are lost at rates of $\gamma_N$ (day$^{-1}$), $\gamma_R$ (day$^{-1}$) and $\gamma_P$ (day$^{-1}$), respectively. In addition, circulating erythrocytes and platelets are actively destroyed at fixed times $\tau_{RS}$ (days) and $\tau_{PS}$ (days), respectively, from their time of entering the circulating compartment.

Let

$$N(t) = \int_{\tau_N}^{+\infty} n(t, a)da, \quad R(t) = \int_{\tau_R}^{\tau_{Rsum}} r(t, a)da, \quad P(t) = \int_{\tau_P}^{\tau_{Psum}} p(t, a)da, \quad (6.86)$$

which are the populations of circulating cells. Here, we set $\tau_{Rsum} = \tau_R + \tau_{RS}$ and $\tau_{Psum} = \tau_P + \tau_{PS}$. Using the above notations, the age-structured model of hematopoiesis is then described by the following partial differential equations [45]:

$$\nabla s(t, a) = -\gamma_S(t)s(t, a) \qquad\qquad\qquad (t > 0, 0 \le a \le \tau_S)$$

$$\frac{dQ}{dt} = 2s(t, \tau_S) - (\beta(Q) + \kappa_N(N) + \kappa_R(R) + \kappa_P(P))Q \quad (t > 0)$$

$$\nabla n(t, a) = \begin{cases} \eta_N(t, a)n(t, a) & (t > 0, 0 \le a \le \tau_N) \\ -\gamma_N(t, a)n(t, a) & (t > 0, a \ge \tau_N) \end{cases} \qquad (6.87)$$

$$\nabla r(t, a) = \begin{cases} \eta_R(t, a)r(t, a) & (t > 0, 0 \le a \le \tau_R) \\ -\gamma_R(t, a)r(t, a) & (t > 0, \tau_R \le a \le \tau_{Rsum}) \end{cases}$$

$$\nabla p(t, a) = \begin{cases} \eta_P(t, a)p(t, a) & (t > 0, 0 \le a \le \tau_P) \\ -\gamma_P(t, a)p(t, a) & (t > 0, \tau_P \le a \le \tau_{Psum}) \end{cases}$$

where $\nabla = \dfrac{\partial}{\partial t} + \dfrac{\partial}{\partial a}$. The boundary conditions at $a = 0$ are given by

$$
\begin{aligned}
s(t, 0) &= \beta(Q(t))Q(t), \\
n(t, 0) &= \kappa_N(N(t))Q(t), \\
r(t, 0) &= \kappa_R(R(t))Q(t), \\
p(t, 0) &= \kappa_P(P(t))Q(t),
\end{aligned}
\qquad (t \geq 0)
\qquad (6.88)
$$

according to the negative feedback loops. Moreover, we have

$$
\lim_{a \to +\infty} n(t, a) = 0. \qquad (6.89)
$$

The initial conditions are

$$
\begin{aligned}
s(0, a) &= g_S(a), \ (0 \leq a \leq \tau_S), \\
Q(0) &= Q_0, \\
n(0, a) &= g_N(a), \ (0 \leq a \leq +\infty), \\
r(0, a) &= g_R(a), \ (0 \leq a \leq \tau_{Rsum}), \\
p(0, a) &= g_P(a), \ (0 \leq a \leq \tau_{Psum}),
\end{aligned}
\qquad (6.90)
$$

where $g_S(a)$, $g_N(a)$, $g_R(a)$ and $g_P(a)$ give the initial population distributions of proliferative phase stem cells and the precursors of neutrophils, erythrocytes, and platelets, respectively.

The negative feedback functions are represented by Hill-type functions [46]:

$$
\kappa_N(N) = f_0 \frac{\theta_1^{s_1}}{\theta_1^{s_1} + N^{s_1}}, \qquad \beta(Q) = k_0 \frac{\theta_2^{s_2}}{\theta_2^{s_2} + Q^{s_2}},
$$

$$
\kappa_R(R) = \frac{\bar{\kappa}_r}{1 + K_r R^{s_3}}, \qquad \kappa_P(P) = \frac{\bar{\kappa}_p}{1 + K_p P^{s_4}}.
\qquad (6.91)
$$

Typical parameter values for the feedback functions are given in Table 6.3.

Equations (6.86)–(6.91) define the initial-boundary value problem for the age-structured model of hematopoietic regulation and are the basis of the following simplified model and analysis.

In hematological modeling, we are mainly interested in the dynamics of circulating blood cell populations $N(t)$, $R(t)$ and $P(t)$. This can be modeled by the delay differential equations obtained from the above age-structured model.

We assume the apoptosis rates $\gamma_N$, $\gamma_R$, and $\gamma_P$ are constant. Using the method of characteristics with Eq. (6.87) and using the boundary conditions (6.88) and (6.89), we obtain the following delay differential equations when $t > \tau_{\max} = \max\{\tau_S, \tau_N, \tau_{Rsum}, \tau_{Psum}\}$:

**Table 6.3** Parameters for the Hill functions (6.91) [45, 46]

| Parameter Name | Value | Unit |
|---|---|---|
| *Function $\beta(Q)$* | | |
| $k_0$ | 8.0 | $day^{-1}$ |
| $\theta_2$ | 0.0826 | $10^6$ cells/kg |
| $s_2$ | 2 | (none) |
| *Function $\kappa_N(N)$* | | |
| $f_0$ | 0.154744 | $day^{-1}$ |
| $\theta_1$ | 0.2942 | $10^8$ cells/kg |
| $s_1$ | 1 | (none) |
| *Function $\kappa_R(R)$* | | |
| $\bar{\kappa}_r$ | 1.23744 | $day^{-1}$ |
| $K_r$ | 0.0382 | $(10^{-11}$ cells/kg$)^{-s_3}$ |
| $s_3$ | 6.96 | $day^{-1}$ |
| *Function $\kappa_P(P)$* | | |
| $\bar{\kappa}_p$ | 0.2802 | $day^{-1}$ |
| $K_p$ | 20.343 | $(10^{10}$ cells/kg$)^{-s_4}$ |
| $s_4$ | 1.29 | $day^{-1}$ |

$$\frac{dQ}{dt} = 2e^{-\tau_S \hat{\gamma}_S(t-\tau_S)}\beta(Q_{\tau_S})Q_{\tau_S} - (\beta(Q) + \kappa_N(N) + \kappa_R(R) + \kappa_P(P))Q,$$

$$\frac{dN}{dt} = -\gamma_N N + e^{\tau_N \hat{\eta}_N(t-\tau_N)}\kappa_N(N_{\tau_N})Q_{\tau_N},$$

$$\frac{dR}{dt} = -\gamma_R R + e^{\tau_R \hat{\eta}_R(t-\tau_R)}\kappa_R(R_{\tau_R})Q_{\tau_R} - e^{-\gamma_R \tau_{RS}}e^{\tau_R \hat{\eta}_R(t-\tau_{Rsum})}\kappa_R(R_{\tau_{Rsum}})R_{\tau_{Rsum}},$$

$$\frac{dP}{dt} = -\gamma_P P + e^{\tau_P \hat{\eta}_P(t-\tau_P)}\kappa_P(P_{\tau_P})Q_{\tau_P} - e^{-\gamma_P \tau_{PS}}e^{\tau_P \hat{\eta}_P(t-\tau_{Psum})}\kappa_P(P_{\tau_{Psum}})Q_{\tau_{Psum}},$$

$$(6.92)$$

where

$$\hat{\gamma}_S(t) = \frac{1}{\tau_S}\int_0^{\tau_S}\gamma_S(t+s)ds, \quad \hat{\eta}_k(t) = \frac{1}{\tau_k}\int_0^{\tau_k}\eta_k(t+s,s)ds, \quad (k = N, R, P).$$

$$(6.93)$$

Here, the subscripts of the dependent variables indicate delayed arguments, i.e., $Q_{\tau_S} = Q(t - \tau_S)$.

The delay differential equations (6.92) determine the dynamic behaviors of the circulating blood cell populations. Here, we note that when $t < \tau_{\max}$, Eq. (6.92) is not equivalent to the original age-structured model (6.87). In this case, the initial conditions (6.90) have to be involved in the dynamical equation. Refer to [45] for detailed discussions.

For hematologically normal individuals, we assume that the apoptosis rate $\gamma_S$ and amplification rates $\eta_k$ $(k = N, R, P)$ are constants, and hence, $\hat{\gamma}_S = \gamma_S, \hat{\eta}_k = \eta_k, (k = N, R, P)$. Thus, we obtain the following delay differential equations:

$$
\begin{cases}
\dfrac{dQ}{dt} = 2e^{-\tau_S \gamma_S} \beta(Q_{\tau_S})Q_{\tau_S} - (\beta(Q) + \kappa_N(N) + \kappa_R(R) + \kappa_P(P))Q, \\[2ex]
\dfrac{dN}{dt} = -\gamma_N N + e^{\tau_N \eta_N} \kappa_N(N_{\tau_N})Q_{\tau_N}, \\[2ex]
\dfrac{dR}{dt} = -\gamma_R R + e^{\tau_R \eta_R}\left( \kappa_R(R_{\tau_R})Q_{\tau_R} - e^{-\gamma_R \tau_{RS}} \kappa_R(R_{\tau_{Rsum}})Q_{\tau_{Rsum}} \right), \\[2ex]
\dfrac{dP}{dt} = -\gamma_P P + e^{\tau_P \eta_P}\left( \kappa_P(P_{\tau_P})Q_{\tau_P} - e^{-\gamma_P \tau_{PS}} \kappa_P(P_{\tau_{Psum}})Q_{\tau_{Psum}} \right).
\end{cases}
\tag{6.94}
$$

The Eq. (6.94) were first presented in [46]. The equations and their various forms have been used to study different types of dynamic hematological diseases [45, 46, 83, 118, 119].

### 6.4.2.3   Modeling the Clinical Treatment

Dynamic hematological diseases show abnormal dynamics in cell counts, mostly due to changes in the control parameters. Hence, clinical treatments intend to alter the control parameters to control the cell count dynamics.

To study the effect of clinical treatments, such as G-CSF and chemotherapy administration, which are known to affect hematopoiesis in the bone marrow, we further divide the amplification/maturation compartment of each cell line into two subcompartments, corresponding to the compartments of amplification and maturation. Let

$$
\tau_k = \tau_{kP} + \tau_{kM}, \quad (k = N, R, P),
\tag{6.95}
$$

where $\tau_{kP}$ are the durations of the amplification stages, and $\tau_{kM}$ are the times for maturation. The amplification rates $\eta_k$, $(k = N, R, P)$ are defined separately in the two stages as:

$$
\eta_k(t, a) = \begin{cases} \eta_{kP}(t) & 0 \le a < \tau_{kP} \\ -\gamma_{kM}(t) & \tau_{kP} \le a \le \tau_k \end{cases} \quad (k = N, R, P).
\tag{6.96}
$$

Here, $\eta_{kP}$ are the amplification rates in the amplification stage, and $\gamma_{kM}$ are the apoptosis rates in the maturation stage; these rates are assumed to be independent of the age $a$. Thus, $\hat{\eta}_k$ defined by (6.93) can be written as

$$
\hat{\eta}_k(t) = \frac{1}{\tau_k} \left[ \int_0^{\tau_{kP}} \eta_{kP}(t + s)ds - \int_{\tau_{kP}}^{\tau_k} \gamma_{kM}(t + s)ds \right].
\tag{6.97}
$$

For hematologically normal individuals whose rates $\eta_{kP}$ and $\gamma_{kM}$ are constants, we have

$$\eta_k = (\eta_{kP}\tau_{kP} - \gamma_{kM}\tau_{kM})/\tau_k, \quad (k = N, R, P). \tag{6.98}$$

G-CSF is a potent inducer of HSC mobilization from the bone marrow into the bloodstream and stimulates the bone marrow to produce granulocytes and stem cells and to release them into the bloodstream. The administration of G-CSF is often used to treat severe neutropenia. The pharmaceutical analogs of naturally occurring G-CSF are called filgrastim and lenograstim. G-CSF perturbs hematopoietic dynamics by decreasing the apoptosis rate of HSCs [120], reducing the apoptosis rate of committed neutrophils [121], and decreasing the neutrophil precursor maturation time [115]. Let $G(t)$ ($\mu$g/ml) be the circulating G-CSF concentration. The effect of G-CSF administration can be modeled by [122]

$$\gamma_S(t) = \gamma_S + \Delta\gamma_S^{\mathrm{gcsf}} \frac{b_s}{G(t) + b_s}, \tag{6.99}$$

$$\gamma_0(t) = \gamma_0 + \Delta\gamma_0^{\mathrm{gcsf}} \frac{b_0}{G(t) + b_0}, \tag{6.100}$$

$$\eta_{NP}(t) = \eta_{NP} + \Delta\eta_{NP}^{\mathrm{gcsf}} \frac{G(t)}{G(t) + c_n}, \tag{6.101}$$

$$\tau_{NM}(t) = \tau_{NM}/V_n(t), \quad V_n(t) = 1 + (V_{\max} - 1)\frac{G(t)}{G(t) + b_v}. \tag{6.102}$$

Here, $\gamma_S$, $\gamma_0$, $\eta_{NP}$ and $\tau_{NM}$ are the baseline values in the absence of G-CSF administration. The pharmacokinetic dynamics of G-CSF concentration can be described by a two-compartmental model consisting of tissue and the circulation system (Fig. 6.13):

$$\frac{dX}{dt} = I(t) + k_T V_B G - k_B X \tag{6.103}$$

$$\frac{dG}{dt} = G_{\mathrm{prod}} + k_B X/V_B - k_T G - (\gamma_G G + \sigma N F(G)). \tag{6.104}$$

Here, $X(t)$ ($\mu$g/kg) denotes the tissue level of G-CSF, $G(t)$ ($\mu$g/ml) denotes the circulating G-CSF concentration, and $I(t)$ denotes a step function representing the injection of exogenous G-CSF into the tissue. The parameters $k_T$ and $k_B$ are exchange rate constants, $V_B$ is the volume of the blood compartment, and $G_{\mathrm{prod}}$ is the fixed G-CSF production rate. The clearance is given by two parts: the degradation of G-CSF at a rate of $\gamma_G$ and the removal of G-CSF from the circulation through saturable clearance $\sigma N F(G)$, where

$$F(G) = \frac{G^2}{k_G + G^2}$$

and $N$ represents the circulating neutrophil number.

**Fig. 6.13** A
two-compartment model
for the subcutaneous
administration of G-CSF.
Here, $G_{\text{prod}}$ is the fixed
G-CSF production rate, and
$I(t)$ is a step function
representing the injection of
exogenous G-CSF into the
tissue. Replotted from [123]

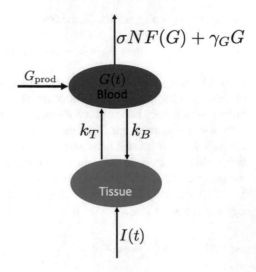

Chemotherapy is a common cancer treatment widely used in various types of cancers. Chemotherapy can cause myelosuppression and excessively low levels of white blood cells, making patients susceptible to infections and sepsis. G-CSF is often used in certain cancer patients to accelerate the recovery of white blood cells and reduce mortality from neutropenia after chemotherapy. To model the effect of chemotherapy, we note that chemotherapy increases apoptosis in both proliferative HSCs and proliferative neutrophil precursors and hence leads to an increase in $\gamma_S$ and a decrease in $\eta_{NP}$ [124]. Denote the active plasma chemotherapy drug concentration as $C(t)$ ($\mu$g/ml). Chemotherapy is often administered periodically, and hence, $C(t)$ can be considered a periodic pulse function. For a relatively short duration of chemotherapy infusion, there is an approximately linear relationship between $C(t)$ and the apoptosis rate [125], so we assume [122, 126]

$$\gamma_S(t) = \gamma_S + \Delta\gamma_S^{\text{chemo}} C(t) \tag{6.105}$$

$$\eta_{NP} = \eta_{NP} - \Delta h_{NP}^{\text{chemo}} C(t). \tag{6.106}$$

Here, $\gamma_S$ and $\eta_{NP}$ are the default apoptosis rate and amplification rate in the absence of chemotherapy, respectively, and $\Delta\gamma_S^{\text{chemo}}$ and $\Delta h_{NP}^{\text{chemo}}$ are constants to quantify the effects of chemotherapy.

Now, we model the pharmacokinetics of chemotherapy. After a single administration of chemotherapy, the active plasma chemotherapy drug concentration consists of an exponentially increasing portion until a maximum value at the end of administration, followed by a falling phase back to zero [127]. Hence, the drug concentration $C(t)$ can be described by

$$C(t) = (I_0/\phi) \times \begin{cases} (1 - e^{-\delta t}), & 0 \le t \le \Delta_c \\ (1 - e^{-\delta\Delta_c})e^{-\delta(t-\Delta_c)}, & \Delta_c < t. \end{cases} \tag{6.107}$$

Here, $\Delta_c$ (days) is the duration of the rising phase (infusion time) of chemotherapy, $I_0$ (mg/(kg $\times$ day)) measures the injection rate of the drug into the plasma, $\phi$ (days$^{-1}$) is the clearance rate of the drug from the body, and $\delta$ (days$^{-1}$) is the effective rate constant of the removal of the chemotherapeutic drug. The total amount of chemotherapy drug administered in the plasma (dosage $D$, mg/kg) is therefore given by

$$D = I_0 \Delta c.$$

Thus, the injection rate is expressed through the infusion time $\Delta_c$ and the dosage $D$ as

$$I_0 = D/\Delta_c.$$

When multiple doses of chemotherapy are delivered in a period $T$, a single dose is administered in the interval $jT \le t \le jT + \Delta_c$. The above formulation can be extended as

$$C(t) = \begin{cases} (I_0/\phi)(1 - e^{-\delta(t-jT)}), & \text{if } jT \le t \le jT + \Delta_c \\ (I_0/\phi)(1 - e^{-\delta\Delta_c})e^{-\delta(t-jT-\Delta c)}, & \text{if } (jT + \Delta c) \le t \le (j+1)T. \end{cases} \quad (6.108)$$

To consider the situation of administrating G-CSF after chemotherapy, we can combine the above equations and write

$$\gamma_S(t) = \gamma_S + \Delta\gamma_S^{\text{chemo}} C(t) + \Delta\gamma_S^{\text{gcsf}} \frac{b_s}{G(t) + b_s}, \quad (6.109)$$

$$\eta_{NP} = \eta_{NP} - \Delta h_{NP}^{\text{chemo}} C(t) + \Delta\eta_{NP}^{\text{gcsf}} \frac{G(t)}{G(t) + c_n}. \quad (6.110)$$

The protocols of drug administration are given by the functions $I(t)$ and $C(t)$. The timing of the cyclic administration of chemotherapy and G-CSF is crucial for the occurrence of neutropenia [122, 126, 128].

### 6.4.3 Origins of Dynamic Hematological Diseases

We have introduced mathematical models for hematopoiesis. From these models, we are able to explore the origins of different patterns of oscillations in dynamic hematological diseases. These results are important to help improve clinical treatment. Here, we briefly introduce related studies on cyclical neutropenia (CN) and cyclical thrombocytopenia (CT). For a more complete review of the field, please refer to [82].

### 6.4.3.1　Cyclical Neutropenia

Although it is a rare disorder, cyclical neutropenia (CN) is probably the most extensively studied periodic hematological disorder due to its interesting dynamics and its clinical and laboratory manifestations. Here, we introduce several models, from simple to sophisticated, that have provided significant insight into the origin and dynamical features of CN. Then, we show how these models have been used to understand and improve the effects of CN treatments.

Since CN patients display oscillations in all cell lines, it was suggested that the oscillation in CN may originate from a loss stability in HSCs [11]; hence we can consider the following delay differential equation:

$$\frac{dQ}{dt} = -(\beta(Q) + \kappa)Q + 2e^{-\gamma_S \tau_S}\beta(Q_{\tau_S})Q_{\tau_S} \tag{6.111}$$

for the resting phase HSC populations. Equation (6.111) is obtained from (6.94) by omitting the cell lines of neutrophils, erythrocytes, and platelets and writing $\kappa$ for the overall differentiation rate of HSCs.

Equation (6.111) is exactly the $G_0$ cell cycle model in Sect. 6.2. There are two steady states: one state corresponds to the phenomena with no cells, which is stable only if it is the only steady state; the other steady state $Q^* > 0$ exists when

$$\kappa < \beta(0)(2e^{-\gamma_S \tau_S} - 1), \tag{6.112}$$

i.e., the differentiation rate is smaller than the net increase rate of HSCs in each cell cycle. The stability of the positive steady state depends on the value of the apoptosis rate $\gamma_S$. When $\gamma_S = 0$, this steady state cannot be destabilized to produce the oscillatory dynamics of cyclical neutropenia. For $\gamma_S > 0$, increases in $\gamma_S$ lead to a decrease in the hematopoietic stem cell numbers, and the steady state is destabilized when a critical value of $\gamma_S$ is reached, $\gamma_S = \gamma_{crit,1}$, at which a supercritical Hopf bifurcation occurs. When $\gamma_S$ is further increased, a reverse bifurcation occurs at a critical value $\gamma_S = \gamma_{crit,2}$, where the positive steady state becomes stable and approaches the zero steady state as $\gamma_S$ further increases. For all values of $\gamma_S$ that satisfy $\gamma_{crit,1} < \gamma_S < \gamma_{crit,2}$, Eq. (6.111) has a periodic solution, whose period is in good agreement with that seen in CN [11]. These results suggest that CN might be related to defects, possibly genetic defects, within the hematopoietic stem cell population that lead to an abnormal apoptotic loss of cells from the proliferative phase of the cell cycle.

Now, we further include the neutrophil cell line in the model. While we neglect the compartments of erythrocytes and platelets, Eq. (6.94) can be reduced to a two-variable delay differential equation model that couples the HSC population model with the neutrophil compartment [39]:

$$\frac{dQ}{dt} = 2e^{-\tau_S \gamma_S} \beta(Q_{\tau_S})Q_{\tau_S} - (\beta(Q) + \kappa_N(N) + \kappa_\delta)Q,$$

(6.113)

$$\frac{dN}{dt} = -\gamma_N N + e^{\tau_N \eta_N} \kappa_N(N_{\tau_N})Q_{\tau_N},$$

where

$$\kappa_N(N) = f_0 \frac{\theta_1^{s_1}}{\theta_1^{s_1} + N^{s_1}}, \qquad \beta(Q) = k_0 \frac{\theta_2^{s_2}}{\theta_2^{s_2} + Q^{s_2}}.$$

(6.114)

Here, $\kappa_\delta = \kappa_R + \kappa_P$ represents the differentiation rate from HSCs to erythrocytes or platelets.

Equation (6.113) has a positive steady state when the condition of (6.112) (with $\kappa = \kappa_N(0) + \kappa_\delta$) is satisfied. Let $(Q^*, N^*)$ be the positive steady state of (6.113); we linearize the equation at the steady state to obtain the linearization equation

$$\begin{cases} \dfrac{dx}{dt} = a_{11}x + b_{12}y + b_1 x_{\tau_S} \\[2mm] \dfrac{dy}{dt} = -\gamma_N y + b_2 x_{\tau_N} + b_3 y_{\tau_N}, \end{cases}$$

(6.115)

where

$$\begin{aligned} a_{11} &= -(\beta(Q^*) + Q^*\beta'(Q^*) + \kappa_N(N^*) + \kappa_\delta), \\ a_{12} &= -Q^*\kappa_N'(N^*), \\ b_1 &= 2e^{-\gamma_S \tau_S}(\beta(Q^*) + Q^*\beta'(Q^*)), \\ b_2 &= e^{\tau_N \eta_N}\kappa_N(N^*), \\ b_3 &= e^{\tau_N \eta_N}\kappa_N'(N^*)Q^*. \end{aligned}$$

Accordingly, the characteristic function for the stability of the steady state $(Q^*, N^*)$ is given by

$$\begin{aligned} h(\lambda) &= \begin{vmatrix} a_{11} - \lambda + b_1 e^{-\lambda \tau_S} & a_{12} \\ b_2 e^{-\lambda \tau_N} & -\gamma_N - \lambda + b_3 e^{-\lambda \tau_N} \end{vmatrix} \\ &= (\lambda - a_{11} - b_1 e^{-\lambda \tau_S})(\lambda + \gamma_N - b_3 e^{-\lambda \tau_N}) - a_{12}b_2 e^{-\lambda \tau_N}. \end{aligned}$$

Bifurcation occurs when there is a pair of imaginary roots in the characteristic equation, i.e.,

$$h(i\omega) = 0.$$

(6.116)

Equation (6.116) with $\omega$ varying in $(0, +\infty)$ defines the conditions for the occurrence of Hopf bifurcation, which can be obtained through numerical methods.

Bifurcation analysis shows two possible origins for CN, due to either an increased apoptosis rate in the stem cell compartment $(\gamma_S)$ or an increase in the apoptosis rate of neutrophil precursors (which leads to a decrease in $\eta_N$); both changes lead to a desta-bilization of the hematopoietic stem cell compartment through a supercritical Hopf bifurcation [39]. These results are in accordance with those of previous modeling studies [94] and agree with experimental data on grey collies. Moreover, numerical analyses show that the Eq. (6.113) have bistability, i.e., the coexistence of a stable steady state and a stable oscillatory solution when $\kappa_\delta$ is large and $\gamma_S$ or $\eta_N$ take values from a certain range [39, 129]. This bistability is essential for understanding the diverse effects of G-CSF treatment on cyclical neutropenia [45, 92].

The above analyses of the simple models have discovered the possible origins of CN. More comprehensive studies of the full model (6.94) with all cell lines can give good estimations of the model parameters that can reproduce the characteristics of CN [83]. The results support the hypothesis obtained from simple models that CN oscillations in neutrophils and platelets can originate from the destabilization of the stem cell compartment. Moreover, to mimic the experimental data, one may need to decrease the rate of HSC differentiation into the neutrophil line and the changes in the apoptosis rate of stem cells in the proliferative phase.

Further numerical studies in [45] confirmed the two possible mechanisms to induce CN, and the periods of the resulting oscillations are insensitive to the changed parameters. In particular, oscillations reminiscent of those in CN patients can be induced by either reducing the neutrophil precursor proliferation rate to 3–15% below the normal value or increasing the HSC apoptosis rate to 40–100% above the normal value.

Furthermore, simulations with changing initial conditions showed that the hematopoietic system possesses multistability over a wide range of parameter values, which include the typical parameter values for a healthy state. In this region with multistability, the hematopoietic system shows the coexistence of a stable steady state along with an oscillatory state. This result is crucial for our understanding of the effects of CN patient treatment. Because of the multistability, patients with CN caused by changes in system parameters may not recover to the healthy state even if the parameters regain their normal values through clinical therapy, for example, through G-CSF treatment. Hence, more complex control strategies of drug adminis-tration are required to eliminate the oscillatory cell counts in CN [45, 92, 129].

### 6.4.3.2 Cyclical Thrombocytopenia

Cyclical thrombocytopenia (CT) is a rare hematological disorder characterized by periodic cycling in platelet counts [89, 130–133]. This disorder is characterized by a variety of symptoms, including purpura, petechiae, epistaxis, gingival bleeding, menorrhagia, and easy bruising. In normal subjects, circulating platelet levels remain relatively stable for years ($150 - 450 \times 10^9$ platelets/L, with an average of $290 \times 10^9$ platelets/L). However, in CT patients, the platelet counts oscillate from very low ($1 \times 10^9$ platelets/L) to normal or very high levels ($2000 \times 10^9$ platelets/L), with periods varying from 13 days to 65 days among different patients [89].

In most CT patients, oscillations appear only in platelets and not in white or red blood cells [89, 90]. There are also rare cases in which some CT patients show statistically significant oscillations in neutrophil counts with the same period (39 days) as platelet oscillation [91]; however, they tested negative for the presence of the ELANE mutation that is characteristic of cyclical neutropenia [93].

The clinical findings suggest that at least two pathways may be associated with CT: immune-mediated platelet destruction (autoimmune CT) and megakaryocyte deficiency and cyclical failure in platelet production (amegakaryocytic CT). Autoimmune CT is thought to be an unusual form of immune thrombocytopenia purpura (ITP) and is more common in females. Amegakaryocytic CT is postulated to be a variant of acquired amegakaryocytic thrombocytopenic purpura and is mainly characterized by the absence of megakaryocytes in the thrombocytopenia phase and an increased megakaryocyte number during thrombocytosis. In autoimmune CT patients, the mean size of megakaryocytes does not change with cyclic variations in the platelet count, while in patients with the amegakaryocytic variety, the number of colony-forming unit-megakaryocyte (CFU-Meg), megakaryocyte number, and cytoplasmic area fluctuate with the platelet cycle [134].

Many studies have tried to understand the origination of CT through mathematical model approaches. An early mathematical model was proposed by [135], in which an age-structured model for the regulation of platelet production was developed. This model considered how thrombopoietin (TPO) affects the transition between megakaryocytes of various ploidy classes and their individual contributions to platelet production. Model simulations reproduce the dynamic characteristics of autoimmune cyclical thromobocytopenia, while the rate of platelet destruction in the circulation is upregulated.

Later, [90] expanded the approach in [135] to incorporate an assumption of megakaryocyte deficiency and cyclical failure in platelet production (amegakaryocytic CT), but they did not suggest that oscillations in platelet levels could be accompanied by oscillations in neutrophil levels. Hence, the authors assumed that the platelet differentiation rate $\kappa_P$ is a constant. Moreover, TPO is the main agent that controls the peripheral platelet regulatory system through the amplification rate $\bar{A}_P$ (the average number of platelets released per megakaryocyte). The rate $\bar{A}_P$ depends on TPO concentration through

$$\bar{A}_P(t) = A_0 e^{\mu \tau_P \bar{T}(t)}, \quad \bar{T}(t) = \frac{1}{\tau_P} \int_{t-\tau_P}^{t} T(t')dt',$$

where $A_0$ denotes the minimal number of platelets produced per megakaryocyte, and $\bar{T}(t)$ represents the average TPO concentration at time $t$. The study in [135] assumed a negative feedback mechanism for the control of platelets based on the total number $(P)$ of circulating platelets and that the negative feedback control is mediated by TPO. These assumptions give

$$\frac{dT}{dt} = \frac{a}{1 + K_p P^{s_4}} - \kappa T.$$

At dynamic equilibrium, we have

$$T = \frac{T_{\max}}{1 + K_P P^{s_4}},$$

where $T_{\max}$ represents the maximum TPO level in the blood. Thus, these arguments and the previous Eq. (6.94) together give the model equation

$$
\begin{cases}
\dfrac{dQ}{dt} = 2e^{-\tau_S \gamma_S} \beta(Q_{\tau_S})Q_{\tau_S} - (\beta(Q) + \kappa_N(N) + \kappa_R(R) + \kappa_P)Q, \\[2mm]
\dfrac{dN}{dt} = -\gamma_N N + e^{\tau_N \eta_N} \kappa_N(N_{\tau_N})Q_{\tau_N}, \\[2mm]
\dfrac{dR}{dt} = -\gamma_R R + e^{\tau_R \eta_R}\left(\kappa_R(R_{\tau_R})Q_{\tau_R} - e^{-\gamma_R \tau_{RS}}\kappa_R(R_{\tau_{Rsum}})Q_{\tau_{Rsum}}\right), \\[2mm]
\dfrac{dP}{dt} = -\gamma_P P + \bar{A}_P(t)\kappa_P \left(Q_{\tau_P} - e^{-\gamma_P \tau_{PS}}Q_{\tau_{Psum}}\right).
\end{cases}
\tag{6.117}
$$

where

$$\beta(Q) = k_0 \frac{\theta_2^{s_2}}{\theta_2^{s_2} + Q^{s_2}}, \quad \kappa_N(N) = f_0 \frac{\theta_1^{s_1}}{\theta_1^{s_1} + N^{s_1}},$$

$$\kappa_R(R) = \frac{\bar{\kappa}_r}{1 + K_r R^{s_3}}, \quad T = \frac{T_{\max}}{1 + K_P P^{s_4}},$$

$$\bar{A}_P(t) = A_0 e^{\mu \tau_P \bar{T}(t)}, \quad \bar{T}(t) = \frac{1}{\tau_P}\int_{t-\tau_P}^{t} T(t')dt'.$$

This model captures the essential features of hematopoiesis and successfully duplicates the characteristics of CT; however, it cannot explain the oscillations in neutrophil counts in CT patients.

Langlois [136] proposed a mathematical model for the regulation of megakaryocyte, platelet, and thrombopoietin dynamics in humans. The model describes megakaryocyte dynamics as an age-structured model and platelet ($P$) and thrombopoietin ($T$) dynamics through their feedback regulation to the stages of mitosis and endomitosis of megakaryocytes. In summary, the model equations are given by

$$\frac{dP}{dt} = \frac{D_0}{\beta_P}m_e(t, \tau_e) - \gamma_P P - \alpha_P \frac{P^{n_P}}{b_P^{n_P} + P^{n_P}}, \tag{6.118}$$

$$\frac{dT}{dt} = T_{\text{prod}} - \gamma_T T - \alpha_T (M_e(t) + k_S \beta_P P)\frac{T^{n_T}}{k_T^{n_T} + T^{n_T}}, \tag{6.119}$$

where

$$m_e(t, a) = V_m k_P Q^* \exp\left[\int_{t-a-\tau_m}^{t-a} \eta_m(T(s))ds\right] \exp\left[\int_{t-a}^{t} \eta_e(T(s))ds\right],$$

and

$$M_e(t) = \int_0^{\tau_e} m_e(t, a)da.$$

The functions $\eta_m(T)$ and $\eta_e(T)$ are given by

$$\eta_m(T) = \eta_m^{\min} + \Delta\eta_m \frac{T}{b_m + T},$$

and

$$\eta_e(T) = \eta_e^{\min} + \Delta\eta_e \frac{T}{b_e + T}.$$

Here, $m_e(t, a)$ is the volume density of megakaryocytes in the endomitosis phase as a function of time $t$ and age $a$, $V_m$ is the volume of a single megakaryocyte of ploidy $2N$ at age $a = 0$, and $\eta_m(T)$ and $\eta_e(T)$ are the proliferation rates of mitosis phase and endomitosis phase megakaryocytes, respectively. The model concluded that the primary change in cyclic thrombocytopenia is the interference with or destruction of the thrombopoietin receptor, with secondary changes in other processes, including the immune-mediated destruction of platelets, megakaryocyte deficiency, and failure in platelet production.

To explain the possible oscillation in neutrophil counts in some CT patients, we may need to keep the assumption that circulating platelets can repress the differentiation of HSCs to platelet precursors, which was discarded in [90]. Moreover, if we further assume that net precursor amplification is regulated by mature blood cell counts so that the proliferation rates are decreasing functions of the corresponding mature cell counts [119, 136], the apoptosis rates of precursor cells are increasing functions of the corresponding mature cell numbers. Hence, let $A_N$ and $A_P$ (here, we omit the red blood cell line) be amplification factors, and the net proliferation of precursors following cell division and apoptosis is

$$A_N(N) = A_{N,\max} e^{-\eta_N \tau_N}, \quad A_P(P) = A_{P,\max} e^{-\eta_P \tau_P}.$$

The apoptosis rates of neutrophil ($\eta_N$) and platelet ($\eta_P$) precursors may depend on the number of mature cells, which are represented as Hill-type functions

$$\eta_N(N) = \bar{\eta}_N \frac{N^{\nu_1}}{\vartheta_1^{\nu_1} + N^{\nu_1}}, \tag{6.120}$$

$$\eta_P(P) = \bar{\eta}_P \frac{P^{\nu_4}}{\vartheta_4^{\nu_4} + P^{\nu_4}}. \tag{6.121}$$

Hence, if we omit the red blood cell line and let $\kappa_R^*$ (days$^{-1}$) represent the differentiation rate of HSCs to erythrocytes, Eq. (6.94) becomes [119]

$$\frac{dS}{dt} = -(\beta(S) + \kappa_N(N) + \kappa_P(P) + \kappa_R^*)S + 2e^{-\gamma_S \tau_S}\beta(S_{\tau_S})S_{\tau_S} \qquad (6.122)$$

$$\frac{dN}{dt} = -\gamma_N N + A_N(N_{\tau_N})\kappa_N(N_{\tau_N})S_{\tau_N} \qquad (6.123)$$

$$\frac{dP}{dt} = -\gamma_P P + A_P(P_{\tau_P})\left(\kappa_P(P_{\tau_P})S_{\tau_P} - e^{-\gamma_P \tau_{PS}}\kappa_P(P_{\tau_{PSum}})S_{\tau_{PSum}}\right), \qquad (6.124)$$

where $\tau_{Psum} = \tau_P + \tau_{PS}$.

To explore the possible oscillation patterns under perturbations to the platelet component, we introduce variations to the parameters $\bar{\kappa}_P$ and $\tau_P$ and examine the amplitudes of platelet counts of the resulting solutions (Fig. 6.14a). From the model simulations, increasing the maturation time $\tau_P$ results in relatively small amplitude oscillations in the platelet counts (Fig. 6.14c, d), while increasing the maximum differentiation rate $\bar{\kappa}_P$ yields larger amplitude oscillations in the platelet counts (Fig. 6.14e). In Fig. 6.14d, e, there are oscillations in both stem cell and neutrophil counts. Specifically, neutrophils show small amplitude oscillations with high nadirs, while both $\tau_P$ and $\bar{\kappa}_P$ increase with respect to normal levels (Fig. 6.14d). Moreover, neutrophil oscillations can display large amplitudes with a low-level nadir (Fig. 6.14e), somehow akin to the dynamics of cyclical neutropenia.

The nadirs of neutrophils are crucial for patients because low neutrophil counts are often associated with severe neutropenia and immune-deficient symptoms that may cause death. To investigate the effects of perturbations in the platelet compartment on the full system, we further explore the dependence of oscillation patterns on all key regulation parameters in the platelet compartment ($\bar{\kappa}_P$, $\bar{\eta}_P$, $\tau_{PS}$, $\tau_P$, $\gamma_P$, $\vartheta_4$ and $A_{P,\max}$). For each parameter set with oscillatory dynamics, we calculate the neutrophil nadirs. Despite perturbations in all 7 parameters, an obvious separation occurs in the neutrophil nadirs at a boundary of approximately $2 \times 10^8$ cells/kg (Fig. 6.14b). These observations suggest a classification of oscillatory dynamics based on the neutrophil nadirs: Pattern 1 when the neutrophil nadir is larger than $2.3 \times 10^8$ cell/kg and Pattern 2 for smaller neutrophil nadirs.

For Pattern 1 oscillations, most cases have large amplitudes in platelet oscillations and small amplitudes in neutrophil oscillations, similar to the dynamics observed in cyclical thrombocytopenia. An example is shown in Fig. 6.14c. Nevertheless, we also observed cases with obvious oscillations in neutrophils, as shown in Fig. 6.14d. Typical cyclical thrombocytopenia patients show significant oscillations in platelets and no oscillations in neutrophils [90], but a small number of patients show oscillations in both platelets and neutrophils [91]. Pattern 1 oscillations in the model simulations reproduce these two types of dynamics: most perturbations in the platelet compartment show typical dynamics with no (or low amplitude) oscillation in neutrophils, and in rare situations, we see oscillations in both platelets and neutrophils. Hence, Pattern 1 oscillations can be further divided into two subclasses, Pattern 1a (typical CT-like) and Pattern 1b (unusual CT-like), for the two types of dynamics.

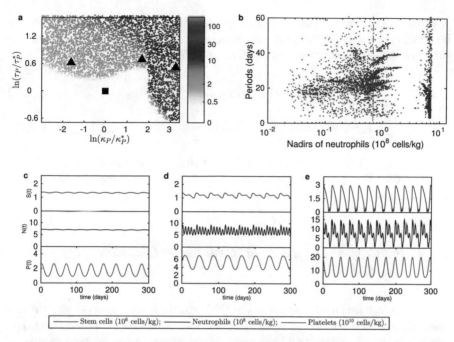

**Fig. 6.14 Simulation results for hematopoietic dynamics** (6.122)–(6.124). **a** Dependence of platelet amplitudes on the parameters $\tau_P$ and $\bar{\kappa}_P$. Each dot represents a parameter set $(\tau_P, \kappa_P)$ shown by the logarithm of the ratio to their default values. The other parameters are the same as those in the normal healthy condition. For each parameter set, the color represents the platelet amplitude (refer to the color bar) for the corresponding solution (with initial conditions from the normal steady state). The color bar shows the platelet amplitudes (in $10^{10}$ cells/kg). The black square marks the default parameter value. The three triangles, from left to right, represent the parameter values corresponding to the solutions in **c-e**, respectively. **b** Characterization of oscillatory dynamics with various parameters $\bar{\kappa}_P$, $\bar{\eta}_P$, $\tau_{PS}$, $\gamma_P$, $\vartheta_4$, and $A_{P,\max}$ varied from 1/30 to 30 times these default values. The dots show the oscillation periods of platelets versus the neutrophil nadirs for different sets of parameters. The vertical dashed line shows the nadir of $0.69 \times 10^8$ cells/kg, corresponding to 1/10 of the normal level. Blue dots show high neutrophil nadirs, and red dots show low neutrophil nadirs, which are separated by a gap (green line) between 2.3 and 4.5 ($\times 10^8$ cells/kg). **c-e** Typical hematopoiesis dynamics with parameters given by black triangles in (A) (from left to right). The bar below **c** through **e** gives the units for the three cell populations. Replotted from [119]

Pattern 2 oscillations show obviously different dynamics, and there are significant oscillations in all cell lines (Fig. 6.14e). In particular, obvious oscillations are observed in the HSC counts, which suggest destabilization in the stem cell compartment. Moreover, many cases show periods of approximately 20 days, and the neutrophil nadirs are lower than 1/10 of the normal level (Fig. 6.14b), which is typical in CN. These results suggest that perturbations in the platelet cell line can induce CN-like behavior, which would be a novel mechanism of inducing CN-like oscillations by increasing the platelet differentiation rate $\bar{\kappa}_P$.

Further discussions of the oscillation patterns in cyclical thrombocytopenia can be found in [119]. In particular, the three oscillation patterns (1a, 1b, and 2) are also seen in a reduced model with a constant neutrophil count ($N(t) \equiv N^*$)

$$\frac{dS}{dt} = -(\beta(S) + \kappa_P(P) + \kappa_0)S + 2e^{-\gamma_S \tau_S}\beta(S_{\tau_S})S_{\tau_S} \tag{6.125}$$

$$\frac{dP}{dt} = -\gamma_P P + A_P(P_{\tau_P})\big(\kappa_P(P_{\tau_P})S_{\tau_P} - e^{-\gamma_P \tau_{PS}}\kappa_P(P_{\tau_{PSum}})S_{\tau_{PSum}}\big), \tag{6.126}$$

where $\kappa_0 = \kappa_N(N^*) + \kappa_R^*$. Hence, the reduced model (6.125)–(6.126) can be essential to understand the oscillation dynamics when there are perturbations in the platelet compartment.

### 6.4.4  Neutrophil Dynamics in Response to Chemotherapy and G-CSF

In addition to permanent changes in the HSC apoptosis rate or neutrophil precursor proliferation rates, chemotherapy can also induce neutropenia [128, 137]. The administration of recombinant hematopoietic cytokines following chemotherapy is frequently used in an effort to circumvent this side effect. G-CSF is now a standard postchemotherapy treatment for neutropenia.

The interval (period) between the repeated administration of chemotherapy is known to have effects on the hematopoietic response [138, 139]. The neutrophil response to G-CSF is highly variable and depends on the timing and protocol of the drug's administration [140–142]. Quantitative systems pharmacology (QSP) models predicted that the timing of cyclic chemotherapy can aggravate the neutropenia nadir in a phenomenon called resonance [126, 143]. In [128], a study of 286 Caucasian children diagnosed with ALL identified resonance in neutrophil counts from lymphoma patients undergoing cyclic cytotoxic chemotherapy and predicted that the length of chemotherapy cycles (with and without G-CSF) has an important influence on the severity of the neutrophil nadir.

To analyze the neutrophil response to chemotherapy and G-CSF, we considered the repeated administration of chemotherapy and G-CSF as periodic perturbations to hematopoiesis and applied the transfer function technique to study the systemic response. In control theory, a transfer function (or system function) is a frequency representation of a dynamical system relating the output or response of a system to the perturbation input or stimulus.

Consider the equation for neutrophil dynamics

$$\frac{dN}{dt} = -\gamma_N N + A_N(t)\kappa_N(N_{\tau_N})Q^*. \tag{6.127}$$

Here, we assume that the HSC number $Q(t) \equiv Q^*$ is a constant, and the amplification factor

$$A_N(t) = e^{\tau_N \hat{\eta}_N(t - \tau_N)}.$$

When chemotherapy is administered with a period $T$, the amplification factor $A_N(t)$ can be expressed as $A_N(t) = A_N + \xi(t)$, where $A_N$ is the normal amplification rate given by (6.98), and $\xi(t)$ is the periodic perturbation (refer to (6.106)).

Let $N(t)$ be the solution of (6.127) under periodic perturbations. Here, we assume that the population of neutrophils is at the steady state (i.e., $N(t) \equiv N^*$) when $t < 0$. When the deviation from the steady state, $y(t) \equiv N(t) - N^*$, is small and of an order of $\varepsilon \ll 1$, it satisfies the linear differentiation delay equation

$$\frac{dy}{dt} = -\gamma_N y - B y_{\tau_N} + \xi(t) \kappa_N (N^*) Q^*, \tag{6.128}$$

where $B = -A_N Q^* \kappa_N'(N^*)$.

The transfer function of (6.128) is defined as the ratio of the Laplace transform of $y(t)$ to that of the input $\xi(t)$ assuming initial conditions of zero. Let $f(s)$ be the Laplace transform of $f(t)$, defined as

$$\hat{f}(s) = \int_0^\infty e^{-st} f(t) dt;$$

the transfer function is given by

$$H(s) = \frac{\hat{y}(s)}{\hat{\xi}(s)} = \frac{\kappa_N(N^*)Q^*}{(s + \gamma_N + Be^{-s\tau_N})}. \tag{6.129}$$

For a linear system at a stable steady state, the response to a periodic input at a frequency of $\omega$ is expressed in terms of the gain $F(\omega)$ of the system and the phase shift $\phi(\omega)$. Both are related to the transfer function $H(s)$ through

$$F(\omega) = |H(i\omega)|, \quad \phi(\omega) = \arg H(i\omega). \tag{6.130}$$

Resonance occurs when $\omega = 2\pi/T$ that maximizes $F(\omega)$, i.e., $\omega = \arg \max F(\omega)$.

To obtain the dependence of the resonant period on the system parameters, we only need to find the minimum value of

$$h(\omega) = |i\omega + \gamma_N + Be^{-i\omega\tau_N}|^2, \tag{6.131}$$

which is given by the solution of

$$h'(\omega) = \omega - B(\gamma_N \tau_N + 1)\sin(\omega\tau_N) - B\omega\tau_N \cos(\omega\tau_N) = 0. \tag{6.132}$$

If the parameters $B$, $\tau_N$, and $\gamma_N$ satisfy

$$B\tau_N(\gamma_N\tau_N + 1) > \frac{\pi}{2}, \tag{6.133}$$

it is easy to verify that

$$h^{(3)}(\omega) > 0 \text{ when } 0 < \omega < \frac{\pi}{2\tau_N},$$

and

$$h'(0) = 0, \ h'(\frac{\pi}{2\tau_N}) < 0, \ h'(\frac{\pi}{\tau_N}) > 0.$$

Therefore, we have $h'(\omega) < 0$ when $0 < 2\tau_N\omega < \pi$, and there is at least one solution of equation $h'(\omega) = 0$ that satisfies $\pi < 2\tau_N\omega < 2\pi$. This solution gives the minimum positive root that corresponds to the dominant resonant frequency.

Now, we can solve (6.132) to obtain the dominant resonant frequency $\omega \in (\frac{\pi}{2\tau_N}, \frac{\pi}{\tau_N})$. To do this, we note the approximations

$$\sin(\omega\tau_N) \approx \pi - \omega\tau_N, \quad \cos(\omega\tau_N) \approx -1;$$

equation (6.132) becomes

$$\omega - B(\gamma_N\tau_N + 1)(\pi - \omega\tau_N) + B\omega\tau_N \approx 0,$$

which yields

$$\omega \approx \pi B\frac{\gamma_N\tau_N + 1}{1 + B\tau_N(\gamma_N\tau_N + 2)}. \tag{6.134}$$

Now, the dominant resonant period is approximately given as

$$T = \frac{2\pi}{\omega} \approx 2\left[\tau_N + \frac{B\tau_N + 1}{B(\tau_N\gamma_N + 1)}\right]. \tag{6.135}$$

From the parameter values for hematologically normal individuals (Table 6.2), we have

$$B\tau_N \gg 1, \quad \tau_N\gamma_N \gg 1. \tag{6.136}$$

Therefore, (6.135) becomes

$$T \approx 2(\tau_N + \gamma_N^{-1}) \tag{6.137}$$

under normal physiological conditions. We also note that the conditions of (6.136) also yield (6.133) and $\omega \approx \pi$ according to (6.134). Thus, when (6.136) is satisfied, the first resonant period is approximated by (6.137).

Despite the simplicity, the approximation of the resonant period given by (6.137) is still valid for models with a gamma distributed delay $\tau_N$ or the two-compartments model with both HSC and neutrophil cell counts [122, 126].

In the approximation (6.137), we note that $\tau_N$ is the transit time from the entire neu-trophil precursor stage (proliferation plus maturation), and $\gamma_N^{-1}$ is the average lifetime of the circulating neutrophils. Thus, the dominant resonant period of chemotherapy is approximately twice the average lifetime of marrow plus the time of circulating neutrophils starting from when they differentiate from HSCs. From [115], cells spend 3-6 days in the mitotic pool under normal physiological conditions, the transit time through the postmitotic pool is between 6 and 8.4 days, and the circulating neutrophil death rate is $1.7 < \gamma_N < 2.4$ days$^{-1}$. Thus, $9 < \tau_N < 14.4$ days, and hence, (6.137) indicates that for hematologically normal individuals, the resonant period in response to chemotherapy is approximately $18.8 < T < 29.7$ days. When repeated chemotherapy is administered in an interval close to the resonant period, it can result in serious side effects due to the large amplitude fluctuations in neutrophil dynam-ics and sometimes, severe neutropenia [126]. These estimations suggest that if the period $T$ of chemotherapy is outside the range $18.8 < T < 29.7$ days, it should be possible to avoid resonance and possible severe neutropenia.

Protocols for the administration of many common chemotherapy agents (such as cisplatin, cyclophosphamide, docetaxel, and paclitaxel) call for a three-week ($T = 21$ days) cycle [144]. The three-week cycle is right within the above estimated resonant period and hence often induces severe neutropenia. G-CSF is frequently used to treat neutropenia induced by chemotherapy [123, 137]. However, the clin-ical administration schedule of G-CSF after chemotherapy is typically determined by trial and error, and the optimal method of giving G-CSF is not clear [145, 146].

In [123], the authors present a delay differential equation model for the regulation of neutrophil production that accounts for the effect of G-CSF. The study found that varying the starting day or the duration of G-CSF treatment can lead to different qualitative responses in neutrophil counts. Further studies showed that the response of neutrophil dynamics to G-CSF is highly variable, depending on the time of G-CSF delivery after chemotherapy at each cycle [126]. In particular, there are specific times in the chemotherapy cycle that G-CSF administration can have positive effects in terms of ameliorating or even eliminating severe neutropenia. However, there are also broad ranges of administration times that can worsen neutropenia after G-CSF administration.

To better understand the effect of G-CSF after chemotherapy, we re-examine equation (6.128), and the periodic perturbation $\xi(t)$ is now given by $\xi(t) = \xi_G(t) - \xi_C(t)$, where $\xi_G(t)$ and $\xi_C(t)$ are the periodic perturbations due to G-CSF and chemotherapy, respectively. Given $\xi_C(t)$, the function $\xi_G(t)$ depends on the protocol of G-CSF delivery, and the relation between these two functions is complicated. Here, we perform the analysis by simply assuming that the input functions satisfy

$$\xi_G(t) = r\xi_C(t - T_e), \tag{6.138}$$

where $r$ is a constant and $T_e$ is the lag time between G-CSF and chemotherapy to become effective in perturbing the amplification rate $A_N$. Because of the postponed response to chemotherapy, $T_e$ is usually not given by the exact day of G-CSF admin-istration.

Taking the Laplace transform of (6.128), we have

$$H(s) = \frac{\hat{y}(s)}{\hat{\xi}_C(s)} = H_1(s) \cdot H_2(s),$$

where

$$H_1(s) = -\frac{\kappa_N(N^*)Q^*}{s + \gamma_N + Be^{-s\tau_N}},$$
$$H_2(s) = 1 - re^{-sT_e}.$$

Thus, the transfer function is now given by

$$F(\omega) = |H(i\omega)| = |H_1(i\omega)| \cdot |H_2(i\omega)|. \tag{6.139}$$

If we have chosen the period $T$ for chemotherapy that coincides with the resonant period of $|H_1(i\omega)|$, the response to G-CSF administration is determined by

$$|H_2(i\omega)|^2 = 1 - 2r\cos(\omega T_e) + r^2. \tag{6.140}$$

This function has a maximum when $T_e = T/2$.

The above analyses show that when we administer chemotherapy in a period $T$ coinciding with the resonant period, we would expect that the worst outcome from G-CSF would occur when G-CSF delivery yielded $T_e = T/2$. This analysis provides a rough understanding of the effect of G-CSF timing in numerical simulations [122].

In [122], a pharmacokinetic model of neutrophil dynamics after chemotherapy and G-CSF was considered (see Sect. 6.4.2)

$$\frac{dQ}{dt} = -(\beta(Q) + \kappa_N(N) + \kappa_\delta)Q + A_Q(t)\beta(Q_{\tau_S})Q_{\tau_S}, \tag{6.141}$$
$$\frac{dN}{dt} = -\gamma_N N + A_N(t)\kappa_N(N_{\tau_N})Q_{\tau_N}, \tag{6.142}$$

where $\kappa_\delta$ represents the combined rate of HSC differentiation into the megakaryocyte and erythrocyte lines, $\beta(Q)$ and $\kappa_N(N)$ are defined in (6.91) , and

$$A_Q(t) = 2\exp\left[-\int_0^{\gamma_S} \gamma_S(t - \tau_S + s)ds\right],$$
$$A_N(t) = \exp\left[\int_0^{\tau_{NP}} \eta_{NP}(t - \tau_N(t) + s)ds - \int_{\tau_{NP}}^{\tau_N(t)} \gamma_0(t - \tau_N(t) + s)ds\right],$$
$$\tau_N(t) = \tau_{NP} + \tau_{NM}(t).$$

Chemotherapy and G-CSF alter the coefficients $\tau_{NM}(t)$, $\gamma_S(t)$ and $\eta_{NP}(t)$ following the equations introduced previously (6.99–6.108). Based on the above kinetic model, a numerical study showed that a single dose of chemotherapy produces damping oscillations in neutrophil levels, and the short-term application of chemotherapy can induce permanent oscillations. Moreover, the neutropenia caused by chemotherapy can be overcome if G-CSF is given early after chemotherapy but can actually be worsened if G-CSF is given later.

The above discussions show that the hematopoietic response to chemotherapy and G-CSF is highly dynamic. The timing of cyclic chemotherapy can worsen neutropenia and the coadministration of G-CSF is not well characterized in the clinical literature. G-CSF is known to affect the neutrophil maturation time in bone marrow, whose detailed dependence is unknown. Further clinical investigations and pharmacokinetic/pharmacodynamic (PK/PD) modeling studies are needed to characterize this important facet of neutrophil regulation [128, 143, 147].

## 6.5  Summary

In this chapter, we introduced mathematical models of stem cell regeneration and their application to understand the origins of dynamic hematological diseases. From the $G_0$ cell cycle model, which is represented by an age-structured model, we can derive a delay differential equation model for the population dynamics of resting phase stem cells. When cell heterogeneities are considered, we introduce an epigenetic state for the cells and derive a differential-integral equation for the heterogeneous stem cell regeneration dynamics. The obtained equation establishes the connections between different scale biological processes, including single-cell transcriptomes, cell population dynamic properties, the cell cycle, and epigenetic modification inheritance, and hence can be applied to the study of various biological processes.

Dynamic hematological diseases are a group of diseases characterized by predictable oscillations in one or more cellular elements of the blood and are often associated with dysregulations in hematopoiesis. In this chapter, we introduced general mathematical models for hematopoiesis, gave a brief survey of how the study (using mathematical models) of dynamic hematological diseases can provide insights into the physiological origin of these diseases, and provided investigators an opportunity to see how to better treat these diseases. These studies show examples of how mathematical biologists in collaboration with hematologists and oncologists can help us better understand the origin and improve the medical treatment of complex diseases.

## References

1. Ehninger, A., Trumpp, A.: The bone marrow stem cell niche grows up: mesenchymal stem cells and macrophages move in. J. Exp. Med. **208**, 421–428 (2011)
2. Trumpp, A., Essers, M., Wilson, A.: Awakening dormant haematopoietic stem cells. Nat. Rev. Immunol. **10**, 201–209 (2010)

3. Lévesque, J.-P., Helwani, F.M., Winkler, I.G.: The endosteal 'osteoblastic' niche and its role in hematopoietic stem cell homing and mobilization. Leukemia **24**, 1979–1992 (2010)
4. Barker, N.N., van de Wetering, M.M., Clevers, H.H.: The intestinal stem cell. Genes. Dev. **22**, 1856–1864 (2008)
5. Leedham, S.J., Brittan, M., McDonald, S.A.C., Wright, N.A.: Intestinal stem cells. J. Cell Mol. Med. **9**, 11–24 (2005)
6. van der Flier, L.G., Clevers, H.: Stem cells, self-renewal, and differentiation in the intestinal epithelium. Annu. Rev. Physiol. **71**, 241–260 (2009)
7. Lander, A.D., Gokoffski, K.K., Wan, F.Y.M., Nie, Q., Calof, A.L.: Cell lineages and the logic of proliferative control. PLoS Biol. **7** (2009)
8. Burns, F.J., Tannock, I.F.: On the existence of a G0-phase in the cell cycle. Cell Prolif. **3**, 321–334 (1970)
9. Dingli, D., Traulsen, A., Pacheco, J.M.: Stochastic dynamics of hematopoietic tumor stem cells. Cell Cycle (Georgetown, Tex) **6**, 461–466 (2007)
10. Hu, G.M., Lee, C.Y., Chen, Y.-Y., Pang, N.N., Tzeng, W.J.: Mathematical model of heterogeneous cancer growth with an autocrine signalling pathway. Cell Prolif. **45**, 445–455 (2012)
11. Mackey, M.C.: Unified hypothesis for the origin of aplastic anemia and periodic hematopoiesis. Blood **51**, 941–956 (1978)
12. Mackey, M.C.: Cell kinetic status of haematopoietic stem cells. Cell Prolif. **34**, 71–83 (2001)
13. Mangel, M., Bonsall, M.B.: Phenotypic evolutionary models in stem cell biology: replacement, quiescence, and variability. PLoS ONE *3* (2008)
14. Mangel, M., Bonsall, M.B.: Stem cell biology is population biology: differentiation of hematopoietic multipotent progenitors to common lymphoid and myeloid progenitors. Theor. Biol. Med. Model. **10**, 5–5 (2012)
15. Rodriguez-Brenes, I., Komarova, N., Wodarz, D.: Evolutionary dynamics of feedback escape and the development of stem-cell–driven cancers. Proc. Natl. Acad. Sci. USA **108**, 18983–18988 (2011)
16. Traulsen, A., Lenaerts, T., Pacheco, J.M., Dingli, D.: On the dynamics of neutral mutations in a mathematical model for a homogeneous stem cell population. J. R. Soc. Interface/the R. Soc. **10**, 20120810–20120810 (2013)
17. Zhou, D., Wu, D., Li, Z., Qian, M., Zhang, M.Q.: Population dynamics of cancer cells with cell state conversions. Quant. Biol. **1**, 201–208 (2013)
18. Chang, H.H., Hemberg, M., Barahona, M., Ingber, D.E., Huang, S.: Transcriptome-wide noise controls lineage choice in mammalian progenitor cells. Nature **453**, 544–547 (2008b)
19. Dykstra, B., Kent, D., Bowie, M., McCaffrey, L., Hamilton, M., Lyons, K., Lee, S.-J., Brinkman, R., Eaves, C.: Long-term propagation of distinct hematopoietic differentiation programs in vivo. Stem Cell **1**, 218–229 (2007)
20. Gibson, T.M., Gersbach, C.A.: Single-molecule analysis of myocyte differentiation reveals bimodal lineage commitment. Integr. Biol. (Camb) **7**, 663–671 (2015)
21. Hayashi, K., de Sousa Lopes, S.M.C., Tang, F., Surani, M.A.: Dynamic equilibrium and heterogeneity of mouse pluripotent stem cells with distinct functional and epigenetic states. Stem Cell **3**, 391–401 (2008)
22. Singer, Z.S., Yong, J., Tischler, J., Hackett, J.A., Altinok, A., Surani, M.A., Cai, L., Elowitz, M.B.: Dynamic heterogeneity and DNA methylation in embryonic stem cells. Mol. Cell **55**, 319–331 (2014)
23. Zernicka-Goetz, M., Morris, S.A., Bruce, A.W.: Making a firm decision: multifaceted regulation of cell fate in the early mouse embryo. Nat. Rev. Genet. **10**, 467–477 (2009)
24. Till, J.E., McCulloch, E.A., Siminovitch, L.: A stochastic model of stem cell proliferation, based on the growth of spleen colony-forming cells. Proc. Natl. Acad. Sci. USA **51**, 29–36 (1964)
25. MacArthur, B.D.: Collective dynamics of stem cell populations. Proc. Natl. Acad. Sci. USA **111**, 3653–3654 (2014)
26. Lei, J., Levin, S.A., Nie, Q.: Mathematical model of adult stem cell regeneration with cross-talk between genetic and epigenetic regulation. Proc. Natl. Acad. Sci. USA **111**, E880–E887 (2014)

27. Situ, Q., Lei, J.: A mathematical model of stem cell regeneration with epigenetic state transitions. MBE **14**, 1379–1397 (2017)
28. Lei, J.: A general mathematical framework for understanding the behavior of heterogeneous stem cell regeneration. J. Theor. Biol. **492** (2020a)
29. Lei, J.: Evolutionary dynamics of cancer: from epigenetic regulation to cell population dynamics—mathematical model framework, applications, and open problems. Sci. China Math. **63**, 411–424 (2020b)
30. Lajtha, L.: On the concept of the cell cycle. J. Cell. Comp. Physiol. **60**, 143 (1963)
31. Quastler, H.: The analysis of cell population kinetics. In: Lamerton, L.F., Fry, R. (eds.) Cell Proliferation. Blackwell Scientific Publications Oxford
32. Ornitz, D.M., Itoh, N. (2001). Fibroblast growth factors. Genome Biol. **2**, reviews 3005.1–3005.12
33. Massague, J.: TGF$\beta$ signalling in context. Nat. Rev. Mol. Cell. Biol. **13**, 616–630 (2012)
34. Nakao, A., Afrakhte, M., Moren, A., Nakayama, T., Christian, J.L., Heuchel, R., Itoh, S., Kawabata, N., Heldin, N.E., Heldin, C.H., tenDijke, P.: Identification of Smad7, a TGF beta-inducible antagonist of TGF-beta signalling. Nature **389**, 631–635 (1997)
35. Massagué, J.: The transforming growth factor-beta family. Annu. Rev. Cell Biol. **6**, 597–641 (1990)
36. Hanahan, D., Weinberg, R.A.: The hallmarks of cancer. Cell **100**, 57–70 (2000)
37. Batlle, E., Massague, J.: Transforming growth factor-$\beta$ signaling in immunity and cancer. Immunity **50**, 924–940 (2019)
38. Horn, L.A., Fousek, K., Palena, C.: Tumor plasticity and resistance to immunotherapy. Trends Cancer **6**, 432–441 (2020)
39. Bernard, S., Bélair, J., Mackey, M.C.: Oscillations in cyclical neutropenia: new evidence based on mathematical modeling. J. Theor. Biol. **223**, 283–298 (2003)
40. Sanchez-Vega, F., Mina, M., Armenia, J., La, K.C., Dimitriadoy, S., Liu, D.L., Kantheti, H.S., Saghafinia, S., Daian, F., Gao, Q., Bailey, M.H., Liang, W.-W., Foltz, S.M., Shmulevich, I., Ding, L., Heins, Z., Gross, B., Zhang, H., Kundra, R., Bahceci, I., Dervishi, L., Dogrusoz, U., Zhou, W., Way, G.P., Greene, C.S., Xiao, Y., Wang, C., Iavarone, A., Berger, A.H., Bivona, T.G., Lazar, A.J., Hammer, G.D., Giordano, T., Kwong, L.N., McArthur, G., Huang, C., Tward, A.D., Frederick, M.J., McCormick, F., Network, T.C.G.A.R., Caesar-Johnson, S.J., Demchok, J.A., Felau, I., Kasapi, M., Ferguson, M.L., Hutter, C.M., Sofia, H.J., Tarnuzzer, R., Wang, Z., Yang, L., Zenklusen, J.C., Zhang, J.J., Chudamani, S., Liu, J., Lolla, L., Naresh, R., Pihl, T., Sun, Q., Wan, Y., Wu, Y., Cho, J., DeFreitas, T., Frazer, S., Gehlenborg, N., Getz, G., Heiman, D.I., Kim, J., Lawrence, M.S., Lin, P., Meier, S., Noble, M.S., Saksena, G., Voet, D., Zhang, H., Bernard, B., Chambwe, N., Dhankani, V., Knijnenburg, T., Kramer, R., Leinonen, K., Liu, Y., Miller, M., Reynolds, S., Thorsson, V., Zhang, W., Akbani, R., Broom, B.M., Hegde, A.M., Ju, Z., Kanchi, R.S., Korkut, A., Li, J., Liang, H., Ling, S., Liu, W., Lu, Y., Mills, G.B., Ng, K.-S., Rao, A., Ryan, M., Wang, J., Weinstein, J.N., Zhang, J., Abeshouse, A., Chakravarty, D., Chatila, W.K., de Bruijn, I., Gao, J., Gross, B.E., Heins, Z.J., La, K., Ladanyi, M., Luna, A., Nissan, M.G., Ochoa, A., Phillips, S.M., Reznik, E., Sander, C., Sheridan, R., Sumer, S.O., Sun, Y., Taylor, B.S., Wang, J., Anur, P., Peto, M., Spellman, P., Benz, C., Stuart, J.M., Wong, C.K., Yau, C., Hayes, D.N., Parker, J.S., Wilkerson, M.D., Ally, A., Balasundaram, M., Bowlby, R., Brooks, D., Carlsen, R., Chuah, E., Dhalla, N., Holt, R., Jones, S.J.M., Kasaian, K., Lee, D., Ma, Y., Marra, M.A., Mayo, M., Moore, R.A., Mungall, A.J., Mungall, K., Robertson, A.G., Sadeghi, S., Schein, J.E., Sipahimalani, P., Tam, A., Thiessen, N., Tse, K., Wong, T., Berger, A.C., Beroukhim, R., Cherniack, A.D., Cibulskis, C., Gabriel, S.B., Gao, G.F., Ha, G., Meyerson, M., Schumacher, S.E., Shih, J., Kucherlapati, M.H., Kucherlapati, R.S., Baylin, S., Cope, L., Danilova, L., Bootwalla, M.S., Lai, P.H., Maglinte, D.T., Van Den Berg, D.J., Weisenberger, D.J., Auman, J.T., Balu, S., Bodenheimer, T., Fan, C., Hoadley, K.A., Hoyle, A.P., Jefferys, S.R., Jones, C.D., Meng, S., Mieczkowski, P.A., Mose, L.E., Perou, A.H., Perou, C.M., Roach, J., Shi, Y., Simons, J.V., Skelly, T., Soloway, M.G., Tan, D., Veluvolu, U., Fan, H., Hinoue, T., Laird, P.W., Shen, H., Zhou, W., Bellair, M., Chang, K., Covington, K., Creighton, C.J., Dinh, H., Doddapaneni, H., Donehower, L.A., Drummond,

J., Gibbs, R.A., Glenn, R., Hale, W., Han, Y., Hu, J., Korchina, V., Lee, S., Lewis, L., Li, W., Liu, X., Morgan, M., Morton, D., Muzny, D., Santibanez, J., Sheth, M., Shinbrot, E., Wang, L., Wang, M., Wheeler, D.A., Xi, L., Zhao, F., Hess, J., Appelbaum, E.L., Bailey, M., Cordes, M.G., Fronick, C.C., Fulton, L.A., Fulton, R.S., Kandoth, C., Mardis, E.R., McLellan, M.D., Miller, C.: Oncogenic signaling pathways in the cancer genome atlas. Cell **173**, 321–337.e10 (2018)

41. Lei, J.: A general mathematical framework for understanding the behavior of heterogeneous stem cell regeneration. J. Theor. Biol. **492** (2020c)

42. Abkowitz, J., Holly, R., Hammond, W.P.: Cyclic hematopoiesis in dogs: studies of erythroid burst forming cells confirm and early stem cell defect. Exp. Hematol. *16*, 941–945 (1988)

43. Boggs, D.R., Boggs, S.S., Saxe, D.F., Gress, L.A., Canfield, D.R.: Hematopoietic Stem Cells with high proliferative potential: assay of their concentraion in marrow by the frequency and duration of cure of W/Wv mice. J. Clin. Investig. **70**, 242–253 (1982)

44. Micklem, H.S., Lennon, J.E., Ansell, J.D., Gray, R.A.: Numbers and dispersion of repopulating hematopoietic cell clones in radiation chimeras as functions of injected cell dose. Exp. Hematol. **15**, 251–257 (1987)

45. Lei, J., Mackey, M.C.: Multistability in an age-structured model of hematopoiesis: cyclical neutropenia. J. Theor. Biol. **270**, 143–153 (2011)

46. Colijn, C., Mackey, M.: A mathematical model of hematopoiesis—I. Periodic chronic myelogenous leukemia. J. Theor. Biol. **237**, 117–132 (2005a)

47. Probst, A.V., Dunleavy, E., Almouzni, G.: Epigenetic inheritance during the cell cycle. Nat. Rev. Mol. Cell. Biol. **10**, 192–206 (2009)

48. Wu, H., Zhang, Y.: Reversing DNA methylation: mechanisms, genomics, and biological functions. Cell **156**, 45–68 (2014)

49. Schepeler, T., Page, M.E., Jensen, K.B.: Heterogeneity and plasticity of epidermal stem cells. Development **141**, 2559–2567 (2014)

50. Takaoka, K., Hamada, H.: Origin of cellular asymmetries in the pre-implantation mouse embryo: a hypothesis. Philos. Trans. R. Soc. Lond. B Biol. Sci. *369* (2014)

51. Tang, F., Barbacioru, C., Wang, Y., Nordman, E., Lee, C., Xu, N., Wang, X., Bodeau, J., Tuch, B.B., Siddiqui, A., Lao, K., Surani, M.A.: mRNA-Seq whole-transcriptome analysis of a single cell. Nat. Methods **6**, 377–382 (2009)

52. Rotem, A., Ram, O., Shoresh, N., Sperling, R.A., Goren, A., Weitz, D.A., Bernstein, B.E.: Single-cell ChIP-seq reveals cell subpopulations defined by chromatin state. Nat. Biotechnol. **33**, 1165–1172 (2015)

53. Smallwood, S.A., Lee, H.J., Angermueller, C., Krueger, F., Saadeh, H., Peat, J., Andrews, S.R., Stegle, O., Reik, W., Kelsey, G.: Single-cell genome-wide bisulfite sequencing for assessing epigenetic heterogeneity. Nat. Methods **11**, 817–820 (2014)

54. Shahrezaei, V., Swain, P.S.: Analytical distributions for stochastic gene expression. Proc. Natl. Acad. Sci. USA **105**, 17256–17261 (2008a)

55. Friedman, N., Cai, L., Xie, X.S.: Linking stochastic dynamics to population distribution: an analytical framework of gene expression. Phys. Rev. Lett. **97** (2006)

56. Arai, F., Stumpf, P.S., Ikushima, Y.M., Hosokawa, K., Roch, A., Lutolf, M.P., Suda, T., MacArthur, B.D.: Machine learning of hematopoietic stem cell divisions from paired daughter cell expression profiles reveals effects of aging on self-renewal. Cell Syst. **11**, 640–652.e5 (2020)

57. Fischer, D.S., Fiedler, A.K., Kernfeld, E.M., Genga, R.M.J., Bastidas-Ponce, A., Bakhti, M., Lickert, H., Hasenauer, J., Maehr, R., Theis, F.J.: Inferring population dynamics from single-cell RNA-sequencing time series data. Nat. Biotechnol. **37**, 461–468 (2019)

58. Woodworth, M.B., Girskis, K.M., Walsh, C.A.: Building a lineage from single cells: genetic techniques for cell lineage tracking. Nat. Rev. Genet. **18**, 230–244 (2017)

59. Notta, F., Zandi, S., Takayama, N., Dobson, S., Gan, O.I., Wilson, G., Kaufmann, K.B., McLeod, J., Laurenti, E., Dunant, C.F., McPherson, J.D., Stein, L.D., Dror, Y., Dick, J.E.: Distinct routes of lineage development reshape the human blood hierarchy across ontogeny. Science *351*, aab2116–aab2116 (2016)

60. Wu, C., Li, B., Lu, R., Koelle, S.J., Yang, Y., Jares, A., Krouse, A.E., Metzger, M., Liang, F., Loré, K., Wu, C.O., Donahue, R.E., Chen, I.S.Y., Weissman, I., Dunbar, C.E.: Clonal tracking of rhesus macaque hematopoiesis highlights a distinct lineage origin for natural killer cells. Stem Cell **14**, 486–499 (2014)

61. Perié, L., Hodgkin, P.D., Naik, S.H., Schumacher, T.N., de Boer, R.J., Duffy, K.R.: Determining lineage pathways from cellular barcoding experiments. Cell Rep. **6**, 617–624 (2014)

62. Bocharov, G., Quiel, J., Luzyanina, T., Alon, H., Chiglintsev, E., Chereshnev, V., Meier-Schellersheim, M., Paul, W.E., Grossman, Z.: Feedback regulation of proliferation versus differentiation rates explains the dependence of CD4 T-cell expansion on precursor number. Proc. Nat. Acad. Sci. **108**, 3318–3323 (2011)

63. Vermeulen, L., Snippert, H.J.: Stem cell dynamics in homeostasis and cancer of the intestine. Nat. Rev. Cancer **14**, 468–480 (2014)

64. Knapp, D.J., Eaves, C.J.: Control of the hematopoietic stem cell state. Nat. Publ. Group **24**, 3–4 (2014)

65. Lugli, E., Dominguez, M.H., Gattinoni, L., Chattopadhyay, P.K., Bolton, D.L., Song, K., Klatt, N.R., Brenchley, J.M., Vaccari, M., Gostick, E., Price, D.A., Waldmann, T.A., Restifo, N.P., Franchini, G., Roederer, M.: Superior T memory stem cell persistence supports long-lived T cell memory. J. Clin. Invest. **123**, 594–599 (2013)

66. D'Urso, A., Brickner, J.H.: Epigenetic transcriptional memory. Curr. Genet. **63**, 435–439 (2017)

67. Su, Y., Wei, W., Robert, L., Xue, M., Tsoi, J., Garcia-Diaz, A., Homet Moreno, B., Kim, J., Ng, R.H., Lee, J.W., Koya, R.C., Comin-Anduix, B., Graeber, T.G., Ribas, A., Heath, J.R.: Single-cell analysis resolves the cell state transition and signaling dynamics associated with melanoma drug-induced resistance. Proc. Natl. Acad. Sci. USA **114**, 13679–13684 (2017)

68. Macaulay, I.C., Svensson, V., Labalette, C., Ferreira, L., Hamey, F., Voet, T., Teichmann, S.A., Cvejic, A.: Single-Cell RNA-Sequencing reveals a continuous spectrum of differentiation in hematopoietic cells. Cell Rep. **14**, 1–13 (2016)

69. Richard, A., Boullu, L., Herbach, U., Bonnafoux, A., Morin, V., Vallin, E., Guillemin, A., Papili Gao, N., Gunawan, R., Cosette, J., Arnaud, O., Kupiec, J.-J., Espinasse, T., Gonin-Giraud, S., Gandrillon, O.: Single-Cell-Based analysis highlights a surge in Cell-to-Cell molecular variability preceding irreversible commitment in a differentiation process. PLoS Biol. **14** (2016)

70. Joost, S., Zeisel, A., Jacob, T., Sun, X., La Manno, G., Lönnerberg, P., Linnarsson, S., Kasper, M.: Single-Cell transcriptomics reveals that differentiation and spatial signatures shape epidermal and hair follicle heterogeneity. Cell Syst. **3**, 221–237.e9 (2016)

71. Gupta, P.B., Fillmore, C.M., Jiang, G., Shapira, S.D., Tao, K., Kuperwasser, C., Lander, E.S.: Stochastic state transitions give rise to phenotypic equilibrium in populations of cancer cells. Cell **146**, 633–644 (2010)

72. Furness, S.G.B., McNagny, K.: Beyond mere markers: functions for CD34 family of sialomucins in hematopoiesis. IR **34**, 13–32 (2006)

73. Chang, H.H., Hemberg, M., Barahona, M., Ingber, D.E., Huang, S.: Transcriptome-wide noise controls lineage choice in mammalian progenitor cells. Nature **453**, 544–547 (2008a)

74. Li, Q., Wennborg, A., Aurell, E., Dekel, E., Zou, J.-Z., Xu, Y., Huang, S., Ernberg, I.: Dynamics inside the cancer cell attractor reveal cell heterogeneity, limits of stability, and escape. Proc. Natl. Acad. Sci. USA **113**, 2672–2677 (2016)

75. Weston, W., Zayas, J., Perez, R., George, J., Jurecic, R.: Dynamic equilibrium of heterogeneous and interconvertible multipotent hematopoietic cell subsets. Sci. Rep. **4**, 5199–5199 (2014)

76. Ren, X., Kang, B., Zhang, Z.: Understanding tumor ecosystems by single-cell sequencing: promises and limitations. Genome Biol. **19**, 211 (2018)

77. Stuart, T., Satija, R.: Integrative single-cell analysis. Nat. Rev. Genet. **20**, 257–272 (2019)

78. Liu, J., Song, Y., Lei, J.: Single-Cell entropy to quantify the cellular order parameter from Single-Cell RNA-Seq data. Biophys. Rev. Lett. **15**, 35–49 (2020)

79. Ye, Y., Yang, Z., Lei, J.: Using single-cell entropy to describe the dynamics of reprogramming and differentiation of induced pluripotent stem cells. Int. J Mod. Phys. B **34**, 2050288(2020)
80. Glass, L., Mackey, M.C.: From Clocks to Chaos-the Rhythms of Life. Princeton University Press, Princeton (1988)
81. Foley, C., Mackey, M.C.: Dynamic hematological disease: a review. J. Math. Biol. **58**, 285–322 (2009b)
82. Dale, D.C., Mackey, M.C.: Understanding, treating and avoiding hematological disease: better medicine through mathematics? Bull. Math. Biol. **77**, 739–757 (2015)
83. Colijn, C., Mackey, M.: A mathematical model of hematopoiesis: II cyclical neutropenia. J. Theor. Biol. **237**, 133–146 (2005b)
84. Hearn, T., Haurie, C., Mackey, M.C.: Cyclical neutropenia and the peripheral control of white blood cell production. J. Theor. Biol. **192**, 167–181 (1998)
85. Haurie, C., Dale, D.C., Mackey, M.C.: Occurrence of periodic oscillations in the differential blood counts of congenital, idiopathic, and cyclical neutropenic patients before and during treatment with G-CSF. Exp. Hematol. **27**, 401–409 (1999)
86. Fortin, P., Mackey, M.C.: Periodic chronic myelogenous leukemia: spectral analysis of blood cell counts and etiological implications. Br. J. Haematol. **104**, 336–345 (1999)
87. Mackey, M.C.: Periodic auto-immune hemolytic anemia: an induced dynamical disease. Bull. Math. Biol. **41**, 829–834 (1979)
88. Mackey, M.C., Glass, L.: Oscillation and chaos in physiological control systems. Science **197**, 287–289 (1977)
89. Swinburne, J.L., Mackey, M.C.: Cyclical thrombocytopenia: characterization by spectral analysis and a review. J. Theor. Med. **2**, 81–91 (2000)
90. Apostu, R., Mackey, M.C.: Understanding cyclical thrombocytopenia: a mathematical modeling approach. J. Theor. Biol. **251**, 297–316 (2008)
91. Langlois, G.P., Arnold, D.M., Potts, J., Leber, B., Dale, D.C., Mackey, M.C.: Cyclic thrombocytopenia with statistically significant neutrophil oscillations. Clin. Case Rep. **19**, 53 (2018)
92. Mackey, M.C.: Periodic hematological disorders: quintessential examples of dynamical diseases. Chaos **30**, 063123 (2020)
93. Horwitz, M., Benson, K.F., Person, R.E., Aprikyan, A.G., Dale, D.C.: Mutations in ELA2, encoding neutrophil elastase, define a 21-day biological clock in cyclic haematopoiesis. Nat. Genet. **23**, 433–436 (1999)
94. Haurie, C., Dale, D.C., Rudnicki, R., Mackey, M.C.: Modeling complex neutrophil dynamics in the grey collie. J. Theor. Biol. **204**, 505–519 (2000)
95. Gill, M., Ockelford, P., Morris, A., Bierre, T., Kyle, C.: Diagnostic handbook-the interpretation of laboratory tests. Diagnostic Medlab, Auckland (2000)
96. Haurie, C., Person, R., Dale, D.C., Mackey, M.C.: Hematopoietic dynamics in grey collies. Exp. Hematol. **27**, 1139–1148 (1999)
97. Beutler, E., Lichtman, M., Coller, B., Kipps, T.: Williams Hematology. McGraw-Hill, New York (1995)
98. Engstrom, K., Lundquist, A., Söderströ, N.: Periodic thrombocytopenia or tidal platelet dysgenesis in man. Scand. J. Haematol. **3**, 290–292 (1966)
99. Ranlov, P., Videbaek, A.: Cyclic haemolytic anaemia synchronous with Pel-Ebstein fever in a case of Hodgkin's disease. Acta Medica Scandinavica **174**, 583–588 (1963)
100. Gordon, R.R., Varadi, S.: Congenital hypoplastic anemia (pure red-cell anemia) with periodic erythroblastopenia. Lancet **1**, 296–299 (1962)
101. Gurney, C.W., Simmons, E.L., Gaston, E.O.: Cyclic erythropoiesis in W/Wv mice following a single small dose of $^{89}$Sr. Exp. Hematol. **9**, 118–122 (1981)
102. Gibson, C.M., Gurney, C.W., Gaston, E.O., Simmons, E.L.: Cyclic erythropoiesis in the S1/S1d mouse. Exp. Hematol. **12**, 343–348 (1984)
103. Gibson, C.M., Gurney, C.W., Simmons, E.L., Gaston, E.O.: Further studies on cyclic erythropoiesis in mice. Exp. Hematol. **13**, 855–860 (1985)
104. Vácha, J., Znoji, V.: Application of a mathematical model of erythropoiesis to the process of recovery after acute X-irradiation of mice. Biofizika **20**, 872–879 (1975)

105. Vácha, J.: Postirradiational oscillations of erythropoiesis in mice. Acta Sc. Nat. Brno. **16**, 1–52 (1982)
106. Orr, J.S., Kirk, J., Gray, K.G., Anderson, J.R.: A study of the interdependence of red cell and bone marrow stem cell populations. Br. J.Haematol. **15**, 23–24 (1968)
107. Munje, C., Copland, M.: Exploring stem cell heterogeneity in Chronic Myeloid Leukemia. Trends Cancer **4**, 167–169 (2018)
108. Melo, J.V.: The diversity of BCR-ABL fusion proteins and their relationship to Leukemia phenotype. Blood **88**, 2375–2384 (1996)
109. Henderson, E.S., Lister, T.A., Greaves, M.F. (eds.): Leukemia. Saunders, Philadelphia (1996)
110. Mackey, M.C., Bélair, J.: Cell replication and control. In: Beuter, A., Glass, L., Mackey, M.C., Titcombe, M.S. (eds.) Nonlinear dynamics in physiology and medicine. Springer, New York (2003)
111. Hoffbrand, A.V., Pettit, J.E., Moss, P.A.: Essential Haematology, 4th edn. Blackwell Science, Milan (2011)
112. Mahaffy, J.M., Bélair, J., Mackey, M.C.: Hematopoietic model with moving boundary condition and state dependent delay: applications in erythropoiesis. J. Theor. Biol. **190**, 135–146 (1998)
113. Bélair, J., Mackey, M.C.: A model for the regulation of mammalian platelet. Ann. N. Y. Acad. Sci. **504**, 280–282 (1987)
114. Adamson, J.W.: The relationship of erythropoietin and iron metabolism to red blood cell production in humans. Semin. Oncol. **2**, 9–15 (1994)
115. Price, T.H., Chatta, G.S., Dale, D.C.: Effect of recombinant granulocyte colony-stimulating factor on neutrophil kinetics in normal young and elderly humans. Blood **88**, 335–340 (1996)
116. Ratajczak, M.Z.M., Ratajczak, J.J., Marlicz, W.W., Pletcher, C.H.C., Machalinski, B.B., Moore, J.J., Hung, H.H., Gewirtz, A.M.A.: Recombinant human thrombopoietin (TPO) stimulates erythropoiesis by inhibiting erythroid progenitor cell apoptosis. Br. J. Haematol. **98**, 8–17 (1997)
117. Pujo-Menjouet, L.: Blood cell dynamics: half of a century of modelling. Math. Model. Nat. Phenom. **11**, 92–115 (2016)
118. Colijn, C., Foley, C., Mackey, M.C.: G-CSF treatment of canine cyclical neutropenia: a comprehensive mathematical model. Exp. Hematol. **35**, 898–907 (2007)
119. Zhuge, C., Mackey, Lei, J.: Origins of oscillation patterns in cyclical thrombocytopenia. J. Theor. Biol. **462**, 432–445 (2018)
120. Merchant, A.A., Singh, A., Matsui, W., Biswal, S.: The redox-sensitive transcription factor Nrf2 regulates murine hematopoietic stem cell survival indepently of ROS levels. Blood **118**, 6572–6579 (2011)
121. Leavey, P.J., Sellins, K.S., Thurman, G., Elzi, D., Hiester, A., Silliman, C.C., Zerbe, G., Cohen, J.J., Ambruso, D.R.: In vivo treatment with granulocyte colony-stimulating factor results in divergent effects on neutrophil functions measured in vitro. Blood **92**, 4366–4374 (1998)
122. Brooks, G., Langlois, G.P., Lei, J., Mackey, M.C.: Neutrophil dynamics after chemotherapy and G-CSF: the role of pharmacokinetics in shaping the response. J. Theor. Biol. **315**, 97–109 (2012)
123. Foley, C., Mackey, M.C.: Mathematical model for G-CSF administration after chemotherapy. J. Theor. Biol. **257**, 27–44 (2009a)
124. Hannun, Y.A.: Apoptosis and the dilemma of cancer chemotherapy. Blood **89**, 1845–1853 (1997)
125. Häcker, S., Karl, S., Mader, I., Cristofanon, S., Schweitzer, T., Krauss, J., Rutkowski, S., Debatin, K.-M., Fulda, S.: Histone deacetylase inhibitors prime medulloblastoma cells for chemotherapy-induced apoptosis by enhancing p53-dependent Bax activation. Oncogene **30**, 2275–2281 (2011)
126. Zhuge, C., Lei, J., Lei, J., Mackey, M.C.: Neutrophil dynamics in response to chemotherapy and G-CSF. J. Theor. Biol. **293**, 111–120 (2012)

127. Fetterly, G.J., Grasela, T.H., Sherman, J.W., Dul, J.L., Grahn, A., Lecomte, D., Fiedler-Kelly, J., Damjanov, N., Fishman, M., Kane, M.P., Rubin, E.H., Tan, A.R.: Pharmacokinetic/pharmacodynamic modeling and simulation of neutropenia during phase I development of liposome-entrapped paclitaxel. Clin. Cancer Res. **14**, 5856–5863 (2008)
128. Mackey, M.C., Glisovic, S., Leclerc, J.-M., Pastore, Y., Krajinovic, M., Craig, M.: The timing of cyclic cytotoxic chemotherapy can worsen neutropenia and neutrophilia. Br. J. Clin. Pharmacol. **13**
129. Ma, S., Zhu, K., Lei, J.: Bistability and state transition of a delay differential equation model of neutrophil dynamics. Int. J. Bifurc. Chaos **25**, 1550017 (2015)
130. Wahlberg, P., Nyman, D., Ekelund, P., Carlsson, S.A., Granlund, H.: Cyclical thrombocytopenia with remission during lynestrenol treatment in a woman. Ann. Clin. Res. **9**, 356–358 (1977)
131. Sekine, T., Takagi, M., Uemura, Y., Mori, Y., Wada, H., Minami, N., Deguchi, K., Shirakawa, S.: A cyclic thrombocytopenia associated with Sjögren syndrome. Rinsho Ketsueki **30**, 1021–1026 (1989)
132. Kosugi, S., Tomiyama, Y., Shiraga, M., Kashiwagi, H., Nakao, H., Kanayama, Y., Kurata, Y., Matsuzawa, Y.: Cyclic thrombocytopenia associated with IgM anti-GPllb-Illa autoantibodies. Br. J. Haematol. **88**, 809–815 (1994)
133. Pavord, S., Sivakumaran, M., Furber, P., Mitchell, V.: Cyclical thrombocytopenia as a rare manifestation of myelodysplatic syndrome. Clin. Lab. Haematol. **18**, 221–223 (1996)
134. Nagasawa, T., Hasegawa, Y., Shimizu, S., Kawashima, Y., Nishimura, S., Suzukawa, K., Mukai, H., Hori, M., Komeno, T., Kojima, H., Ninomiya, H., Tahara, T., Abe, T.: Serum thrombopoietin level is mainly regulated by megakaryocyte mass rather than platelet mass in human subjects. Br. J. Haematol. **101**, 242–244 (1998)
135. Santillan, M., Mahaffy, J.M., Bélair, J., Mackey, M.C.: Regulation of platelet production: the normal response to perturbation and cyclical platelet disease. J. Theor. Biol. **206**, 585–603 (2000)
136. Langlois, G.P., Craig, M., Humphries, A.R., Mackey, M.C., Mahaffy, J.M., Bélair, J., Moulin, T., Sinclair, S.R., Wang, L.: Normal and pathological dynamics of platelets in humans. J. Math. Biol. **75**, 1411–1462 (2017)
137. Crawford, J., Dale, D.C., Lyman, G.H.: Chemotherapy-induced: risks, consequences, and new directions for its management. Cancer **100**, 228–237 (2003)
138. Thatcher, N., Girling, D.J., Hopwood, P., Sambrook, R.J., Qian, W., Stephens, R.J.: Improving survival without reducing quality of life in small-cell lung cancer patients by increasing the dose-intensity of chemotherapy with granulocyte colony-stimulating factor support: results of a British Medical Research Council Multicenter Randomized Trial. Medical Research Council Lung Cancer Working Party. J. Clin. Oncol. **18**, 395–404 (2000)
139. Tjan-Heijnen, V.C.G., Wagener, D.J.T., Postmus, P.E.: An analysis of chemotherapy dose and dose-intensity in small-cell lung cancer: lessons to be drawn. Ann. Oncol. **13**, 1519–1530 (2002)
140. Morstyn, G., Campbell, L., Lieschke, G., Layton, J., Maher, D., O'Connor, M., Green, M., Sheridan, W., Vincent, M., Alton, K., Souza, L., McGrath, K., Fox, R.: Treatment of chemotherapy-induced neutropenia by subcutaneously administered granulocyte colony-stimulating factor with optimization of dose and duration of therapy. J. Clin. Oncol. **7**, 1554–1562 (1989)
141. Meishenberg, B.R., Davis, T.A., Melaragno, A.J., Stead, R., Monroy, R.L.: A comparison of therapeutic schedules for administering granulocyte colony-stimulating factor to nonhuman primates after high-dose chemotherapy. Blood **79**, 2267–2272 (1992)
142. Butler, R.D., Waites, T.M., Lamar, R.E., Hainsworth, J.D., Greco, F.A., Johnson, D.H.: Timing of G-CSF administration during intensive chemotherapy for breast cancer (abstract). Am. Soc. Clin. Oncol. **11**, 1411 (1992)
143. Craig, M., Humphries, A.R., Nekka, F., Bélair, J., Li, J., Mackey, M.C.: Neutrophil dynamics during concurrent chemotherapy and G-CSF administration: mathematical modelling guides dose optimisation to minimise neutropenia. J. Theor. Biol. **385**, 77–89 (2015)

144. Skeel, R.T., (ed.): Handbook of Cancer Chemotherapy. Lippincott Williams & Wilkins (2007)
145. Clark, O.A., Lyman, G.H., Castro, A.A., Clark, L.G., Djulbegovic, B.: Colony-stimulating factors for chemotherapy-induced febrile neutrophenia: a meta-analysis of randomized controlled trails. J. Clin. Oncol. **23**, 4198–4214 (2005)
146. Bennett, C.L., Weeks, J.A., Somerfield, M.R., et al.: Use of hematopoietic colony-stimulating factors: comparision of the 1994 and 1997 American Society of Clinical Oncology surveys regarding ASCO clinical practice guidelines. J. Clin. Oncol. **17**, 3676–3681 (1999)
147. Craig, M., Humphries, A.R., Mackey, M.C.: A mathematical model of granulopoiesis incorporating the negative feedback dynamics and Kinetics of G-CSF/Neutrophil binding and internalization. Bull. Math. Biol. **78**, 2304–2357 (2016)

# Chapter 7
# Mathematical Models of Morphogen Gradients and Growth Control

*Science may be described as the art of systematic oversimplification.*

—Karl Popper

## 7.1 Introduction

In developmental biology, it is a fundamental question to understand how cells with distinct phenotypes and organs are generated from a single cell or a small number of cells. In 1969, Wolpert proposed the concept of *positional information* in which cells acquire positional identities as in a coordinate system and then interpret their positions to give rise to spatial patterns [1]. In the last 60 years, substantial experimental evidence has shown that the positional information of cells strongly depends on various types of signaling molecules known as *morphogens* (aka ligands). During embryonic development, morphogens are synthesized at a localized site and disperse (e.g., through diffusion) from their production site to bind to cell receptors on cell membranes. These processes result in different levels of receptor occupancy at different cell locations, leading to different levels of downstream signals for different cells. The spatial concentration gradient of morphogen-receptor complexes (aka *signaling gradients*) induces spatially graded differences in cell signaling. These signaling gradients provide positional information for the global patterns across the target tissue [1, 2]. An important issue in the positional information theory is to understand how cells know precisely and reliably where they are [3].

The concept of morphogens is at the center of positional information [4, 5]. Morphogens are signaling molecules that can diffuse and act over several cell diameters to induce concentration-dependent cellular responses. Different control strategies are involved in this process due to the complexity and diversity of different types of organ development. 'Wet' experiments alone are usually insufficient for understanding the complex machinery used for morphogen-mediated patterning and growth control. Mathematical modeling approaches play critical roles in delineating

J. Lei, *Systems Biology*, Lecture Notes on Mathematical Modelling in the Life Sciences, https://doi.org/10.1007/978-3-030-73033-8_7

these complex processes, as well as the principles behind them, providing an influx of fresh ideas and novel methods to biology [6–8]. In this chapter, we present mathematical models of morphogen gradient formation, patterning robustness, and growth control. This chapter focuses mainly on morphogen dynamics and growth driven by localized sources of morphogens. Spontaneous patterning and some other aspects of patterning studies can be found in other review papers [8–11].

In this chapter, we take the formation of morphogen gradients in the *Drosophila* wing imaginal disc as an example to illustrate the mathematical models. We first begin with an introduction to the biological background, followed by mathematical models for morphogen gradient formation, which are formulated by reaction-diffusion equations. Next, we discuss the robustness of morphogen gradients and introduce analysis methods. Finally, we consider morphogen-mediated growth control models that are closely related to scaling solutions and moving boundary problems.

## 7.2   Morphogen-Mediated Pattern Formation

Morphogens are signaling protein molecules synthesized at a localized site, some of which disperse from their production site, bind to cell receptors along the way, and result in different receptor occupancies at different cell locations [11]. The spatial concentration gradient of morphogen-receptor complexes (aka *signaling gradients*) induces spatially graded differences in cell signaling, which in turn gives rise to different cell fates, visual tissue patterns and organs during development.

Morphogen-mediated pattern formation can be illustrated by a simple nonbiological system: the patterning of the French flag (Fig. 7.1), which has a simple pattern of one-third blue, one-third white, and one-third red in one direction, similar to a line of cells with three different fates. In this system, each cell has the potential to become blue, white, or red. One of the simplest ways of achieving this patterning is that the cells acquire positional information through the spatial gradient of the morphogen. If there is a source of morphogen at one end and a sink at the other

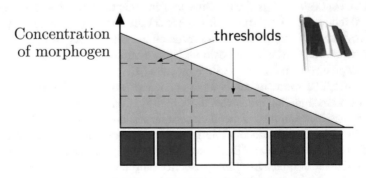

**Fig. 7.1 The French flag model of pattern formation**. Replotted from [12]

end, the morphogen diffuses along the line of cells, leading to graded concentrations in space, with different positional information for each spatial region. If the cells respond to the *threshold concentrations* of the morphogen—for example, becoming blue, white, or red in accordance with high, medium, or low levels of morphogen concentrations—the line of cells then develops to become a French flag (Fig. 7.1).

Much of the experimental evidence for morphogens has been observed in various systems, such as the anterior-posterior patterning of the vertebrate limb, the dorsal-ventral patterning of the vertebrate neural tube, the axes in Xenopus, and the leg and wing imaginal discs of *Drosophila*. Known morphogens include sonic hedgehog (shh), which provides a graded signal for pattern formation in the ventral neural tube [13]; the transforming growth factor beta (TGF-$\beta$) family involved in dorsal-ventral patterning [14, 15]; retinoic acid, which stimulates the growth of the posterior end of some organisms [16]; and decapentaplegic (Dpp) and wingless (Wg), which regulate patterning and growth in *Drosophila* leg and wing imaginal discs [17–20]. More examples of the specific biological functions of various morphogens can be found in other reviews [2, 5, 21, 22].

Here, we take decapentaplegic (Dpp) as an example to illustrate how morphogens play roles in tissue development. Dpp is a key morphogen involved in the development of the fruit fly *Drosophila melanogaster* and is the first validated secreted morphogen [17]. Dpp is known to be necessary for the correct patterning and development of the early *Drosophila* embryo and the fifteen imaginal discs, which are tissues that become limbs and other organs and structures in the adult fly. Dpp also plays a role in regulating the growth and size of tissues. Flies with mutations in Dpp fail to form these structures correctly. Studies of Dpp in *Drosophila* have led to a greater understanding of the function and importance of their homologs in vertebrates such as humans [5, 23].

Dpp is found in early embryos and at the wing imaginal discs of fruit flies [24]. During embryonic development, Dpp is uniformly expressed on the dorsal side of the embryo and establishes a sharp concentration gradient. In the imaginal discs, Dpp is strongly expressed in a narrow stripe of cells down the middle of the disc that marks the border between the anterior and posterior sides, the anteroposterior axis. Dpp diffuses from this stripe towards the edges of the tissues, forming a gradient as expected of a morphogen (Fig. 7.2). In the fly wing, Dpp expression is regulated by the transcription factor Engrailed (En) and its downstream gene hedgehog (Hh). Engrailed is mostly expressed in the posterior region but not in the anterior region and activates the short-range signaling factor hedgehog. Hedgehog signaling instructs neighboring cells to express Dpp, but Dpp expression is also repressed by En. As a result, Dpp is only produced in a narrow stripe of cells immediately adjacent to but not within the posterior half of the tissue [25]. Dpp is produced at this anterior/posterior border and then diffuses out to the edges of the tissue, forming a spatial concentration gradient. In Drosophila, the Dpp receptor is formed by two proteins, Thickveins (Tkv) and Punt [26]; genetic tools suggest that in vivo, Dpp binds strongly to the type I receptor Tkv but not to the type II receptor Punt [27]. When a cell receives a Dpp signal, the receptors are able to activate an intracellular protein called mothers against Dpp (mad) by phosphorylation. Activated Mad is able to bind to DNA and

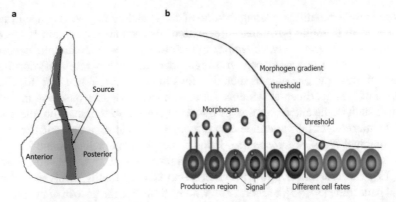

**Fig. 7.2  Positional information and the Dpp gradient shape. a** The Dpp gradient expanded in the wing imaginal disc. **b** Morphogen concentration as a function of the distance to the source. Different target genes are activated in accordance with spatial concentration-dependent signaling

act as a transcription factor to affect the expression of different downstream genes including Spalt (Sal) and Potomotorblind (Omb) in response to Dpp signaling. The expression patterns of the Hh, Sal, and Omb genes determine the location of veins in the *Drosophila* wing [24].

## 7.3   Mathematical Models of Morphogen Gradient Formation

Various mathematical models have been proposed for studying the formation of morphogen gradients. Most models consist of morphogen diffusion and the interactions among morphogens and cellular molecules. Hence, the reaction-diffusion equations introduced in Sect. 2.5 are basic formulations for modeling the morphogen gradient system [6, 11, 28]. Here, we focus on the formation of long-range morphogen gradients (for example, the Dpp gradient in the *Drosophila* wing imaginal disc shown in Fig. 7.2a) and introduce mathematical models developed from studies of morphogen gradient formation, robustness, and growth control.

### 7.3.1   Ligand Diffusion in Extracellular Space

In the Dpp gradient formation, Dpp is mostly produced at the anterior/posterior border and then diffuses to the edges of the wing imaginal disc. Because of the symmetry, we simplify the wing imaginal disc as a one-dimensional space and only consider the Dpp gradient formation along the direction perpendicular to the anterior-posterior

**a** **b**

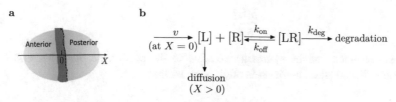

**Fig. 7.3 A simple diffusion model of morphogen gradient formation. a** Illustration of the distribution of Dpp in the *Drosophila* wing imaginal disc. **b** The model with diffusion, reversible binding, and degradation. Replotted from [30]

axis (Fig. 7.3a). Hence, let $X$ represent the distance from the source of morphogen production and $0 \leq X \leq X_{max}$ be half of the imaginal disc. In the case of *Drosophila* wing imaginal discs, we usually have $X_{max} = 100\ \mu m$ (approximately 40 cells, and the diameter of each cell is approximately 2.5 $\mu m$) [29].

A simple biologically possible model is shown in Fig. 7.3b. In this model, diffusion ligands are synthesized at a local production region with a rate specified by the function $V(X)$ (s$^{-1}$). The ligands reversibly bind to receptors to form signaling complexes, and the formation and dissociation rates of the ligand-receptor complexes are $k_{on}$ (Ms$^{-1}$ with M being the unit of concentration) and $k_{off}$ (s$^{-1}$), respectively. The signaling complexes are endocytosed and degraded at a rate of $k_{deg}$ (s$^{-1}$). These behaviors lead to the following one-dimensional reaction-diffusion equation of *ligand extracellular diffusion* (LED) (model B in [30]):

**LED:**

$$\frac{\partial[L]}{\partial t} = D\frac{\partial^2[L]}{\partial X^2} - k_{on}[L](R_{tot} - [LR]) + k_{off}[LR] + V(X), \quad (7.1)$$

$$\frac{\partial[LR]}{\partial t} = k_{on}[L](R_{tot} - [LR]) - k_{off}[LR] - k_{deg}[LR] \quad (7.2)$$

$$(0 < X < X_{max}, t > 0)$$

Here, [L] and [LR] represent the concentrations of ligands and signaling complexes, respectively, and the total receptor concentration is assumed to be a constant, $R_{tot} = [R] + [LR]$; hence, ($R_{tot} - [LR]$) represents the concentration of free receptors. The diffusion is assumed to be governed by Fick's second law, with $D$ being the diffusion coefficient. The localized synthesis site is often represented by a narrow region of finite width $0 \leq X \leq X_{min}$. Thus, the production rate $V(X)$ is given in terms of the Heaviside unit step function $H(z)$ (please refer to [31, 32]):

$$V(X) = \frac{v}{X_{min}}H(X_{min} - X), \qquad H(z) \equiv \begin{cases} 0, & (z < 0) \\ 1, & (z \geq 0) \end{cases}. \quad (7.3)$$

Here, $v$ is the synthesis rate per unit length.

To define the boundary condition, the no-flux condition is often assumed at one side of morphogen production

$$\frac{\partial [L]}{\partial X} = 0 \quad \text{at} \quad X = 0 \tag{7.4}$$

as a consequence of the symmetry relative to the border. At the other end of the imaginal disc, a sink condition is often used so that:

$$[L] = 0 \quad \text{at} \quad X = X_{\text{max}}. \tag{7.5}$$

For simplicity, we often consider the situation of a point source, where the ligand is only produced at $X = 0$. To this end, we can take the limit $X_{\text{min}} \to 0$ in (7.3); therefore, the production rate $V(X)$ is given as a delta function

$$V(X) = v\delta(X). \tag{7.6}$$

We note that the boundary condition (7.4) is not valid in this case, and Eq. (7.1) becomes

$$\frac{\partial [L]}{\partial t} = -k_{\text{on}}[L](R_{\text{tot}} - [LR]) + k_{\text{off}}[LR] + v \tag{7.7}$$

at $X = 0$. Hence, for the point course model, we have

**LED$_0$:**
$$\frac{\partial [L]}{\partial t} = D\frac{\partial^2 [L]}{\partial X^2} - k_{\text{on}}[L](R_{\text{tot}} - [LR]) + k_{\text{off}}[LR], \tag{7.8}$$

$$\frac{\partial [LR]}{\partial t} = k_{\text{on}}[L](R_{\text{tot}} - [LR]) - k_{\text{off}}[LR] - k_{\text{deg}}[LR]. \tag{7.9}$$

We can set the initial condition as the situation before the production of free ligands,

$$[L] = [LR] = 0, \quad t = 0, \ X \in [0, X_{\text{max}}]. \tag{7.10}$$

Experiments show that in the *Drosophila* wing imaginal disc, extracellular Dpp is unstable and almost completely degraded in 3 h [19]. If we assume that the concentration of Dpp drops to 5% of the original level in 3 h, then

$$e^{-k_{\text{deg}} \times 3h} \approx 0.05,$$

which gives $k_{\text{deg}} \approx 2.77 \times 10^{-4} \text{ s}^{-1}$. Usually, we have $k_{\text{deg}} > 10^{-4} \text{ s}^{-1}$.

Based on the above equations, the evolution of the Dpp gradient and signal complexes can be obtained numerically. Figure 7.4 shows the evolution of signal complexes [LR] under either small ($v/R_{\text{tot}} = 5 \times 10^{-5} \text{ s}^{-1}$) or large ($v/R_{\text{tot}} = 5 \times 10^{-4} \text{ s}^{-1}$) ligand production rates. The signaling gradient depends on the model parameters. When the Dpp production rate is small, the signaling concentration exponentially decays with the distance, which is in agreement with experimental observations [33, 34]. However, when the Dpp production rate is large, all receptors within the region are bound to the ligands and hence cannot form a biologically meaningful signaling gradient.

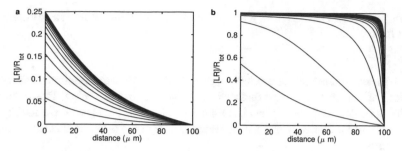

**Fig. 7.4  Signaling gradient based on the ligand extracellular diffusion model with point source LED$_0$.** The parameters are $D = 10\,\mu m^2 s^{-1}$, $X_{max} = 100\,\mu m$, $k_{on}R_{tot} = 0.01\,s^{-1}$, $k_{off} = 10^{-6}\,s^{-1}$, and $k_{deg} = 2 \times 10^{-4}\,s^{-1}$; **a.** $v/R_{tot} = 5 \times 10^{-5}\,s^{-1}$, $\psi = 10$, $\beta = 0.25$; **b.** $v/R_{tot} = 5 \times 10^{-4}\,s^{-1}$, $\psi = 10$, $\beta = 2.5$. Here, $\psi$ and $\beta$ refer to (7.13). Replotted from [30]

### 7.3.1.1  Steady-State Morphogen Gradient

To analyze how the steady state signaling gradient depends on the model parameters, we take the point source model LED$_0$ and solve the steady-state gradient. In Eqs. (7.8)–(7.9), let the derivative with $t$ be zero, and we have the following boundary problem of the ordinary differential equation:

$$\begin{cases} D\dfrac{\partial^2[L]}{\partial X^2} - k_{deg}[LR] = 0 \\ [LR] = R_{tot}\dfrac{[L]}{[L] + k_m}, \quad k_m = \dfrac{k_{off} + k_{deg}}{k_{on}}, \\ [LR]|_{X=0} = v/R_{tot}, \quad [L]|_{X=max} = 0. \end{cases} \tag{7.11}$$

We introduce the nondimensional variables

$$a = \frac{[L]}{k_m}, \quad b = \frac{[LR]}{R_{tot}}, \quad x = \frac{X}{X_{max}}, \quad \psi = \frac{k_{deg}X_{max}^2}{D}\frac{R_{tot}}{k_m}, \quad \beta = \frac{v}{R_{tot}k_{deg}}$$

and obtain the nondimensional boundary value problem

$$\begin{cases} \dfrac{\partial^2 a}{\partial x} - \psi\dfrac{a}{1+a} = 0, \\ a(0) = \dfrac{\beta}{1-\beta}, \quad a(1) = 0. \end{cases} \tag{7.12}$$

Here, $a$ and $b$ are the nondimensional concentrations of ligands and signaling complexes, respectively, and satisfy the relation

$$b = \frac{a}{1+a}.$$

Hence, we always have $b(0) = \beta < 1$. Therefore, to obtain a stable steady state, we must have

$$\beta = v/(R_{tot}k_{deg}) < 1,$$

which means that the ligand production rate is smaller than the maximum rate of the binding of ligands to receptors and their removal due to degradation. This condition is biologically reasonable. When the ligand production rate is too large for ligands to be removed effectively, ligands will accumulate in the tissue so that all receptors are saturated.

From (7.12), the steady state gradient of the $LED_0$ model depends on only two dimensionless parameters, the ligand *synthesis-to-degradation ratio* $\beta$, and the *effective on rate* $\psi$, which are defined as

$$\beta = \frac{v}{R_{tot}k_{deg}}, \quad \psi = \frac{X_{max}^2 k_{deg}}{D} \frac{k_{on}R_{tot}}{(k_{off} + k_{deg})}. \tag{7.13}$$

Applying the method of upper and lower solutions [35], the boundary value problem (7.12) has a unique positive solution when $\beta < 1$ [36]. Nevertheless, when a narrow region of ligand production is explicitly considered (the LED model (7.1)–(7.3)), the system always has a positive steady-state morphogen gradient, and there is no restriction on the synthesis-to-degradation ratio for the existence of a steady-state gradient [31]. Moreover, the solution satisfies $a(x) > 0$ and $a'(x) < 0$ for $0 < x < 1$.

### 7.3.1.2   Upper and Lower Solution Method

Consider a second-order boundary value problem

$$\begin{cases} u'' - f(u, x) = 0, & (0 < x < 1) \\ (au + bu')|_{x=0} = h_0 \\ (cu + du')|_{x=1} = h_1 \end{cases} \tag{7.14}$$

A general form of the upper and lower solution method is given below. Let $H_p = C^p([0, 1], \mathbb{R})$ represent the set of all $p^{th}$-order derivative functions defined on the interval $[0, 1]$. Let $L : H_2 \mapsto H_0^k$ and $f : H_2 \mapsto H_0^k$ be linear and nonlinear operators, respectively, mapping $H_2$ to $H_0^k$ ($k \geq 1$). Consider an equation defined on $H_2$,

$$Lu + f(u) = 0. \tag{7.15}$$

The upper and lower solution method establishes a condition for the existence of a solution $u \in H_2$ for (7.15).

A function $u_0 \in H_2$ is said to be an upper solution of (7.15) if

$$Lu_0 + f(u_0) \leq 0, \tag{7.16}$$

and a function $v_0 \in H_2$ is said to be a lower solution of (7.15) if

$$Lv_0 + f(v_0) \geq 0. \tag{7.17}$$

Here, for any $a, b \in H_0{}^k$, $a \geq b$ means $a(x) \geq b(x)$ for any $x \in [0, 1]$.

An operator $T : H_0{}^k \mapsto H_0$ is a negative operator if $Ta \leq 0$ for any $a \in H_0{}^k$. The theorem below gives the upper and lower method [35].

**Upper and low solution method**: Consider Eq. (7.15), if the following conditions are satisfied:

(1) there exists an upper solution $u_0$ and a lower solution $v_0$ of (7.15), and $u_0 \geq v_0$;
(2) there exist constants $\lambda_i > 0$ $(i = 1, \ldots, k)$, so that

$$f_i(\varphi_1) - f_i(\varphi_2) > -\lambda_i(\varphi_1 - \varphi_2)$$

for any function $\varphi_1, \varphi_2$ satisfying

$$u_0 \geq \varphi_1 \geq \varphi_2 \geq v_0;$$

(3) define an operator $\Lambda : H_2 \mapsto H_2{}^k$ as $\Lambda = (\lambda_1 u, \ldots, \lambda_k u)$, the inverse operator $(L - \Lambda)^{-1}$ is a negative operator,

Equation (7.15) has at least one solution $u(x) \in H_2$, and $v_0 \leq u \leq u_0$.

When we apply the above theorem to the boundary value problem (7.14), the linear operator $L : H_2 \mapsto H_2{}^3$ is defined as

$$Lu = \left( \frac{d^2}{dx^2}u, \ \pm(a + b\frac{d}{dx})u\Big|_{x=0}, \ \pm(c + d\frac{d}{dx})u\Big|_{x=1} \right),$$

and the nonlinear operator $f : H_2 \mapsto H_0^3$ is defined as

$$f(u) = (-f(u(x), x), \pm h_0, \pm h_1).$$

Now, consider the boundary value problem (7.12); there exists an upper solution (refer to (7.34))

$$\bar{a}(x) = \frac{\beta}{1 - \beta}(1 - x)$$

and a lower solution (refer to (7.34))

$$\underline{a}(x) = \frac{\beta \sinh(\sqrt{\psi}(1 - x))}{\sinh \sqrt{\psi}}.$$

Hence, there exists a solution $\underline{a}(x) \le a(x) \le \bar{a}(x)$. Moreover, the solution is unique [36].

### 7.3.1.3    The Method of Energy Integration

The boundary value problem (7.12) can be solved through the method of energy integration. Let $a_0 = \beta/(1 - \beta)$ for simplicity. Multiply both sides of (7.12) by $a'(x)$ and integrate the equation from 0 to $x$. We obtain

$$\frac{1}{2}a'(x)^2 - \frac{1}{2}a'(0)^2 - \psi(F(a(x)) - F(a_0)) = 0,$$

where $F$ is the potential function

$$F(a) = a - \ln(1 + a).$$

The boundary condition $a'(0)$ is later determined by (7.19). Therefore, we have a first-order differential equation for $a(x)$ (note that $a'(x) < 0$)

$$a'(x) = -\sqrt{a'(0)^2 + 2\psi(F(a) - F(a_0))}, \quad a(1) = 0.$$

Hence, the boundary value problem can be solved through an implicit function between $x$ and $a(x)$ as

$$x = \int_{a(x)}^{a_0} \frac{da}{\sqrt{a'(0)^2 + 2\psi(F(a) - F(a_0))}}. \tag{7.18}$$

Here, the constant $a'(0)$ is determined by the boundary condition $a(1) = 0$, i.e., solving the following implicit equation:

$$1 = \int_0^{a_0} \frac{da}{\sqrt{a'(0)^2 + 2\psi(F(a) - F(a_0))}}. \tag{7.19}$$

The implicit function (7.18) gives the steady state gradient of the boundary problem (7.12). However, the explicit form cannot be obtained from (7.18). To further identify the dependence of the morphogen gradient on the model parameters, we apply the methods of approximation analysis to find the approximation solutions under extreme conditions of either high or low ligand production rates.

### 7.3.1.4    High Ligand Synthesis Rate

First, we consider the case with a high ligand synthesis rate, i.e., $v$ is very large so that $\beta \approx 1$. In this case, we have $a \gg 1$ within most of the region $0 < x < 1$, and

Eq. (7.12) can be approximated as

$$a''(x) - \psi = 0, \quad a(0) = \frac{\beta}{1-\beta}, \quad a(1) = 0. \tag{7.20}$$

The solution is given by

$$a(x) = \frac{\beta}{1-\beta}(1-x) - \frac{1}{2}\psi x(1-x). \tag{7.21}$$

Hence, the nondimensional signaling concentration is

$$b(x) = \frac{\frac{\beta}{1-\beta}(1-x) - \frac{1}{2}\psi x(1-x)}{1 + \frac{\beta}{1-\beta}(1-x) - \frac{1}{2}\psi x(1-x)}. \tag{7.22}$$

When $\beta \approx 1$, we have

$$b(x) \approx \frac{\beta(1-x)}{1-\beta x} = 1 - \frac{1-\beta}{1-\beta x}. \tag{7.23}$$

From (7.23), we have $b(x) \approx 1$ except for the narrow region close to $x = 1$, i.e., most receptors are saturated. For example, if $\beta = 0.99$, we have $[LR] > 0.9R_{\text{tot}}$ within $0 < X < 0.9X_{\text{max}}$. In this case, it is unlikely to have a biologically useful gradient to yield multiple cell fates. Consequently, we come to the following conclusion: to obtain a biological gradient, the ligand synthesis rate should not be too high.

### 7.3.1.5 Low Ligand Synthesis Rate

Next, we consider the case with a low ligand synthesis rate ($\beta \ll 1$); hence, the ligand concentration is very low so that $a(x) < a(0) \ll 1$. Therewith, Eq. (7.12) is approximated as

$$a''(x) - \psi a = 0, \quad a(0) = \beta, \quad a(1) = 0, \tag{7.24}$$

which yield the solution

$$a(x) = \frac{\beta \sinh(\sqrt{\psi}(1-x))}{\sinh\sqrt{\psi}}. \tag{7.25}$$

The nondimensional signaling concentration is

$$b(x) = \frac{\frac{\beta \sinh(\sqrt{\psi}(1-x))}{\sinh\sqrt{\psi}}}{1 + \frac{\beta \sinh(\sqrt{\psi}(1-x))}{\sinh\sqrt{\psi}}}. \tag{7.26}$$

When $\beta \ll 1$, we have approximately

$$b(x) = \frac{\beta \sinh(\sqrt{\psi}(1-x))}{\sinh\sqrt{\psi}}. \tag{7.27}$$

In particular,

$$b(x)/b(1) = \sinh(\sqrt{\psi}(1-x)).$$

The relative signaling concentration decays exponentially, in agreement with experimental observations [33].

### 7.3.1.6  Linear Stability of the Steady State

To investigate the stability of the steady-state solution, we introduce nondimensional parameters

$$\theta = \frac{D}{X_{\max}^2(k_{\text{off}} + k_{\text{deg}})}, \quad h_0 = \frac{X_{\max}^2 k_{\text{on}} R_{\text{tot}}}{D}, \quad f_0 = \frac{X_{\max}^2 k_{\text{off}} R_{\text{tot}}}{D k_m}$$

and a nondimensional time

$$t' = (k_{\text{off}} + k_{\text{deg}})t;$$

Equations (7.8)–(7.9) become (here, we write $t$ for the nondimensional time for simplicity)

$$\begin{cases} \dfrac{\partial a}{\partial t} = \theta(\dfrac{\partial a^2}{\partial x^2} - h_0 a(1-b) + f_0 b) \\ \dfrac{\partial b}{\partial t} = a(1-b) - b \end{cases} \tag{7.28}$$

Consider small perturbations from the steady state in the form

$$\{a(x,t), b(x,t)\} = \{a(x), b(x)\} + e^{-\lambda t}\{\hat{a}(x), \hat{b}(x)\}. \tag{7.29}$$

After linearization, Eq. (7.28) becomes

$$\begin{cases} -\lambda\hat{a} = \theta(\hat{a}'' + h_0 a(x)\hat{b} - h_0(1-b(x))\hat{a} + f_0\hat{b}) \\ -\lambda\hat{b} = -a(x)\hat{b} - (1-b(x))\hat{a}. \end{cases} \tag{7.30}$$

Thus,

$$\hat{b} = \frac{1-b(x)}{\lambda - a(x)}\hat{a},$$

and $\hat{a}$ satisfies a nonlinear eigenvalue problem

$$\hat{a}'' + q(x; \lambda)\hat{a} = 0, \quad \hat{a}(0) = \hat{a}(1) = 0, \tag{7.31}$$

where

$$q(x; \lambda) = \frac{\lambda}{\theta} + \frac{(1 - b(x))(f_0 - h_0\lambda + 2h_0 a(x))}{\lambda - a(x)}.$$

We can prove that all eigenvalues of the eigenvalue problem (7.31) are positive, and hence, the steady-state gradient is asymptotically stable [37].

### 7.3.1.7 Biological Gradient

The above analyses show that a biologically useful gradient (i.e., able to broadly distribute patterning information over the entire field of cells) can be produced only when $\beta < 1$ and the combination of $\beta$ and $\psi$ takes values from a particular region [30]. For biological gradients, the resulting signaling gradient exponentially decays from the production border toward the other end, and the characteristic decay length is represented by $\sqrt{\psi}$. In the *Drosophila* wing imaginal disc, the decay length of Dpp is approximately 20 $\mu$m, which is approximately $X_{max}/5$, and hence $\sqrt{\psi} \approx 5$, i.e., $\psi \approx 25$. Model simulations also confirm the analytic conclusions (Fig. 7.4).

In the above analyses, we obtained estimations when the ligand production rate was either very high or very low. For the general case, there is no simple method to obtain the approximation of the signaling gradient $b(x)$. Nevertheless, we can obtain the estimation of the upper and lower bonds of the signaling gradient. Consider the relation

$$0 < \frac{a}{1 + a} < a$$

and take the upper and lower bounds. We have the following boundary value problems:

$$\bar{a}''(x) = 0, \quad \bar{a}(0) = \frac{\beta}{1 - \beta}, \quad \bar{a}(1) = 0 \tag{7.32}$$

and

$$\underline{a}''(x) - \psi\underline{a} = 0, \quad \underline{a}(0) = \frac{\beta}{1 - \beta}, \quad \underline{a}(1) = 0. \tag{7.33}$$

The corresponding solutions are

$$\bar{a}(x) = \frac{\beta}{1 - \beta}(1 - x), \quad \underline{a}(x) = \frac{\beta}{1 - \beta}\frac{\sinh(\sqrt{\psi}(1 - s))}{\sinh\sqrt{\psi}}. \tag{7.34}$$

From the upper and lower solution method, the solution of (7.12) $a(x)$ satisfies

$$\underline{a}(x) \le a(x) \le \bar{a}(x).$$

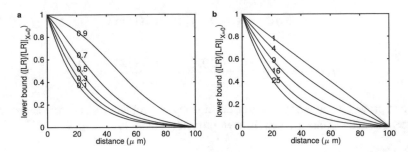

**Fig. 7.5   The lower bounds of the relative signaling gradient versus various values of $\psi$ and $\beta$. a $\psi = 25$ with various $\beta$ values. b $\beta = 0.2$ with various $\psi$ values. Replotted from [30]**

Hence, we obtain an estimation of $a(x)$, and the upper and lower bounds of the signaling gradient are given by

$$\frac{\underline{a}(x)}{1 + \underline{a}(x)} \leq b(x) \leq \frac{\bar{a}(x)}{1 + \bar{a}(x)}. \tag{7.35}$$

The dependencies of the relative lower bound with various values of $\psi$ and $\beta$ are shown in Fig. 7.5.

The basic model (7.1)–(7.2) has been refined in many studies for more realistic situations, including the distributed synthesis of receptors [31, 38], the endocytosis and exocytosis of receptors and signaling complexes [30, 38], and extensions to two- or three-dimensional diffusion [39]. These improvements make little difference to the conclusion of morphogen gradient formation but can be more difficult in terms of mathematical analysis.

Despite the formation of biological gradients, numerical simulations shown that the signaling gradients produced from the above diffusion mechanism are not robust enough with respect to changes in the system parameters, as small changes in the ligand production rate can cause substantial changes in the gradient shape [30]. In contrast, embryonic patterning is usually highly robust, resisting not only substantial changes in the expression level of individual genes but also fluctuating environmental conditions. These results suggest that additional biological processes must be at work to ensure such robustness, leading to further model improvements.

## 7.3.2   Self-enhanced Ligand Degradation

In [40], a mechanism of self-enhanced ligand degradation was proposed to enhance the robustness of morphogen gradients (Fig. 7.6). In this case, the degradation of morphogens is formulated as a nonlinear function of its own concentration instead of a linear function by first-order degradation. A simple aspect of *self-enhanced ligand degradation* (SED) can be seen from the following diffusion equation with a

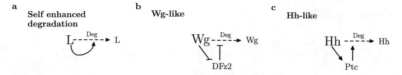

**Fig. 7.6 Self-enhanced ligand degradation. a** Illustration of self-enhanced degradation. **b** Wg-like class self-enhanced degradation. Morphogen signaling represses receptor expression, while receptors stabilize morphogens. **c** Hh-like class self-enhanced degradation. Morphogens activate receptor expression, while receptors enhance morphogen degradation. Replotted from [40]

power-law degradation profile

**SED**:
$$\frac{\partial[L]}{\partial t} = D\frac{\partial^2[L]}{\partial X^2} - \alpha[L]^n + v\delta(X). \tag{7.36}$$

In the case of linear degradation ($n = 1$), the steady-state gradient is exponentially decaying (here, we assume the approximation $X_{max} \rightarrow +\infty$) (also see [34])

$$[L] = [L]_0 e^{-X/\Delta_d}, \quad [L]_0 = v/\alpha, \quad \Delta_d = \sqrt{D/\alpha}. \tag{7.37}$$

It is obvious that the steady-state concentration is linearly dependent on the ligand production rate; thus, the morphogen gradient linearly tracks the changes in the production rate when there are external perturbations to ligand synthesis.

In the case of nonlinear degradation ($n > 1$), the steady state gradient is power-law decaying:

$$[L] = \frac{[L]_0}{(X/\varepsilon + 1)^m}, \quad [L]_0 = \sqrt[n]{v/\alpha}, \quad m = \frac{2}{n-1}, \quad \varepsilon = \sqrt{\frac{Dm(m+1)}{\alpha L_0^{n-1}}}. \tag{7.38}$$

Now, the steady state concentration depends on the ligand production rate in a sublinear way, and hence, the morphogen gradient can buffer against the fluctuations in the production rate.

Biologically, self-enhanced ligand degradation can be achieved through a morphogen network composed of morphogen signaling-regulated (enhanced or repressed) receptor expression and receptor-mediated ligand degradation (often through a protease). Two types of regulation were proposed in [40]: the Wingless (Wg)-like class, in which morphogen signaling represses the receptor and the receptor stabilizes the morphogen, and the Hedgehog(Hh)-like class, in which morphogen signaling activates receptor expression and the receptor enhances morphogen degradation (Fig. 7.6). A general formulation is given by the following set of reaction-diffusion equations [40]:

$$\frac{\partial [L]}{\partial t} = D\frac{\partial^2 [L]}{\partial X^2} - k_+^1 [L][R] + k_-^1 [LR] - a_1 [PR][L] \tag{7.39}$$
$$- a_2 [P][L] - a_3 [L] + V(X),$$

$$\frac{\partial [LR]}{\partial t} = k_+^1 [L][R] - k_-^1 [LR] - a_4 [LR], \tag{7.40}$$

**SED':**
$$\frac{\partial [PR]}{\partial t} = k_+^2 [R][P] - k_-^2 [PR], \tag{7.41}$$

$$\frac{\partial [R]}{\partial t} = \eta_{r1}\frac{K_a^m}{K_a^m + [LR]^m} + \eta_{r2}\frac{[LR]^n}{K_b^n + [LR]^n} - k_+^2 [R][P] \tag{7.42}$$
$$+ k_-^2 [PR] - k_+^1 [R][L] + k_-^1 [LR] - \alpha_5 [R] + \rho\alpha_4 [LR],$$

$$[P] = P_{\text{tot}} - [PR], \tag{7.43}$$

where [L], [R] and [P] denote the concentrations of the ligand, receptor, and protease, respectively, and the complexes are denoted by their constituents. Different types of Wg-like or Hh-like regulation can be defined by adjusting the parameters for protease-mediated ligand degradation ($a_1$ and $a_2$) and for receptor expression ($\eta_{r1}$ and $\eta_{r2}$). Please refer to [40] for further discussions of the model.

### 7.3.3   Non-receptor-Mediated Ligand Transport

Morphogen gradient formation may be modulated by interactions with heparan sulfate proteoglycans (HSPGs) and other extracellular proteins that tether morphogens to the cell surface [41–46]. Such nonsignaling entities are called *non-receptors* because they bind with morphogens in a way similar to receptors, but the resulting complexes do not signal cell fate decisions. The non-receptors are often members of the HSPG family. Proteoglycans are abundant components of the cell surface and extracellular matrix. They consist of a transmembrane core protein and one or more long unbranched disaccharide chains. These chains differ by the nature of repeating disaccharides. Heparan sulfate (HS), an example of this type of chain, contains alternating N-acetyl glucosamine (GlcNAc) and glucuronic or iduronic acid residues. [47] provided a good introduction of how HSPGs affect the stability and distribution of extracellular gradients. Many experiments have shown that non-receptors play an important role in the formation and robustness of morphogen gradients, such as Dpp, Wg, and Hh [43, 46, 48, 49].

    The basic model in Sect. 7.3.1 can be extended to include the binding of morphogens with non-receptors, as illustrated in Fig. 7.7. In this model, we consider a fixed concentration $N_{\text{tot}}$ of proteoglycan-type non-diffusive non-receptors and introduce a set of similar activities for the non-receptor sites. These assumptions yield the following set of reaction-diffusion equations of *non-receptor-mediated ligand extracellular diffusion* (N-LED), described in [50]:

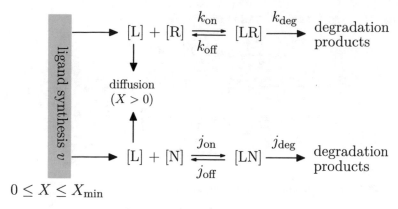

$$0 \leq X \leq X_{\min}$$

**Fig. 7.7 Illustration of the non-receptor-mediated morphogen gradient formation.** The model includes free ligands [L], receptors [R], ligand-receptor complexes [LR], non-receptors [N], and ligand-non-receptor complexes [LN]. The ligands are synthesized locally over a few cells, diffuse away from the source and bind to both receptors and non-receptors on the cell membrane. The complexes are endocytosed and degraded in the cell interior. Replotted from [28]

**N-LED**:

$$\frac{\partial [\mathrm{L}]}{\partial t} = D\frac{\partial^2 [\mathrm{L}]}{\partial X^2} - k_{\mathrm{on}}[\mathrm{L}](R_{\mathrm{tot}} - [\mathrm{LR}]) + k_{\mathrm{off}}[\mathrm{LR}] \quad (7.44)$$
$$- j_{\mathrm{on}}[\mathrm{L}](N_{\mathrm{tot}} - [\mathrm{LN}]) + j_{\mathrm{off}}[\mathrm{LN}] + V(X),$$

$$\frac{\partial [\mathrm{LR}]}{\partial t} = k_{\mathrm{on}}[\mathrm{L}](R_{\mathrm{tot}} - [\mathrm{LR}]) - (k_{\mathrm{off}} + k_{\mathrm{deg}})[\mathrm{LR}], \quad (7.45)$$

$$\frac{\partial [\mathrm{LN}]}{\partial t} = j_{\mathrm{on}}[\mathrm{L}](N_{\mathrm{tot}} - [\mathrm{LN}]) - (j_{\mathrm{off}} + j_{\mathrm{deg}})[\mathrm{LN}]. \quad (7.46)$$

Here, [LN] represents the concentration of ligand-non-receptor complexes, $N_{\mathrm{tot}} = [\mathrm{N}] + [\mathrm{LN}]$, and the other variables are the same as those in the above LED model. Similar to the discussions in Sect. 7.3.1, the ligand production region is specified at $0 \leq X \leq X_{\min}$, and the ligand production rate $V(X)$ is taken as a delta function $V(X) = v_0\delta(X)$ for a point source ($X_{\min} = 0$) or a step function $V(X) = (v_0/X_{\min})H(X_{\min} - X)$ if $X_{\min} > 0$.

At the steady state, we have a nonlinear boundary value problem

$$\begin{cases} D\dfrac{\partial^2 [\mathrm{L}]}{\partial X^2} - k_{\mathrm{deg}}[\mathrm{LR}] - j_{\mathrm{off}}[\mathrm{LN}] + V(X) = 0, \\ \dfrac{\partial [\mathrm{L}]}{\partial X}\bigg|_{X=0} = 0, \quad [\mathrm{L}]|_{X=X_{\max}} = 0, \end{cases} \quad (7.47)$$

where

$$\frac{[\mathrm{LR}]}{R_{\mathrm{tot}}} = \frac{[\mathrm{L}]}{[\mathrm{L}] + k_m}, \quad k_m = \frac{k_{\mathrm{off}} + k_{\mathrm{deg}}}{k_{\mathrm{on}}}, \quad \frac{[\mathrm{LN}]}{N_{\mathrm{tot}}} = \frac{[\mathrm{L}]}{[\mathrm{L}] + j_m}, \quad j_m = \frac{j_{\mathrm{off}} + j_{\mathrm{deg}}}{j_{\mathrm{on}}}.$$

Introduce the nondimensional variables

$$a = \frac{[L]}{k_m}, \quad b = \frac{[LR]}{R_{tot}}, \quad c = \frac{[LN]}{R_{tot}},$$

$$\lambda_R^2 = \frac{X_{max}^2 k_{deg} R_{tot}}{Dk_m}, \quad \lambda_N^2 = \frac{X_{max}^2 j_{deg} N_{tot}}{Dj_m}, \quad \lambda^2 = \lambda_R^2 + \lambda_N^2,$$

$$\gamma = \frac{k_m}{j_m}, \quad p = \frac{\lambda_R^2}{\lambda^2}, \quad d = \frac{X_{min}}{X_{max}},$$

$$v(x) = \frac{X_{max}^2}{Dk_m} V(X_{max}x) = \left(\frac{v}{d}\right) H(d - x), \quad v = \frac{v_0 X_{max}}{Dk_m}.$$

We have a nondimensional boundary value problem

$$a''(x) - \lambda^2 \left(\frac{p}{1+a} + \frac{1-p}{1+\gamma a}\right) a + v(x) = 0, \quad a'(0) = a(1) = 0, \qquad (7.48)$$

and the relative signaling concentration is

$$b(x) = \frac{a(x)}{1 + a(x)}. \qquad (7.49)$$

Here, $v$ is the normalized synthesis rate.

The existence, uniqueness, and linear stability of the steady-state gradient for the N-LED model were analytically studied in [50]. Similarly to the case without a non-receptor, when $X_{min} = 0$ (point source), the system has a unique steady-state gradient if the synthesis-to-degradation ratio, now defined as

$$\beta = \frac{v_0}{R_{tot} k_{deg} + N_{tot} j_{deg}}, \qquad (7.50)$$

satisfies $\beta < 1$. When $X_{min} > 0$ (narrow production region), the system always has a unique steady-state gradient. Furthermore, the steady-state gradient is linearly stable in either of these situations. The presence of non-receptors should reduce the number of morphogens available for binding to receptors and thereby inhibit cell signaling. This aspect was analytically proven in [50]: for sufficiently low morphogen synthesis rates, the presence of non-diffusive non-receptors generally lowers the normalized concentration level of both free ligand and ligand-receptor concentrations at each point of the solution domain, reduces the steepness of the negative slope, and increases the convexity of the concentrations.

Applying the method of energy integration to the boundary value problem (7.48), we obtain the steady-state solution that is given by the implicit function as [36]

$$x = \begin{cases} \displaystyle\int_a^{a_0} \frac{du}{\sqrt{2\lambda^2(E(u) - E(a_0)) - 2v(u - a_0)}}, & (0 \le x < d), \\[4mm] \displaystyle d + \int_a^{a_d} \frac{du}{\sqrt{2\lambda^2 E(u) + s_1^2}}, & (d \le x \le 1), \end{cases} \tag{7.51}$$

where

$$E(u) = p(u - \ln(1 + u)) + \frac{1 - p}{\gamma^2}(\gamma u - \ln(1 + \gamma u)).$$

The unknown constants $a_0 \equiv a(0)$, $a_d \equiv a(d)$ and $s_1 \equiv a'(1)$ are determined by $a(1) = 0$ and the continuity of $a(x)$ and $a'(x)$ at $x = d$, i.e.,

$$1 - d = \int_0^{a_d} \frac{du}{\sqrt{2\lambda^2 E(u) + s_1^2}}, \tag{7.52}$$

$$d = \int_{a_d}^{a_0} \frac{du}{\sqrt{2\lambda^2(E(u) - E(u_0)) - 2v(u - a_0)}}, \tag{7.53}$$

$$a_d - a_0 = -(2\lambda^2 E(a_0) + s_1^2). \tag{7.54}$$

Equations (7.51)–(7.54) together give the unique steady-state gradient of the N-LED model. The steady-state gradient is determined by the nondimensional parameters $\lambda, p, \gamma$, and $v$.

Although the system always has a steady-state gradient when $X_{\min} > 0$, not all gradient shapes are biologically useful, i.e., some gradients may not be multi-fate gradients that can broadly distribute patterning information over the entire field of cells. In [36], the concept of a multi-fate gradient was defined mathematically based on the following three aspects of the gradient profile:

1. the slope of the normalized signaling gradient should not be too steep, e.g., $|b'(x)/b(x)| < \delta^{-1}$ for some $\delta > 0$ over the signaling region;
2. the concentration of the patterning signal should not be too low in the vicinity of the ligand production region, e.g., $b(d) \ge \theta$ with $0 < \theta < 1$;
3. the slope of the normalized free ligand concentration at the farther end of the tissue should be a small value, e.g., $|a'(1)| \ll \varepsilon a_\theta$ for some $\varepsilon \ll 1$, and $a_\theta = [a]_{b=\theta} = \theta/(1 - \theta)$.

For example, the parameters $\delta = 0.05, \theta = 0.1$ and $\varepsilon = 0.002$ give reasonable citations for the biologically useful multi-fate gradient.

From the nondimensional equation (7.48), the steady-state gradient mainly depends on four parameters [36]: the normalized ligand synthesis rate $v$, the ratio of the saturation levels of receptors to non-receptors $\gamma$, the ratio of degradation flux of receptors to non-receptors $p$, and the total degradation flux of receptors and non-receptors $\lambda$. For a sufficiently small ligand synthesis rate, we expect $0 \le a(x) \ll 1$ and $0 \le \gamma a(x) \ll 1$. In this case, a leading term approximate solution of the steady-state problem (7.48) is determined by (here, we consider the case $d > 0$)

$$\underline{a}'' - \lambda^2 \underline{a} + (v/d)H(d - x) = 0, \quad \underline{a}'(0) = \underline{a}(1) = 0. \qquad (7.55)$$

This boundary value problem gives the relative morphogen gradient

$$\underline{a}(x) = \begin{cases} \dfrac{v(\cosh(\lambda) - \cosh(\lambda x)\cosh(\lambda(1 - d)))}{d\lambda^2 \cosh(\lambda)}, & (0 \leq x \leq d), \\ \dfrac{v(\sinh(\lambda d)\sinh(\lambda(1 - x)))}{d\lambda^2 \cosh(\lambda)}, & (d \leq x \leq 1). \end{cases} \qquad (7.56)$$

When the ligand synthesis rate is sufficiently large so that $vd$ is large compared to $\max\{1, \lambda^2, \lambda^2/\gamma\}$, the leading term approximation $\bar{a}(x)$ for the steady-state solution of (7.48) is determined by the boundary value problem

$$\bar{a}''(x) + v(x) = 0, \quad \bar{a}'(0) = \bar{a}(1) = 0, \qquad (7.57)$$

which gives

$$\bar{a}(x) = \begin{cases} v(1 - \tfrac{1}{2}d) - \tfrac{1}{2}vx^2, & (0 \leq x \leq d), \\ v(1 - x), & (d \leq x \leq 1). \end{cases} \qquad (7.58)$$

However, from

$$b(x) = \frac{a(x)}{1 + a(x)},$$

it is not difficult to show that when the ligand synthesis rate is large enough, the signaling gradient $b(x)$ is not a multi-fate gradient according to the above definition [36]. Further discussions revealed that when $\gamma$ is small enough, a multi-fate gradient can be achieved for a suitable level of the ligand synthesis rate $v$. These results outline the restrictions to produce a biologically acceptable multi-fate gradient following the N-LED model. For further discussions on the gradients based on the N-LED model, please refer to [36].

Most HSPGs are static components in the extracellular matrix and therefore are considered to be non-diffusible. However, these non-diffusible non-receptors can transport ligands through a "bucket brigade" pathway to form long-range signals [51–53] (restricted diffusion model in [46]) (Fig. 7.8). With these bucket brigade transports, ligands move across the tissue in a manner similar to diffusion through the random work of the heparan sulfate (HS) chains; hence, they can mathematically be described by diffusion equations [51]. Thus, we have a set of reaction-diffusion equations in which both [L] and [LN] are allowed to diffuse ($N_D$-LED, N-LED with diffusion in non-receptors) [53]:

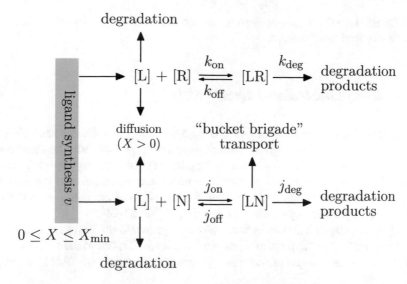

**Fig. 7.8 The model of morphogen gradient formation with "bucket brigade" transport through membrane-associated non-receptors and possible fast degradation of free ligands.** Replotted from [28]

$$\frac{\partial[L]}{\partial t} = D_L \frac{\partial^2[L]}{\partial X^2} - k_{on}[L][R] + k_{off}[LR] \tag{7.59}$$
$$- j_{on}[L](N_{tot} - [LN]) + j_{off}[LN] - k_{deg,L}[L] + V(X),$$

**N$_D$-LED:** $$\frac{\partial[R]}{\partial t} = \omega_R([LR]) - k_{on}[L][R] + k_{off}[LR] - k_{deg,R}[R], \tag{7.60}$$

$$\frac{\partial[LR]}{\partial t} = k_{on}[L][R] - (k_{off} + k_{deg,LR})[LR], \tag{7.61}$$

$$\frac{\partial[LN]}{\partial t} = D_{LN}\frac{\partial^2[LN]}{\partial X^2} + j_{on}(N_{tot} - [LN]) - (j_{off} + j_{deg})[L]. \tag{7.62}$$

Here, the synthesis and degradation of receptors are explicitly included in the equations, and the feedback from the signaling molecule to receptor synthesis is given by the function $\omega_R$. Furthermore, the degradation of free ligands, with a degradation rate of $k_{deg,L}$, is also included. Discussions in [53] showed that if the morphogen is rapidly turned over ($k_{deg,L} \gg j_{on}N_{tot}$) but is protected against degradation while binding to the non-receptors ($j_{deg} = 0$), the system has a unique steady-state solution, and the solution is linearly stable.

In several studies of the formation of the BMP gradient in *Drosophila* or zebrafish embryos [38, 54–56], the binding of morphogens to diffusible non-receptors can result in the sharp enhancement of the signaling gradients. Mathematically, the BMP gradient models are similar to the above equations; however, the boundary conditions are different since the tissue geometries are modeled as closed circles in the patterning of dorsal-ventral development.

## 7.3.4   Receptor-Mediated Transcytosis

In addition to the formation of long-distance gradients by free diffusion and/or bucket brigade transportation through non-receptors, morphogens can also be transported with the help of receptors, a process termed transcytosis, i.e., by repeated rounds of morphogen binding to cell surface receptors, internalization into the cell and subsequent externalization, and release of the ligand from the receptor on the cell surface, e.g., dynamin-dependent endocytosis [29, 33, 57–59].

In dynamin-dependent endocytosis, the free ligands do not move away from the source, but diffuse in the form of signaling complexes. Similar to the LED model in Sect. 7.3.1, a simple model of such *receptor-mediated transcytosis* (RMT) is formulated as [30]

$$\textbf{RMT:} \qquad \frac{\partial [L]}{\partial t} = V(X) - k_{on}[L](R_{tot} - [LR]) + k_{off}[LR], \qquad (7.63)$$

$$\frac{\partial [LR]}{\partial t} = D\frac{\partial^2 [LR]}{\partial X^2} + k_{on}[L](R_{tot} - [LR]) - k_{off}[LR] - k_{deg}[LR]. \quad (7.64)$$

At the steady state, we have

$$\begin{cases} D\dfrac{\partial^2 [LR]}{\partial X^2} - k_{deg}[LR] + V(X) = 0 \\ \dfrac{\partial [LR]}{\partial X}\bigg|_{X=0} = 0, \quad [LR]|_{X=X_{max}} = 0. \end{cases} \qquad (7.65)$$

Similar to the previous discussion, the steady-state solution exponentially decays to form a multi-fate gradient, and the decay length is given by $1/\sqrt{k_{deg}/D}$. In the *Drosophila* wing imaginal disc, the decay length is approximately 20 $\mu$m, which gives $D \approx 400k_{deg}$. Since $k_{deg} \approx 10^{-4}$ s$^{-1}$, we have $D \approx 0.04$ $\mu$m$^2$s$^{-1}$. This diffusion coefficient is too small to form the gradient in a reasonable timeframe [30].

There has been a long debate regarding whether a long-range morphogen gradient is formed by morphogen diffusion or by non-diffusive mechanisms [30, 59–61]. In [34], the kinetic parameters of two key morphogens, decapentaplegic (Dpp) and wingless (Wg), during the development of the *Drosophila* wing were studied quantitatively. Dpp forms a longer-range gradient than Wg. The kinetic parameters suggest that dynamin-dependent endocytosis may be required for the gradient formation of Dpp but not for that of Wg. However, the dynamin-dependent endocytosis of Dpp can be the result of ligands binding to either receptors or non-receptors, which yields different types of diffusive ligand-receptor formulations (the RMT model (7.63)–(7.64)) and ligand-non-receptor formulations (the $N_D$-LED model (7.59)–(7.62)).

The receptor plays different roles in these two models. In the receptor-mediated transcytosis model, the receptor is essential for endocytosis, re-secretion, and thus the transport of ligands, whereas in the ligand extracellular diffusion model, receptors merely modulate ligand distribution by binding to the ligand at the cell surface for the purpose of internalization and signaling. Furthermore, experimental monitors for the Dpp gradient in wing discs containing receptor gain-of-function and loss-of-function clones refute the receptor-mediated transcytosis model for Dpp gradient formation while supporting the restricted extracellular diffusion model in which the majority of Dpp is not bound to the receptor Thickveins [27].

## 7.3.5 Other Models

In the previous models, we mainly discuss gradient formations in which morphogens are synthesized at a localized site and transported to farther tissue cells at the other side to form a long-range signaling gradient. The formulation of these gradients is usually modeled by reaction-diffusion equations with a source term on one side and a sink on the other side, and hence, the signaling gradient at the steady state is driven by the boundary conditions (the source term). In most models, the positional information is assumed to be decoded by the concentration of signaling molecules at the steady state, which is described by a set of nonlinear boundary value problems. However, alternative decoding strategies are also possible, including the pre-steady state readout [62], the temporal derivatives of the morphogen concentration [63], the dynamics of morphogen signaling [5, 64], and the integration of signals from multiple morphogens [65, 66].

There are many other biological problems that yield alternative forms of mathematical models. In the dorsal-ventral patterning of *Drosophila* embryonic development, the morphogens Dpp and short gastrulation (Sog) are synthesized in different regions, diffuse to the whole tissue, and then interact with each other to pattern tissue development [67, 68]. A mathematical model of such Dpp/Sog patterning was developed in [38]. In this model, the *Drosophila* embryo was represented by a ring for the dorsal-ventral cross-section, Dpp was only produced in the upper part (dorsal region), and Sog was only produced in the lower part (ventral region). Hence, there is no boundary condition in this model, and the steady-state gradient is regulated by the interactions between the two morphogens. A similar computational model, but in three spatial dimensions, was developed for the BMP gradients in the dorsal-ventral patterning of the zebrafish embryo [56].

In addition to the mechanisms of patterning with signaling gradients, other developmental patterning strategies include patterning with activator-inhibitor systems (i.e., Turing's theory), genetic oscillations in neighboring cells, and mechanical deformations. Please refer to [6, 8, 9] for reviews of the mathematical models and computational approaches of these patterning strategies.

## 7.4   Robustness of Morphogen Gradients

Robustness—a phenotypic trait of the absence or low level of variation in phenotype in the face of genetic and/or environmental perturbation—has become a commonly used term in biological studies [69]. Embryonic pattering is usually highly robust, resisting not only substantial changes in the expression level of individual genes but also fluctuating environmental conditions (e.g., unseasonal heat waves). Exploring the robustness of morphogen gradient formation and identifying ways in which robust patterning is produced have become major research topics in the study of pattern formation [3, 70, 71]. Most works have focused on parametric robustness, i.e., insensitivity to parameter values [36, 53, 55, 67, 70, 72–77]. Some investigators have also focused on the "precision" of morphogen gradients, i.e., the natural variation among individuals in a population [33, 78–81]. Some studies have drawn attention to the effects of noise in morphogen gradients, especially the precision of patterning boundaries [33, 82–84].

### 7.4.1   Definition of Robustness

The precise definition of robustness is often ambiguous, despite its common use in biological studies. When defining robustness, it is important to specify which trait is robust to which perturbation and to provide a quantification for measuring the robustness [69]. The robustness of a system to external or internal variations is often quantified by the sensitivity coefficient, which corresponds to the fold change in the output of interest in response to a given fold change in a particular input [85].

For the morphogen gradient formations described by a set of reaction-diffusion equations with suitable boundary conditions, the signaling gradient is described by the solution of the resulting boundary value problem at the steady state. Here, we introduce general ways to define robustness. To this end, we always assume that the boundary value problem has a unique solution, denoted by $\mathrm{Sig}(X; p)$, and the signal concentration at position $X$ is dependent on the parameter value $p$. Therefore, the sensitivity coefficient of the signal with respect to changes in $p$ can be defined as

$$S_{\mathrm{Sig}, p}(X) = \left| \frac{\partial \ln \mathrm{Sig}}{\partial \ln p} \right| = \left| \frac{p}{\mathrm{Sig}} \frac{\partial \mathrm{Sig}(X; p)}{\partial p} \right|. \tag{7.66}$$

This sensitivity coefficient is defined based on the robustness in many studies, such as [40, 51, 53, 57].

In experiments, a natural pattering phenotype is used to measure the location $X$ of cells with a particular cell type, which is determined by the corresponding signaling level Sig. Thus, similar to the discussion in [75], the corresponding sensitivity coefficient is given by

$$S_{X,p}(\text{Sig}) = \left| \frac{p}{X} \frac{\partial X(\text{Sig}, p)}{\partial p} \right|. \tag{7.67}$$

When the cell type is well defined by a signaling threshold, the robustness of patterning can be measured by the sensitivity coefficient at the signaling boundary, i.e., $S_{X,p}$, with the signaling concentration taken as the threshold value. Otherwise, the robustness of patterning formation with respect to the parameter $p$ can be defined as the mean sensitivity over the signaling region

$$R_p = \frac{1}{\text{Sig}_1 - \text{Sig}_0} \int_{\text{Sig}_0}^{\text{Sig}_1} \left| \frac{p}{X} \frac{\partial X(\text{Sig}, p)}{\partial p} \right| d\,\text{Sig}. \tag{7.68}$$

The sensitivity coefficient defined by (7.67) is a good quantity to measure the variation with respect to small changes in the input parameter, i.e., perturbations in system parameters. However, in biological systems, mutations in genes can yield significant changes to certain parameters, such as the protein synthesis rates [76]. To measure the robustness with respect to these significant changes, the root mean square of cell displacement after parameter changes would be a meaningful measurement of robustness and is formulated as [36]

$$R_{p \to p'} = \frac{1}{|X(\text{Sig}_1, p) - X(\text{Sig}_0, p)|} \sqrt{\frac{1}{\text{Sig}_1 - \text{Sig}_0} \int_{\text{Sig}_0}^{\text{Sig}_1} (\Delta X)^2 d\,\text{Sig}}, \tag{7.69}$$

where $\Delta X = X(\text{Sig}, p') - X(\text{Sig}, p)$ represents the change in cell location with a given signaling concentration Sig when $p$ is changed to $p'$.

The above definitions are straightforward for parametric robustness in which changes in parameters often shift the signaling gradient in a single direction. However, the boundary between different cell fates is well defined by the concentration threshold, whereas noise perturbations in cell-to-cell variations can often mix up the cell fate boundary and induce a "salt-and-pepper" transition zone. In this situation, a sharpness index is often defined for the robustness of patterning formation in the context of noise perturbations [75, 84, 86].

## 7.4.2 Robustness with Respect to Perturbations in Morphogen Production

The robustness usually depends on the system parameters in a highly nonlinear way, and hence, it is difficult to perform mathematical analysis. Here, we consider the robustness with respect to perturbations in morphogen production to show the methods in the study of the robustness of the morphogen gradient.

### 7.4.2.1   Self-Enhanced Clearance

First, we discuss a well-accepted strategy for achieving parametric robustness in morphogen gradients, the principle of self-enhanced clearance [40]. In Sect. 7.3.2, a simple model was introduced to demonstrate how a morphogen's stimulation of its own degradation can provide a way to build a robust gradient. From Eq. (7.36), in the case of linear degradation ($n = 1$), the sensitivity of the ligand concentration at $X$ with respect to the concentration at the synthesis site is

$$S_{X,L_0}|_{n=1} = \left| \frac{L_0}{[L]} \frac{\partial [L]}{\partial L_0} \right| = 1. \tag{7.70}$$

When there is self-enhanced clearance ($n > 1$), the sensitivity coefficient is

$$S_{X,L_0}|_{n>1} = 1 - \frac{X/\varepsilon}{X/\varepsilon + 1}. \tag{7.71}$$

Thus, self-enhanced clearance (increasing $n$) tends to decrease the sensitivity of the ligand concentration to variations in morphogen synthesis, hence improving the robustness. Please refer to [70] for a review of the strategy of self-enhanced clearance.

### 7.4.2.2   Non-receptor-Mediated Morphogen Degradation

A possible means of self-enhanced clearance can be the consequence of enhancing ligand degradation through nonsignaling receptors (non-receptors). Analytical studies in [36] provided a theoretical basis for how a robust signaling gradient can be achieved by substantial binding of the signaling morphogen to non-receptors and the degradation of the resulting complexes at a sufficiently rapid rate.

Here, we consider the N-LED model, which includes the reversible binding and unbinding of ligands with non-receptors and the rapid degradation of the resulting complexes. To analyze the robustness of morphogen gradients, we consider the normalized concentration of the signaling morphogen-receptor complexes given by

$$\begin{cases} a''(x) - \lambda^2 \left( \dfrac{p}{1+a} + \dfrac{1-p}{1+\gamma a} \right) a + v(x) = 0, \quad a'(0) = a(1) = 0 \\ \\ b(x) = \dfrac{a(x)}{1+a(x)}. \end{cases} \tag{7.72}$$

We have shown that the boundary value problem (7.72) has a unique solution. There are five dimensionless parameters in the nondimensional equations: the relative width of the ligand production region $d$, the normalized ligand synthesis rate $v$, the ratio of the saturation level of receptors to the saturation level of non-receptors $\gamma$, the ratio of the degradation fluxes of the receptor to the degradation fluxes of the non-receptor $p$, and the sum of these fluxes $\lambda^2$. Here, $d$ is a geometric parameter and is often fixed in studies (for example, $d = 0.06$ in the case of the *Drosophila* wing imaginal disc, corresponding to the width of 12 $\mu$m for the production region compared with the total width of 200 $\mu$m); hence, the signaling gradient and robustness depend on the other four parameters: $v$, $\gamma$, $p$, and $\lambda$. Here, we note that $p = 1$ corresponds to the situation without non-receptors. Given the nonlinear boundary value problem (7.72), the dependence of the signaling concentration on parameters is not obvious, and hence, it is not straightforward to calculate the robustness.

Based on the unique solution of (7.72), the robustness measures the sensitivity dependence between the solution and the system parameters. Applying the robustness defined by (7.69), when there is a 2-fold change in the synthesis rate ($v \rightarrow 2v$), the corresponding robustness $R_v$ is defined as the root mean square of cell displacement (7.69) over the signaling region [36]

$$R_v = \frac{1}{x(b_{1/5}) - d} \sqrt{\frac{1}{b_{4/5} - b_{1/5}} \int_{b_{1/5}}^{b_{4/5}} (\Delta x)^2 db}, \qquad (7.73)$$

where $b_s = sb(d)$, $x(b_{1/5})$ means the location when $b(x) = b_{1/5}$, and

$$\Delta x = x(b; 2v) - x(b; v)$$

represents the displacement between positions with the same level of signaling complex $b$ for gradients with synthesis rates $v$ and $2v$ (Fig. 7.9). In general, the displacement $\Delta x$ depends on the normalized signaling gradients for the two different

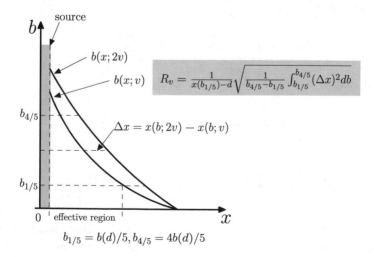

$$b_{1/5} = b(d)/5, \, b_{4/5} = 4b(d)/5$$

**Fig. 7.9 Illustration of the robustness with respect to a 2- fold change in the synthesis rate.** Replotted from [28]

ligand synthesis rates $v$ and $2v$. Since these gradients themselves also depend on the parameters $p$, $\gamma$, and $\lambda$, we indicate these dependencies by writing $R_v(p, \gamma, \lambda)$ as a continuous function of the parameters. The system is considered to have good robustness if $R < 0.2$.

For the case of a low ligand synthesis rate so that $v/\lambda^2 \ll 1$, we have $a(x) \ll 1$, and $b(x) \approx a(x)$. The gradient $a(x; v)$ (here, we specify the dependence on $v$) is approximately given by

$$a''(x; v) - \lambda^2 a(x; v) + v(x) = 0, \quad a'(0) = a(1) = 0,$$

$$a(x; v) = vK(x; \lambda), \quad K(x; \lambda) = \frac{\sinh(\lambda d) \sinh(\lambda(1 - x))}{\lambda^2 \cosh(\lambda)}, \quad (d \le x \le 1).$$

Hence, $a(x; v)$ is proportional to $v$, and $a(x; 2v) = 2a(x; v)$. Moreover,

$$x_v(a) - d = \int_a^{a_d} \frac{du}{\sqrt{s_1(v)^2 + \lambda^2 u^2}}, \quad (x > d)$$

$$= \frac{1}{\lambda} \ln \left( \frac{\lambda a_d(v) + \sqrt{s_1(v)^2 + \lambda^2(a_d(v))^2}}{\lambda a + \sqrt{s_1(v)^2 + \lambda^2 a^2}} \right),$$

where $s_1(v) = a'(1)$. From the second condition of the multi-fate gradient, $s_1(v)/a = O(\varepsilon) \ll 1$ over the signaling region. Hence,

$$x_v(a) - d \approx \frac{1}{\lambda} \ln \frac{a_d(v)}{a}.$$

Thus, note $a_d(2v) = 2a_d(v)$,

$$\Delta x = x_{2v}(a) - x_v(a) = \frac{1}{\lambda} \ln \frac{a_d(2v)}{a_d(v)} = \frac{\ln 2}{\lambda}.$$

Correspondingly, from (7.73) and $b \approx a$, the robustness

$$R_v(p, \gamma, \lambda) \approx \frac{\Delta x}{x_v(a_{1/5}) - d} = \frac{\ln 2}{\ln r},$$

where

$$r = \frac{\lambda a_d(v) + \sqrt{s_1(v)^2 + \lambda^2 a_d(v)^2}}{\lambda a_{1/5}(v) + \sqrt{s_1(v)^2 + \lambda^2 a_{1/5}(v)^2}} \lesssim \frac{a_d(v)}{a_{1/5}(v)} = 5.$$

Hence, the robustness $R_v(p, \gamma, \lambda) \gtrsim \frac{\ln 2}{\ln 5} = 0.43067 \ldots$. Thus, we conclude that in the range of low ligand synthesis rates, multi-fate gradients do not have good robustness.

In general, it is difficult to write the explicit function $R_v(p, \gamma, \lambda)$. We solve the boundary value problem (7.72) through the method of energy integration to obtain

$$
x = \begin{cases}
\displaystyle\int_a^{a_0} \frac{du}{\sqrt{2\lambda^2(E(u) - E(a_0)) - 2(v/d)(u - a_0)}}, & (0 \leq x < d) \\[18pt]
\displaystyle d + \int_a^{a_d} \frac{du}{\sqrt{2\lambda^2 E(u) + (a'(1))^2}}, & (d \leq x \leq 1)
\end{cases}
\tag{7.74}
$$

where

$$
E(u) = p(u - \ln(1 + u)) + \frac{1 - p}{\gamma^2}(\gamma u - \ln(1 + \gamma u)),
\tag{7.75}
$$

and $a_0 = a(0)$, $a_d = a(d)$, and $a'(1)$ are constants determined by the boundary conditions. While we focus on the multi-fate signaling gradients, we have $a'(1)^2 \ll 2\lambda^2 E(u)$, and hence approximately [36]

$$
x(a) \approx d + \frac{1}{\lambda} \int_a^{a_d} \frac{du}{\sqrt{2E(u)}}, \quad (a_{1/5} \leq a \leq a_d),
\tag{7.76}
$$

and therewith

$$
\Delta x \approx \frac{1}{\sqrt{2}\lambda} \int_{a_d(v)}^{a_d(2v)} \frac{du}{\sqrt{E(u)}},
\tag{7.77}
$$

where $a_d(v)$ and $a_d(2v)$ represent $a(d; v)$ and $a(d; 2v)$, respectively. This allows us to simplify the robustness measure $R_v(p, \gamma, \lambda)$ as

$$
R_v^{\mathrm{msg}}(p, \gamma, \lambda) \approx \frac{\Delta x}{x(a_{1/5}) - d} = \frac{\displaystyle\int_{a_d(v)}^{a_d(2v)} \frac{du}{\sqrt{E(u)}}}{\displaystyle\int_{\frac{a_d(v)}{5 + 4a_d(v)}}^{a_d(v)} \frac{du}{\sqrt{E(u)}}}
\tag{7.78}
$$

for multi-fate signaling gradients. Moreover, we can prove that $a_d(2v) > 2a_d(v)$ [36], and hence the robustness $R_v^{\mathrm{msg}}(p, \gamma, \lambda)$ has a lower bound defined by a function $J(p, \gamma)$:

$$
J(p, \gamma) = \min_{\xi > 0} \frac{\displaystyle\int_\xi^{2\xi} \frac{du}{\sqrt{E(u)}}}{\displaystyle\int_{\frac{\xi}{5 + 4\xi}}^{\xi} \frac{du}{\sqrt{E(u)}}}, \quad E(u) = \int_0^u a\left(\frac{p}{1 + a} + \frac{1 - p}{1 + \gamma a}\right) da.
\tag{7.79}
$$

In particular, when $p = 1$ (the case without non-receptors), we have

$$
E(u) = u - \ln(1 + u)
$$

and $J(1, \gamma) > 0.35$. These results indicate that without non-receptors, the system always has poor robustness (biologically acceptable robustness is often taken to be $R < 0.2$). This analysis provides a theoretical basis for the numerical simulation studies in [32] of the nonexistence of robust multi-fate gradients without non-receptors among simulations conducted for $2^{20}$ random sets of parameter values.

In the opposite situation, when $p = 0$ (the case without receptors) and $\gamma = 0$, we have $E(u) = \frac{1}{2}u^2$, and the boundary value problem (7.72) can be solved explicitly, which gives

$$R_v^{\text{msg}} = \frac{\ln 2}{\ln(5 + 4a_d)}, \quad a_d = \frac{v \sinh(\lambda d) \sinh(\lambda(1 - d))}{d\lambda^2 \cosh(\lambda)}. \tag{7.80}$$

Hence, good robustness can be achieved when $a_d > 27/4$, i.e., when the ligand synthesis rate $v$ is large enough. Despite the biologically non-realistic situation $p = 0$, based on the continuous dependence of the robustness $R$ on the four parameters $p$, $\gamma$, $\lambda$, and $v$, good robust multi-fate gradients are possible if both $p$ and $\gamma$ are small values and $v$ is large enough. Biologically, these conditions are met by

(1) a receptor degradative flux that is sufficiently low relative to the non-receptor, and
(2) a synthesis rate of free ligands that is sufficiently high but not high enough to saturate the available receptors in signaling cells.

For detailed discussions, please refer to [36].


### 7.4.2.3 "Bucket Brigade" Transportation Through Non-receptors

Now, we consider the $N_D$-LED model (7.59)–(7.62), in which the formation of a morphogen gradient is formed by "bucket brigade" transportation so that free ligands move long distances with the help of non-receptors.

Similar to the previous discussions, we introduce nondimensional quantities

$$l = \frac{[L]}{j_{\text{off}} N_{\text{tot}}/k_{\text{deg,L}}}, \quad r = \frac{[R]}{R_{\text{tot}}}, \quad u = \frac{[LR]}{R_{\text{tot}} k_{\text{on}} j_{\text{off}} N_{\text{tot}}/(k_{\text{deg,L}} k_{\text{deg,LR}})}, \quad \omega = \frac{[LN]}{N_{\text{tot}}},$$

$$t = k_{\text{deg,L}} T, \quad x = \frac{X}{X_{\text{max}}}, \quad d = \frac{X_{\text{min}}}{X_{\text{max}}}, \quad k(u) = \frac{\omega_R([LR])}{k_{\text{deg,R}} R_{\text{tot}}},$$

$$\varepsilon = \frac{j_{\text{on}} N_{\text{tot}}}{k_{\text{deg,L}}}, \quad \lambda^2 = \frac{j_{\text{off}} X_{\text{max}}^2}{D_{\text{LN}}}, \quad \delta_r = \frac{k_{\text{deg,R}}}{k_{\text{deg,L}}}, \quad \delta_u = \frac{k_{\text{deg,LR}}}{k_{\text{deg,L}}},$$

$$\alpha = \frac{k_{\text{off}}}{k_{\text{deg,LR}}}, \quad \gamma = \frac{k_{\text{on}}, R_{\text{tot}}}{j_{\text{on}}, N_{\text{tot}}}, \quad \eta = \frac{v_0 j_{\text{on}}}{k_{\text{deg,L}} j_{\text{off}}},$$

$$\theta_l = \frac{D_L}{X_{max}^2 k_{deg,L}}, \quad \theta_w = \frac{D_{LN}}{X_{max}^2 k_{deg,L}},$$

where $R_{tot} = \omega(0)/k_{deg,R}$ represents the concentration of unbound receptors at the steady state prior to the onset of ligand production. The normalized ligand production rate is given by

$$v(x) = (\eta/\varepsilon)H(d - x).$$

The normalized steady-state gradient is given by the following boundary value problem [53]:

$$\theta_l \frac{\partial^2 l}{\partial x^2} - (l - w) - \varepsilon (l(1 - w) - \gamma(\alpha u - l r)) + v(x) = 0, \qquad (7.81)$$

$$\frac{\partial^2 w}{\partial x^2} - \lambda^2(w - \varepsilon l(1 - w)) = 0, \qquad (7.82)$$

where $0 < x < 1$, and the functions $l$ and $r$ depend on $u$ through

$$l = (\alpha + 1)u/r, \quad r = k(u) - \mu u. \qquad (7.83)$$

Here, $l$ and $r$ are normalized ligand and ligand-non-receptor complex concentrations, respectively, and $u$ is the normalized signaling (ligand-receptor complex) concentration. The boundary conditions are

$$w'(0) = l'(0) = w(1) = l(1) = 0, \quad ()' = \frac{d()}{dx}. \qquad (7.84)$$

The nonlinear function $k(u)$ represents the feedback from the signal concentration to the synthesis of receptors. We often consider non-positive feedbacks so that

$$k(0) = 1, k'(u) \le 0, k(u) > 0, \quad \forall u \ge 0. \qquad (7.85)$$

To study the robustness with respect to the ligand synthesis rate, we need to investigate how the solution $u(x)$ of the above boundary value problem depends on the synthesis rate $\eta$. In [53], we have proved, through the upper and lower solution method, that if $\varepsilon$ is small enough, a unique combination of biologically acceptable gradients $\{w, (x), l(x), u(x), r(x)\}$ exists, i.e., the gradients satisfy (7.81)–(7.83), and the restrictions

$$0 \le w(x) \le 1, \ l(x) \ge 0, \ r(x) \ge 0, \ u(x) \ge 0, \quad (0 \le x \le 1). \qquad (7.86)$$

When $\varepsilon \ll 1$ and $\theta_l \ll 1$, the first-order approximation of the acceptable gradients at the steady state can be calculated explicitly to give [53]

$$w(x) \approx \begin{cases} \dfrac{\eta}{1+\eta} - a\cosh(\lambda\sqrt{1+\eta}x), & (0 \le x \le d) \\ \dfrac{w_d \sinh(\lambda(1-x))}{\sinh(\lambda(1-d))}, & (d \le x \le 1) \end{cases} \tag{7.87}$$

where

$$w_d = \frac{\eta}{1+\eta}\left[1 + \frac{1}{1+\eta}\coth(\lambda(1-d))\coth(\lambda d\sqrt{1+\eta})\right]^{-1}, \tag{7.88}$$

$$a = \frac{\coth(\lambda(1-d))}{\sqrt{1+\eta}\sinh(\lambda\sqrt{1+\eta}d)}w_d, \tag{7.89}$$

and

$$l(x) \approx \begin{cases} \dfrac{\eta}{\varepsilon} + \dfrac{\eta}{1+\eta} - a\cosh(\lambda\sqrt{1+\eta}x) - \dfrac{\eta}{\varepsilon}\dfrac{\cosh\frac{1-d}{\sqrt{\theta_l}}}{\cosh\frac{1}{\sqrt{\theta_l}}}\cosh(x/\sqrt{\theta_l}), & (0 < x \le d) \\ \dfrac{w_d\sinh(\lambda(1-x))}{\sinh(\lambda(1-d))} + \dfrac{\eta}{\varepsilon}\dfrac{\sinh\frac{d}{\sqrt{\theta_l}}}{\cosh\frac{1}{\sqrt{\theta_l}}}\sinh((1-x)/\sqrt{\theta_l}), & (d \le x < 1). \end{cases} \tag{7.90}$$

Consequently, the signaling gradient $u(x)$ is given implicitly by

$$u(x) = \frac{k(u(x))l(x)}{(\alpha+1)+\mu l(x)}. \tag{7.91}$$

The robustness of the signaling gradient with respect to changes in ligand production, $R(\eta, \eta')$, is defined as the relative change in the signaling gradient when the parameter $\eta$ is changed to $\eta'$

$$R(\eta, \eta') = \frac{1}{\Delta\eta/\eta}\frac{1}{1-d}\int_d^1 \frac{\Delta u(x)}{u(x;\eta)}dx; \tag{7.92}$$

here, we denote the signaling concentration as $u(x;\eta)$ to specify the dependence on the ligand synthesis rate, and

$$\Delta\eta = |\eta - \eta'|, \quad \Delta u = |u(x;\eta) - u(x;\eta')|.$$

Based on these approximations, [53] discussed the robustness $R(\eta, \eta')$ through the above approximation solution. Specifically, when $\eta' > \eta$, the robustness index satisfies

$$R_0(\eta, \eta') < \frac{A(\eta)}{\sqrt{1+\eta}}\left(1 + \frac{\coth(\lambda d\sqrt{1+\eta})}{\sqrt{1+\eta}}\right) + \frac{B(\eta)}{\sqrt{\eta}},$$

where $A(\eta)$ and $B(\eta)$ are bounded for all $\eta > 0$. Thus, we have $R(\eta, \eta') = O(\eta^{-1/2})$ $(\eta' > \eta)$ when $\eta$ is sufficiently large. Here, $\eta' > \eta$ indicates robustness with respect to the increase in the ligand synthesis rate. This result suggests that a robust gradient can be achieved when the ligand synthesis rate is large enough. Biologically, the conditions $\varepsilon \ll 1, \theta_l \ll 1$ and $\eta \gg 1$ are met if (1) the free ligand is rapidly turned over and (2) the ligand synthesis rate is large enough so that non-receptors in the ligand production region have high occupancy.

Cell membrane nonsignaling receptors (such HSPGs) are important for the formation of morphogen gradients. A series of studies have shown that the presence of non-receptors is favorable for robust gradients [36, 42, 51, 53, 87]. The desired robust morphogen gradient with respect to substantial perturbations of the morphogen synthesis rate is achievable through two different mechanisms that involve regulation by non-receptors [53].

Mechanism 1: Substantial (reversible) binding of slowly turned over morphogen molecules with membrane-bound non-receptors, with the resulting nonsignaling complexes degrading at a sufficiently rapid rate.

Mechanism 2: Fast binding of rapidly turned over free morphogen molecules with non-receptors so that the nonsignaling complexes move downstream through the bucket brigade process.

However, the above two mechanisms fail to generate good robustness with respect to perturbations in the receptor synthesis rate due to the restriction of the kinetic and diffusive resistances in molecule transportation [51, 75]. As a consequence of this restriction, there is a trade-off between robustness with respect to the receptor synthesis rate and the signaling length scale [75]. Hence, some other mechanisms must be involved to achieve robustness with respect to multiple perturbations, including nonlocal feedback [88, 89] and noise effects [84].

# 7.5 Morphogen-Mediated Growth Control

Aside from patterning, how morphogen gradients regulate tissue growth is also a critically important question in morphogen systems [5, 17, 90–92]. For example, the *Drosophila* wing disc begins with approximately 40 cells and reaches a size of 50,000 cells in late third instar larvae [93]. Experimental studies on the development of the *Drosophila* wing disc indicate that growth will not occur without morphogens; however, cell proliferation is found to be spatially uniform in the wing disc, even with a spatially graded morphogen [94]. This peculiar result raises the questions of how the spatially inhomogeneous morphogen gradient is translated into uniform growth and how the wing disc maintains a robust final size at the end of development.

The morphogen gradient is closely related to tissue growth. To determine how the inhomogeneous morphogen gradient is translated into uniform growth and how the wing disc maintains a robust final size at the end, a number of potential growth control models have been proposed, including the temporal dynamics model [5, 95, 96], the slope model [97], and the mechanical feedback model [98–100]. In the slope model, the relative slope of the signal controls cell division; in the temporal dynamics model, growth is regulated by the percentage of increase in the signal over time; and in the mechanical feedback model, the mechanical stress induces cell proliferation, while the final size of the tissue is controlled by mechanical compression. However, the mechanisms for morphogen-mediated growth control remain controversial, as none of them can explain all significant biological data.

The mathematical models of growth control are often associated with scale invariance and moving boundary problems. Here, we give the basic principles of the models. For details, please refer to [28].

The phenomenon of scaling has been observed in many biological systems [65, 101–104], including the *Drosophila* wing imaginal disc [5]. The scaling of the morphogen gradient with tissue size is essential for ensuring a body plan of reproducible proportions. The scale invariance of the Dpp gradient was experimentally observed in mutants of the insulin pathway, which affects the wing disc size [19].

When there is a scale invariance of morphogen gradients, the cells divide and proliferate according to morphogen signaling to maintain the growing tissue size. In mathematical models, the growing tissue is represented by a moving boundary. A simple framework to study the scale invariance of a Turing system with ligand [L] and regulatory molecules [E] was proposed in [105]:

$$\frac{\partial[L]}{\partial t} = \frac{\partial}{\partial X}(D_0 + D_1[E]))\frac{\partial}{\partial X}[L] + F([L]), \tag{7.93}$$

$$\frac{\partial[E]}{\partial t} = D_E\frac{\partial^2}{\partial X^2}[E] + v_E. \tag{7.94}$$

Here, $X \in [-X_{max}, X_{max}]$ denotes the region of space occupied by the developing system. In this model, morphogen diffusion is affected by regulatory molecules so that the diffusion rate depends on [E], the concentration of regulatory molecules. The boundary conditions at $X = X_{max}$ are

$$\frac{\partial[L]}{\partial X} = 0 \text{ and } - D_E\frac{\partial[E]}{\partial X} = h[E], \tag{7.95}$$

and the boundary conditions at $X = -X_{max}$ are

$$\frac{\partial[L]}{\partial X} = 0 \text{ and } D_E\frac{\partial[E]}{\partial X} = h[E]. \tag{7.96}$$

The perfect scaling property of the morphogen gradient with the growing tissue size is defined as a scaling solution so that the morphogen gradient can be described by a time-independent function $F(x)$ such that

$$[L](X, t) = [L](0, t)F\left(\frac{X}{X_{max}(t)}\right), \quad X \in [0, X_{max}(t)], \tag{7.97}$$

where $X_{max}(t)$ represents the boundary of the growing tissue.

In models (7.93)–(7.96), the key condition for perfect scale invariance for the gradient [L] is that the morphogen diffusion coefficient $(D_0 + D_1[E])$ is proportional to the tissue size $X_{\max}^2$ [105].

In embryo development, the motion field created by tissue growth may affect the dynamics of the morphogen gradient in the growing domain [106]. Tissue growth control and the scaling of a morphogen gradient are coupled through advection in mathematical models. A generic model for morphogen systems in a growing domain can be modeled based on the Reynolds transport theorem [107]

$$\frac{\partial [L]}{\partial t} + \frac{\partial (v[L])}{\partial X} = D_L \frac{\partial^2 [L]}{\partial X^2}, \quad \text{on } X \in (0, X_{\max}(t)), \tag{7.98}$$

where $v$ denotes the growth field, and the domain expands linearly with time, so that

$$X_{\max}(t) = X_{\max}(0) + v_g t. \tag{7.99}$$

Accordingly, the boundary condition is given by

$$\frac{\partial [L]}{\partial X} = -j_{\text{in}}, \quad \text{at } X = 0; \quad \frac{\partial [L]}{\partial X} = 0, \quad \text{at } X = X_{\max}(t). \tag{7.100}$$

For a uniformly growing domain, the local growth rate $\partial v / \partial X$ is defined as

$$\frac{\partial v}{\partial X} = \frac{v_g}{X_{\max}(t)}. \tag{7.101}$$

In this model, morphogens spread, and an advection term is included since the morphogens may be attached to cells during tissue development. This inclusion is based on the fact that in the *Drosophila* wing disc, at least 97% of morphogens have been found to be either internalized or absorbed by cells [19, 61]. Because of the advection term, the dynamics are different from the previous models in that the morphogen gradient does not reach a steady state within the physiological time scale.

Equation (7.98) can be rewritten in the following form:

$$\frac{\partial [L]}{\partial t} + v \frac{\partial ([L])}{\partial X} = D_L \frac{\partial^2 [L]}{\partial X^2} - [L] \frac{\partial v}{\partial X}. \tag{7.102}$$

In [107], the last term is called the dilution term, which is critical for perfect scaling in a uniformly growing domain. Simulations in [107] have shown that the morphogen gradient has perfect scaling of form

$$[L](X, t) = [L](0, t) \exp\left(-\frac{X}{\lambda(t)}\right)$$

when there is no diffusion ($D_L = 0$). This result indicates that the impact of advective transport is essential for scale invariance.

## 7.6 Summary

Remarkable progress has been made during the past 20 years in understanding morphogen-mediated patterning due to the rapid advancement of experimental techniques and, most importantly, the involvement of mathematical modeling in recent years. The major contribution of mathematical models is in the areas of morphogen gradient formation and the mechanisms of controlling the gradient. In this chapter, we introduced basic principles to model the progress of morphogen gradient formation and presented methods of analyzing the steady-state morphogen gradient and robustness. These results have helped us to better understand the growth control of embryo development and pattern formation. It has become increasingly clear that the interplay between experiments and mathematics is critical to establish general theories that can make sense of the mechanistic complexity of organismal development driven by morphogens [7, 108].

It is not possible to include the complete discussion of the related studies in this lecture. Advanced models need to include more elements in the pathways of morphogen gradient signaling and tissue growth, such as downstream signaling pathways that regulate gene expression [109], the growth Hippo signaling pathway [110–112], the temporal integration of morphogen signals for noise reduction [3, 62], cell-to-cell contacts that modulate growth [99], multiple morphogens for improving robustness [65, 81, 113], cooperative feedback loops [56], the role of tissue geometry in pattern formation [114], and noise in morphogen gradients [82–84, 115]. Moreover, there are many mathematical challenges and unanswered questions that remain to be addressed in future studies. Many characteristics, challenges, and questions associated with the complexity of morphogen systems will require new mathematical and computational tools, likely leading to an emerging research area: mathematical and computational morphogenesis.

## References

1. Wolpert, L.: Positional information and the spatial pattern of cellular differentiation. J. Theor. Biol. **25**, 1–47 (1969)
2. Wolpert, L.: Positional information and patterning revisited. J. Theor. Biol. **269**, 359–365 (2011)
3. Lander, A.D.: How cells know where they are. Science **339**, 923–927 (2013)
4. Briscoe, J., Small, S.: Morphogen rules: design principles of gradient-mediated embryo patterning. Development **142**, 3996–4009 (2015)
5. Wartlick, O., Mumcu, P., Jülicher, F., Gonzalez-Gaitan, M.: Understanding morphogenetic growth control–lessons from flies. Nat. Rev. Mol. Cell Biol. **12**, 594–604 (2011a)
6. Kondo, S., Miura, T.: Reaction-diffusion model as a framework for understanding biological pattern formation. Science **329**, 1616–1620 (2010)
7. Lander, A.D.: Pattern, growth, and control. Cell **144**, 955–969 (2011)
8. Morelli, L.G.L., Uriu, K.K., Ares, S.S., Oates, A.C.A.: Computational approaches to developmental patterning. Science **336**, 187–191 (2012)
9. Maini, P.K., Woolley, T.E., Baker, R.E., Gaffney, E.A., Lee, S.S.: Turing's model for biological pattern formation and the robustness problem. Interf. Focus **2**, 487–496 (2012)

10. Umulis, D.M., Othmer, H.G.: The role of mathematical models in understanding pattern formation in developmental biology. Bull. Math. Biol. **77**, 817–845 (2015)
11. Wartlick, O., Kicheva, A., González-Gaitán, M.: Morphogen gradient formation. Cold Spring Harb. Perspect. Biol. **1**, a001255–a001255 (2009)
12. Wolpert, L., Tickle, C.: Principles of Development. Academic Press, Oxford (2010)
13. Dessaud, E., Ribes, V., Balaskas, N., Yang, L.L., Pierani, A., Kicheva, A., Novitch, B.G., Briscoe, J., Sasai, N.: Dynamic assignment and maintenance of positional identity in the ventral neural tube by the morphogen sonic hedgehog. PLoS Biol. **8**, (2010)
14. Smith, J.: Forming and interpreting gradients in the early Xenopus embryo. Cold Spring Harb Perspect Biol **1**, (2009)
15. Smith, J., Hagemann, A., Saka, Y., Williams, P.: Understanding how morphogens work. Philos. Trans. R. Soc. London B Biol. Sci. **363**, 1387–1392 (2008)
16. Kam, R.K.T., Deng, Y., Chen, Y., Zhao, H.: Retinoic acid synthesis and functions in early embryonic development. Cell Biosci. **2**, 11 (2012)
17. Affolter, M., Basler, K.: The decapentaplegic morphogen gradient: from pattern formation to growth regulation. Nat. Rev. Genet. **8**, 663–674 (2007)
18. Nellen, D., Burke, R., Struhl, G., Basler, K.: Direct and long-range action of a Dpp morphogen gradient. Cell **85**, 357–368 (1996)
19. Teleman, A.A., Cohen, S.M.: Dpp gradient formation in the Drosophila wing imaginal disc. Cell **103**, 971–980 (2000)
20. Theisen, H., Syed, A., Nguyen, B.T., Lukacsovich, T., Purcell, J., Srivastava, G.P., Iron, D., Gaudenz, K., Nie, Q., Wan, F.Y.M., Waterman, M.L., Marsh, J.L.: Wingless directly represses DPP morphogen expression via an Armadillo/TCF/Brinker complex. PLoS ONE **2**, e142–e510 (2007)
21. Gurdon, J.B., Bourillot, P.Y.: Morphogen gradient interpretation. Nature **413**, 797–803 (2001)
22. Tabata, T.: Genetics of morphogen gradients. Nat. Rev. Genet. **2**, 620–630 (2001)
23. Kicheva, A., González-Gaitán, M.: The decapentaplegic morphogen gradient: a precise definition. Curr. Opinion Cell Biol. **20**, 137–143 (2008)
24. Crozatier, M., Glise, B., Vincent, A.: Patterns in evolution: veins of the Drosophila wing. Trends Genet. **20**, 498–505 (2004)
25. Zecca, M., Basler, K., Struhl, G.: Sequential organizing activities of engrailed, hedgehog and decapentaplegic in the Drosophila wing. Development **121**, 2265–2278 (1995)
26. Nellen, D., Affolter, M., Basler, K.: Receptor serine/threonine kinases implicated in the control of Drosophila body pattern by decapentaplegic. Cell **78**, 225–237 (1994)
27. Schwank, G., Dalessi, S., Yang, S.-F., Yagi, R., de Lachapelle, A.M., Affolter, M., Bergmann, S., Basler, K.: Formation of the long range Dpp morphogen gradient. PLoS Biol. **9**, (2011)
28. Lei, J., Lo, W.-C., Nie, Q.: Mathematical models of morphogen dynamics and growth control. Annals Math Sci. Appl. **1**, 427–471 (2016)
29. Entchev, E.V., Schwabedissen, A., González-Gaitán, M.: Gradient formation of the TGF-beta homolog Dpp. Cell **103**, 981–991 (2000)
30. Lander, A.D., Nie, Q., Wan, F.Y.M.: Do morphogen gradients arise by diffusion? Dev. Cell **2**, 785–796 (2002)
31. Lander, A., Nie, Q., Wan, F.: Spatially distributed morphogen production and morphogen gradient formation. Math. Biosci. Eng **2**, 239–262 (2005a)
32. Lander, A., Wan, F., Elledge, H., Mizutani, C., Bier, E., Nie, Q.: (2005b). Diverse paths to morphogen gradient robustness. submitted for publicaiton
33. Bollenbach, T., Pantazis, P., Kicheva, A., Bokel, C., Gonzalez-Gaitan, M., Julicher, F.: Precision of the Dpp gradient. Development **135**, 1137–1146 (2008)
34. Kicheva, A., Pantazis, P., Bollenbach, T., Kalaidzidis, Y., Bittig, T., Jülicher, F., González-Gaitán, M.: Kinetics of morphogen gradient formation. Science **315**, 521–525 (2007)
35. Sattinger, D.H.: Monotone methods in nonlinear elliptic and parabolic boundary value problems. Indiana Univ. Math. J. **21**, 979–1000 (1972)
36. Lei, J., Wan, F.Y.M., Lander, A., Nie, Q.: Robustness of signaling gradient in drosophila wing imaginal disc. Discrete Contin. Dyn. Syst. B **16**, 835–866 (2011)

37. Lou, Y., Nie, Q., Wan, F.Y.M.: Nonlinear Eigenvalue problems in the stability analysis of morphogen gradients. Studies Appl. Math. **113**, 183–215 (2004)
38. Lou, Y., Nie, Q., Wan, F.Y.M.: Effects of Sog on Dpp-receptor binding. SIAM J. Appl. Math. **65**, 1748–1771 (2005)
39. Vergas, B.: Leaky boundaries and morphogen gradietns. PhD thesis, Department of Mathematics, University of California, Irvine (2006)
40. Eldar, A., Rosin, D., Shilo, B.-Z., Barkai, N.: Self-enhanced ligand degradation underlies robustness of morphogen gradients. Dev. Cell **5**, 635–646 (2003)
41. Akiyama, S., Takahashi, S., Kimura, T., Ishimori, K., Morishima, I., Nishikawa, Y., Fujisawa, T.: Conformational landscape of cytochrome c folding studied by microsecond-resolved small-angle x-ray scattering. Proc. Natl. Acad. Sci. USA **99**, 1329–1334 (2002)
42. Belenkaya, T.Y., Han, C., Yan, D., Opoka, R.J., Khodoun, M., Liu, H., Lin, X.: Drosophila Dpp morphogen movement is independent of dynamin-mediated endocytosis but regulated by the glypican members of heparan sulfate proteoglycans. Cell **119**, 231–244 (2004)
43. Häcker, U., Nybakken, K., Perrimon, N.: Heparan sulphate proteoglycans: the sweet side of development. Nat. Rev. Mol. Cell Biol. **6**, 530–541 (2005)
44. Kirkpatrick, C.A., Dimitroff, B.D., Rawson, J.M., Selleck, S.B.: Spatial regulation of wingless morphogen distribution and signaling by Dally-like protein. Dev. Cell **7**, 513–523 (2004)
45. The, I., Bellaiche, Y., Perrimon, N.: Hedgehog movement is regulated through tout velu-dependent synthesis of a heparan sulfate proteoglycan. Mol. Cell **4**, 633–639 (1999)
46. Yan, D., Lin, X.: Shaping morphogen gradients by proteoglycans. Cold Spring Harb Perspect. Biol. **1**, a002493–a002493 (2009)
47. Kirkpatrick, C.A., Selleck, S.B.: Heparan sulfate proteoglycans at a glance. J. Cell Sci. **120**, 1829–1832 (2007)
48. Lin, X.: Functions of heparan sulfate proteoglycans in cell signaling during development. Development **131**, 6009–6021 (2004)
49. Matsuo, I., Kimura-Yoshida, C.: Extracellular distribution of diffusible growth factors controlled by heparan sulfate proteoglycans during mammalian embryogenesis. Philos. Trans. R. Soc. London B Biol. Sci. **369**, 20130545 (2014)
50. Lander, A.D., Nie, Q., Wan, F.Y.M.: Membrane-associated non-receptors and morphogen gradients. Bull. Math. Biol. **69**, 33–54 (2007)
51. Lei, J., Song, Y.: Mathematical model of the formation of morphogen gradients through membrane-associated non-receptors. Bull. Math. Biol. **72**, 805–829 (2010)
52. Lei, J.: Mathematical model of the Dpp gradient formation in drosophila wing imaginal disc. Chinese Sci. Bull. **55**, 984–991 (2010)
53. Lei, J., Wang, D., Song, Y., Nie, Q., Wan, F.Y.M.: Robustness of Morphogen gradients with "bucket brigade" transport through membrane-associated non-receptors. Discrete Contin. Dyn. Syst. B **18**, 721–739 (2012)
54. Mizutani, C.M., Nie, Q., Wan, F.Y., Zhang, Y.-T., Vilmos, P., Sousa-Neves, R., Bier, E., Marsh, J.L., Lander, A.D.: Formation of the BMP activity gradient in the Drosophila embryo. Dev. Cell **8**, 915–924 (2005)
55. Shimmi, O., Umulis, D., Othmer, H., Oconnor, M.: Facilitated transport of a Dpp/Scw heterodimer by Sog/Tsg leads to robust patterning of the blastoderm embryo. Cell **120**, 873–886 (2005)
56. Zhang, Y.-T., Lander, A.D., Nie, Q.: Computational analysis of BMP gradients in dorsal-ventral patterning of the zebrafish embryo. J. Theor. Biol. **248**, 579–589 (2007)
57. Bollenbach, T., Kruse, K., Pantazis, P., González-Gaitán, M., Jülicher, F.: Robust formation of morphogen gradients. Phys. Rev. Lett. **94**, 4 (2005)
58. Bollenbach, T., Kruse, K., Pantazis, P., González-Gaitán, M., Jülicher, F.: Morphogen transport in epithelia. Phys. Rev. E **75**, 16 (2007)
59. Pfeiffer, S., Vincent, J.-P.: Signalling at a distance: transport of wingless in the embryonic epidermis of Drosophila. Semin. Cell Dev. Biol. **10**, 303–309 (1999)
60. Kerszberg, M., Wolpert, L.: Mechanisms for positional signalling by morphogen transport: a theoretical study. J. Theor. Biol. **191**, 103–114 (1998)

61. Zhou, S., Zhou, S.S., Lo, W.-C.W., Lo, W.-C., Suhalim, J.L.J., Suhalim, J.L., Digman, M.A.M., Digman, M.A., Gratton, E.E., Gratton, E., Nie, Q.Q., Nie, Q., Lander, A.D.A., Lander, A.D.: Free extracellular diffusion creates the Dpp morphogen gradient of the Drosophila wing disc. Curr. Biol. **22**, 668–675 (2012)

62. Bergmann, S., Sandler, O., Sberro, H., Shnider, S., Schejter, E., Shilo, B.-Z., Barkai, N.: Pre-steady-state decoding of the bicoid morphogen gradient. PLoS Biol. **5**, (2007)

63. Richards, D.M., Saunders, T.E.: Spatiotemporal analysis of different mechanisms for interpreting morphogen gradients. Biophy. J. **108**, 2061–2073 (2015)

64. Wartlick, O., Mumcu, P., Jülicher, F., Gonaález-Gaitán, M.: Response to comment on dynamics of Dpp signaling and proliferation control. Science **335**, 401 (2012)

65. McHale, P., Rappel, W.-J., Levine, H.: Embryonic pattern scaling achieved by oppositely directed morphogen gradients. Phys. Biol. **3**, 107–120 (2006)

66. Morishita, Y., Iwasa, Y.: Optimal placement of multiple morphogen sources. Phys. Rev. E **77**, (2008)

67. Eldar, A., Dorfman, R., Weiss, D., Ashe, H., Shilo, B.-Z., Barkai, N.: Robustness of the BMP morphogen gradient in Drosophila embryonic patterning. Nature **419**, 304–8 (2002)

68. Shilo, B.-Z., Haskel-Ittah, M., Ben-Zvi, D., Schejter, E.D., Barkai, N.: Creating gradients by morphogen shuttling. Trends Genet. **29**, 339–347 (2013)

69. Félix, M.-A., Barkoulas, M.: Pervasive robustness in biological systems. Nat. Rev. Genet. **16**, 483–496 (2015)

70. Lander, A.D., Lo, W.-C., Nie, Q., Wan, F.Y.M.: The measure of success: constraints, objectives, and tradeoffs in morphogen-mediated patterning. Cold Spring Harb. Perspect. Biol. **1**, a002022–a002022 (2009b)

71. Umulis, D., O'Connor, M.B., Othmer, H.G.: Robustness of embryonic spatial patterning in Drosophila melanogaster. Curr. Top Dev. Biol. **81**, 65–111 (2008)

72. von Dassow, G., Meir, E., Munro, E.M., Odell, G.M.: The segment polarity network is a robust developmental module. Nature **406**, 188–192 (2000)

73. von Dassow, G., Odell, G.M.: Design and constraints of theDrosophila segment polarity module: robust spatial patterning emerges from intertwined cell state switches. J. Exp. Zool. **294**, 179–215 (2002)

74. Lander, A.D.: Morpheus unbound: reimagining the morphogen gradient. Cell **128**, 245–256 (2007)

75. Lo, W.-C., Zhou, S., Wan, F.Y.M., Lander, A.D., Nie, Q.: Robust and precise morphogen-mediated patterning: trade-offs, constraints and mechanisms. J. R. Soc. Interf. **12**, 20141041 (2015)

76. Morimura, S., Maves, L., Chen, Y., Hoffmann, F.M.: decapentaplegic overexpression affects Drosophila wing and leg imaginal disc development and wingless expression. Dev. Biol. **177**, 136–151 (1996)

77. White, R., Nie, Q., Lander, A.D., Schilling, T.: Complex regulation of cyp26a1 creates a robust retinoic acid gradient in the zebrafish embryo. PLoS Biol. **5**, (2007)

78. Emberly, E.: Optimizing the readout of morphogen gradients. Phys. Rev. E **77**, (2008)

79. Houchmandzadeh, B., Wieschaus, E., Leibler, S.: Establishment of developmental precision and proportions in the early Drosophila embryo. Nature **415**, 798–802 (2002)

80. Gregor, T., Tank, D.W., Wieschaus, E.F., Bialek, W.: Probing the limits to positional information. Cell **130**, 153–164 (2007)

81. Tostevin, F., ten Wold, P.R., Howard, M.: Fundamental limits to position determination by concentration gradients. PLoS Comput. Biol. **3**, (2007)

82. He, F., Ren, J., Wang, W., Ma, J.: Evaluating the Drosophila bicoid morphogen gradient system through dissecting the noise in transcriptional bursts. Bioinformatics **28**, 970–975 (2012)

83. Holloway, D.M., Lopes, F.J.P., da Fontoura Costa, L., Travençolo, B.A.N., Golyandina, N., Usevich, K., Spirov, A.V.: Gene expression noise in spatial patterning: hunchback promoter structure affects noise amplitude and distribution in Drosophila segmentation. PLoS Comput. Biol. **7**, e1001069–e1001069 (2010)

84. Zhang, L., Radtke, K., Zheng, L., Cai, A.Q., Schilling, T.F., Nie, Q.: Noise drives sharpening of gene expression boundaries in the zebrafish hindbrain. Mol. Syst. Biol. **8**, 613 (2012)
85. Reeves, G.T., Fraser, S.E.: Biological systems from an engineer's point of view. PLoS Biol. **7**, e21–e21 (2009)
86. Lopes, F.J.P., Vieira, F.M.C., Holloway, D.M., Bisch, P.M., Spirov, A.V.: Spatial bistability generates hunchback expression sharpness in the Drosophila embryo. PLoS Comput. Biol. **4**, (2008)
87. Akiyama, T., Kamimura, K., Firkus, C., Takeo, S., Shimmi, O., Nakato, H.: Dally regulates Dpp morphogen gradient formation by stabilizing Dpp on the cell surface. Dev. Biol. **313**, 408–419 (2008)
88. Kushner, T., Simonyan, A., Wan, F.Y.M.: A new approach to feedback for robust signaling gradients. Studies Appl. Math. **133**, 18–51 (2014)
89. Simonyan, A., Wan, F.Y.: Transient feedback and robust signaling gradients. Int. J. Numer. Anal. Model. **13**, 179–204 (2016)
90. Buchmann, A., Alber, M., Zartman, J.J.: Sizing it up: the mechanical feedback hypothesis of organ growth regulation. Semin. Cell Dev. Biol. **35**, 73–81 (2014)
91. Day, S.J., Lawrence, P., a.: Measuring dimensions: the regulation of size and shape. Development **127**, 2977–2987 (2000)
92. Schwank, G., Basler, K.: Regulation of organ growth by morphogen gradients. Cold Spring Harb. Perspect. Biol. **2**, (2010)
93. Lawrence, P.A., Morata, G.: The early development of mesothoracic compartments in Drosophila. An analysis of cell lineage and fate mapping and an assessment of methods. Dev. Biol. **56**, 40–51 (1977)
94. Martin, F.A., Morata, G.: Compartments and the control of growth in the Drosophila wing imaginal disc. Development **133**, 4421–4426 (2006)
95. Averbukh, I., Ben-Zvi, D., Mishra, S., Barkai, N.: Scaling morphogen gradients during tissue growth by a cell division rule. Development **141**, 2150–2156 (2014)
96. Wartlick, O., Mumcu, P., Kicheva, A., Bittig, T., Seum, C., Jülicher, F., González-Gaitán, M.: Dynamics of Dpp signaling and proliferation control. Science **331**, 1154–1159 (2011b)
97. Rogulja, D., Irvine, K.D.: Regulation of cell proliferation by a morphogen gradient. Cell **123**, 449–461 (2005)
98. Aegerter-Wilmsen, T., Aegerter, C.M., Hafen, E., Basler, K.: Model for the regulation of size in the wing imaginal disc of Drosophila. Mech. Dev. **124**, 318–326 (2007)
99. Aegerter-Wilmsen, T., Heimlicher, M.B., Smith, a. C., de Reuille, P. B., Smith, R. S., Aegerter, C. M. and Basler, K.: Integrating force-sensing and signaling pathways in a model for the regulation of wing imaginal disc size. J. Cell Sci. **125**, 3221–31 (2012)
100. Shraiman, B.I.: Mechanical feedback as a possible regulator of tissue growth. Proc. Natl. Acad. Sci. USA **102**, 3318–23 (2005)
101. Gregor, T., Bialek, W., de Ruyter van Steveninck, R.R., Tank, D.W., Wieschaus, E.F.: Diffusion and scaling during early embryonic pattern formation. Proc Natl Acad Sci USA **102**, 18403–18407 (2005)
102. Lauschke, V.M., Tsiairis, C.D., Francois, P., Aulehla, A.: Scaling of embryonic patterning based on phase-gradient encoding. Nature **493**, 101–105 (2013)
103. Umulis, D.M., Shimmi, O., O'Connor, M.B., Othmer, H.G.: Organism-scale modeling of early Drosophila patterning via bone morphogenetic proteins. Dev. Cell **18**, 260–274 (2010)
104. Umulis, D.M., Othmer, H.G.: Mechanisms of scaling in pattern formation. Development **140**, 4830–43 (2013)
105. Othmer, H.G., Pate, E.: Scale-invariance in reaction-diffusion models of spatial pattern formation. Proc. Natl. Acad. Sci. USA **77**, 4180–4184 (1980)
106. Crampin, E.J., Gaffney, E., a. and Maini, P. K., : Reaction and diffusion on growing domains: scenarios for robust pattern formation. Bull. Math. Biol. **61**, 1093–1120 (1999)
107. Fried, P., Iber, D.: Dynamic scaling of morphogen gradients on growing domains. Nat. Commun. **5**, 5077–77 (2013)
108. Freeman, M.: Morphogen gradients, in theory. Dev. Cell **2**, 689–690 (2002)

109. Shvartsman, S.Y., Baker, R.E.: Mathematical models of morphogen gradients and their effects on gene expression. Dev. Biol. **1**, 715–730 (2012)
110. Buttitta, L., a., Edgar, B. a.: How size is controlled: from Hippos to Yorkies. Nat. Cell Biol. **9**, 1225–1227 (2007)
111. Cho, E., Irvine, K.D.: Action of fat, four-jointed, dachsous and dachs in distal-to-proximal wing signaling. Development **131**, 4489–4500 (2004)
112. Thompson, B.J., Cohen, S.M.: The Hippo pathway regulates the bantam microRNA to control cell proliferation and apoptosis in Drosophila. Cell **126**, 767–774 (2006)
113. Hardway, H., Mukhopadhyay, B., Burke, T., Hichman, T.J., Forman, R.: Modeling the precision and robustness of hunchback border during drosophila embryonic development. J. Theor. Biol. **254**, 390–9 (2008)
114. Umulis, D.M., Othmer, H.G.: The importance of geometry in mathematical models of developing systems. Curr. Opinion Genet. Dev. **22**, 547–52 (2012)
115. Arias, A.M., Hayward, P.: Filtering transcriptional noise during development: concepts and mechanisms. Nat. Rev. Genet. **7**, 34–44 (2005)

# Index

Printed in the United States
by Baker & Taylor Publisher Services